Selected Titles in This Series

(Continued in the back of this publication)

Degree 16 Standard L-function of $GSp(2) \times GSp(2)$

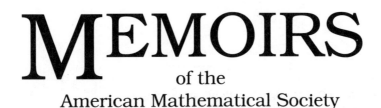

MEMOIRS
of the
American Mathematical Society

Number 588

Degree 16 Standard L-function of $GSp(2) \times GSp(2)$

Dihua Jiang

September 1996 • Volume 123 • Number 588 (third of 4 numbers) • ISSN 0065-9266

American Mathematical Society
Providence, Rhode Island

1991 *Mathematics Subject Classification.*
Primary 11F70.

Library of Congress Cataloging-in-Publication Data

Jiang, Dihua.
  Degree 16 standard L-function of $GSp(2) \times GSp(2)$ / Dihua Jiang.
    p.   cm. – (Memoirs of the American Mathematical Society, ISSN 0065-9266; no. 588)
  "September 1996, volume 123, number 588 (third of 4 numbers)."
  Includes bibliographical references.
  ISBN 0-8218-0476-6 (alk. paper)
  1. Automorphic functions. 2. L-functions. I. Title. II. Series.
QA3.A57   no. 588
[QA353.A9]
510 s–dc20
[515′.98]   96-21897
                                                                                   CIP

Memoirs of the American Mathematical Society

This journal is devoted entirely to research in pure and applied mathematics.

Subscription information. The 1996 subscription begins with Number 568 and consists of six mailings, each containing one or more numbers. Subscription prices for 1996 are $391 list, $313 institutional member. A late charge of 10% of the subscription price will be imposed on orders received from nonmembers after January 1 of the subscription year. Subscribers outside the United States and India must pay a postage surcharge of $25; subscribers in India must pay a postage surcharge of $43. Expedited delivery to destinations in North America $30; elsewhere $92. Each number may be ordered separately; *please specify number* when ordering an individual number. For prices and titles of recently released numbers, see the New Publications sections of the *Notices of the American Mathematical Society*.

Back number information. For back issues see the *AMS Catalog of Publications*.

Subscriptions and orders should be addressed to the American Mathematical Society, P. O. Box 5904, Boston, MA 02206-5904. *All orders must be accompanied by payment.* Other correspondence should be addressed to Box 6248, Providence, RI 02940-6248.

Memoirs of the American Mathematical Society is published bimonthly (each volume consisting usually of more than one number) by the American Mathematical Society at 201 Charles Street, Providence, RI 02904-2213. Periodicals postage paid at Providence, Rhode Island. Postmaster: Send address changes to Memoirs, American Mathematical Society, P. O. Box 6248, Providence, RI 02940-6248.

Contents

ABSTRACT. By the doubling method, we have constructed a global integral of Rankin-Selberg type, which has been proved to represent the degree 16 standard L-function of $GSp(2) \times GSp(2)$, where $GSp(2)$ is the rank two group of symplectic similitudes. After our determination of the location and the degree of the possible poles of a family of Eisenstein series, which is involved in the global convolution, and our establishing of two first term identities in the sense of Kudla-Rallis, the analytic properties of the global integral are completely determined. The local theory of Rankin-Selberg convolution is also developed, but over the archimedean field, the local theory has not been completed as yet. This part together with the applications of the L-function to automorphic representation theory and number theory will be included in our future works.

CHAPTER 1

Introduction

0.1. The Notion of Automorphic L-functions. The notion of automorphic
L-functions, which was introduced by R. Langlands in 1960's, generalizes that of zeta
functions of global fields (number fields or function fields) and Hecke L-functions associated to classical modular forms. By conjectured Reciprocity Laws, L-functions
attached to Galois representations, algebraic varieties, (co)homology groups, motives, and automorphic forms should become special cases of automorphic L-functions.
Let us recall the notion of automorphic L-functions, which can be found in [**Lan3**],
[**Bor1**],[**GeSh**], and [**ArGe**].

For the sake of simplicity, we assume that F is a number field and G is a connected
reductive algebraic group splitting over F. As usual, we let $G_{\mathbb{A}}$ be the adele group of
G and ^LG the L-group associated to G. We can assume that the L-group ^LG is the
complex dual group of G since we only consider the groups splitting over F.

Given an irreducible automorphic cuspidal representation $\pi = \otimes_v \pi_v$ of $G_{\mathbb{A}}$ and a
finite dimensional representation r of the dual group ^LG, it is known that for almost
every finite place v, π_v is unramified. When π_v is unramified, it determines via the
Satake isomorphism a semi-simple conjugacy class τ_v in the local group ^LG_v and the
local Langlands L-factor is defined by

$$L(s, \pi_v, r_v) = [det(I - r_v(t_v)q_v^{-s})]^{-1}$$

where q_v is the cardinality of the residue class field of F_v and $r_v = r \circ \eta_v$ is the
corresponding representation of ^LG_v, the L-group of G as a group over F_v, obtained
by composing r with the natural homomorphism $\eta_v : {}^LG_v \to {}^LG$.

Given the data (G, π, r), there exists a global L-function introduced by Langlands
as follows, which is as usual called the automorphic L-function attached to the data
(G, π, r),

$$L^S(s, \pi, r) = \prod_{v \notin S} L(s, \pi_v, r_v)$$

where S is such a finite set of places of F that if a place v of F is not in S, then
v is finite and π_v is unramified. Langlands proved that the automorphic L-function
$L^S(s, \pi, r)$ converges absolutely for $Re(s)$ large and made the following conjecture on
the analytic properties of the L-function.

1

CONJECTURE 0.1.1 (Langlands). *Let (G, π, r) be the data defined above. Then the Euler product $L^S(s, \pi, r)$ continues to a meromorphic function on the whole complex plane which has only finitely many poles and satisfies a functional equation relating its value at s to its value at $1 - s$.*

Toward the proof of the Langlands conjecture as stated above, two major methods have been developed in recent years: the Langlands-Shahidi method and the Rankin-Selberg method. Both of these two methods give the proof of the Langlands conjecture in many cases, which are listed in Shahidi's list [Shh1] and Rallis' list [Ral4], respectively. It should be mentioned that the Rankin-Selberg method can potentially prove more analytic properties of L-functions than that predicted by the Langlands conjecture as stated above. For example, the precise location and degree of the poles of L-functions may be determined by Rankin-Selberg method.

From the representation theoretical point of view, the automorphic L-function $L^S(s, \pi, r)$ is one of the transcendental invariants of the automorphic representations π. One could understand the significance of the L-function $L^S(s, \pi, r)$ in the following special version of Langlands' principle of functoriality.

CONJECTURE 0.1.2 (Langlands). *Let G and G' be two reductive groups splitting over F, and $^L G$ and $^L G'$ the corresponding complex dual groups. Let ρ be any L-homomorphism from $^L G$ to $^L G'$. Then for any automorphic representation π of $G_{\mathbb{A}}$, there is an automorphic representation π' of $G'_{\mathbb{A}}$ such that for almost every finite place v, the map ρ takes the Satake parameter of π_v to the Satake parameter of π'_v. Moreover,*

$$L^S(s, \pi, r \circ \rho) = L^S(s, \pi', r)$$

for any finite dimension representation r of $^L G'$, where S is a finite set of places such that if a place v is not in S, v is finite and is a place where both π_v and π'_v are unramified.

The general version of Langlands' principle of functoriality may be found in many places, such as [Art3], [Bor], [GeSh], and [ArGe].

One natural question is: How much of the representation π can be determined by the analytic properties of the associated L-function $L^S(s, \pi, r)$? There are many interesting and important problems related to this question.

For example, when $G = GL(n)$, there is the famous Strong Multiplicity One Theorem established by I. Piatetski-Shapiro and J. Shalika, respectively, which says that if two irreducible cuspidal automorphic representations π_1 and π_2 of $GL(n, \mathbb{A})$ have the same L-functions, then $\pi_1 \cong \pi_2$ as automorphic representations of $GL(n, \mathbb{A})$. On the other hand, when G is not of $GL(n)$ type, following the work of J. Labesse and Langlands (on $SL(2)$) [LaLa], the phenomenon of L-indistinguishability may occur. In other words, there may exist two irreducible cuspidal automorphic representations π_1 and π_2, such that π_1 and π_2 have the same L-functions (i.e. $\pi_{1,v}$ is equivalent to

$\pi_{2,v}$ for almost every place v), but π_1 is not equivalent to π_2 globally. This leads to the notion of L-packets of automorphic representations, which gives a partition of the set of equivalence classes of irreducible automorphic representations of $G_{\mathbb{A}}$. One may notice that the Langlands' principle of functoriality gives a lifting from the set of L-packets of one group to the set of L-packets of the other group.

Further explanation about the study of automorphic L-functions can be found in S. Gelbart and F. Shahidi's book [**GeSh**].

0.2. Theta Correspondences. Besides automorphic L-functions, theta series is also one of the three major tools to study automorphic forms and representations. (The other major tool is the trace formula.)

In his fundamental paper [**Wei1**], A. Weil gave a representation theoretical interpretation for the classical Jacobi theta functions. Based on Weil's work, one can view the theta series as certain types of automorphic functions on a metaplectic group. More importantly, theta series gives rise to a typical map called Theta Lifting, which takes representations of one group to representations of the other group, if these two groups form a dual reductive pair in the sense of R. Howe [**How**]. We are going to describe more precisely the theta lifting associated to the dual reductive pair (Sp, O).

Let $(V, (,))$ be a m-dimensional non-degenerate quadratic vector space over the field F with Witt index $r = \frac{m}{2}$ and $H = O(V) = O(r,r)$ the group of all isometries of $(V, (,))$. Let $(W, <,>)$ be a $2n$-dimensional non-degenerate symplectic vector space over F and $G = Sp(n) = Sp(W)$. Then (G, H) forms a dual reductive pair in $Sp(V \otimes W)$ in the sense of Howe. Let $G(\mathbb{A})$ (resp. $H(\mathbb{A})$) denote the adele group of G (resp. H). For a fixed non-trivial additive character $\psi : \mathbb{A}/F \longrightarrow \mathbb{C}^{\times}$, let $\omega = \omega_{\psi}$ denote the Weil representation of the group $G(\mathbb{A}) \times \mathbb{H}(\mathbb{A})$ on the space $\mathcal{S}(V(\mathbb{A})^{\varkappa})$ of Schwartz-Bruhat functions on $V(\mathbb{A})^{\varkappa}$. For $\varphi \in \mathcal{S}(V(\mathbb{A})^{\varkappa})$, $g \in G(\mathbb{A})$ and $h \in H(\mathbb{A})$, define the theta function

$$\theta(g, h; \varphi) = \sum_{x \in V(F)^n} \omega(g)\varphi(h^{-1}x) = \sum_{w \in W(F)^r} \omega(h)\hat{\varphi}(wg),$$

where $\hat{\varphi} \in \mathcal{S}(W(\mathbb{A})^{\backprime})$, which is another model of the Weil representation ω obtained by taking partial Fourier transformation from the model $\mathcal{S}(\mathcal{V}(\mathbb{A})^{\varkappa})$ of ω.

Since this theta function is moderately increasing when restricted to a Siegel domain with respect to either the variable g or the variable h, one can integrate it against any cusp form on $G(\mathbb{A})$ or $H(\mathbb{A})$. That is, one has

$$\theta(\varphi, \phi)(g) := \int_{H(F)\backslash H(\mathbb{A})} \theta(g, h; \varphi)\phi(h)dh \qquad (1)$$

for any cusp form ϕ on $H(\mathbb{A})$ and

$$\theta(\varphi, \phi')(h) := \int_{G(F)\backslash G(\mathbb{A})} \theta(g, h; \varphi)\phi'(g)dg \qquad (2)$$

for any cusp form ϕ' on $G(\mathbb{A})$, where dh, dg is an invariant measure on $H(F)\backslash H(\mathbb{A})$, on $G(F)\backslash G(\mathbb{A})$, respectively, normalized to have total volume one. It is easy to see that both of the integrals are well defined and $\theta(\varphi,\phi)(g)$ and $\theta(\varphi,\phi')(h)$ are automorphic functions on $H(F)\backslash H(\mathbb{A})$, on $G(F)\backslash G(\mathbb{A})$, respectively.

For $H = O(V) = O(r,r)$, we define

$$\theta_r(\pi) := \{\theta(\varphi,f) \ : \ \varphi \in \mathcal{S}(\mathcal{V}(\mathbb{A})^{\varkappa}) \text{ and } \mho \in \pi\} \tag{3}$$

for an irreducible cuspidal automorphic representation π of $G(\mathbb{A})$. As usual, $\theta_r(\pi)$ is called the theta lifting of an irreducible cuspidal automorphic representation π from the group $G(\mathbb{A})$ to a representation of the group $H(\mathbb{A})$ (when G and H form a dual reductive pair). Of course, we can define the theta lifting from the group H to the group G by means of the other theta integral defined as above.

It is well known that under the assumption of local Howe duality principle for the dual reductive pair (G, H) for all places v of the field F, if $\theta_r(\pi)$ is nonzero, then $\theta_r(\pi)$ is irreducible. For example, see §7 in [**KuRa2**] or [**Gel3**] for the detailed discussion about the irreducibility of $\theta_r(\pi)$ and the relation between the theta correspondences and the Howe correspondences in general. The following are two fundamental problems in the study of theta correspondences in general:

(a) When are these theta liftings nonzero?
(b) If they are nonzero, are they cuspidal?

A beautiful description of the cuspidality of the theta liftings is given by Rallis' theory of 'tower of theta liftings' [**Ral1**] and the nonvanishing of the theta liftings is detected by the existence of poles or nonvanishing of the associated L-functions at certain point, following the work of J.-L. Waldspurger [**Wald**], of I. Piatetski-Shapiro [**PS1**], and more recently of S. Kudla and S. Rallis [**KuRa2**].

According to Rallis [**Ral1**], one may consider the theta liftings associated to a tower of dual reductive pairs as follows:

$$
\begin{array}{c}
O(2n, 2n) \\
\vdots \\
\nearrow \quad O(r+1, r+1) \\
Sp(n) \quad \rightarrow \quad O(r, r) \\
\searrow \quad O(r-1, r-1) \\
\vdots \\
O(2, 2)
\end{array}
$$

Then a cusp form π of $Sp(n, \mathbb{A})$ may have nonzero theta lifting $\theta_k(\pi)$ to the space of slowly increasing automorphic functions on $O(k, k; \mathbb{A})$ for different level k in the tower. More precisely, Rallis' theorem says that $\theta_k(\pi)$ can be cuspidal only for one level k in the tower and $\theta_k(\pi)$ can be cuspidal if and only if $\theta_k(\pi) \neq 0$ but $\theta_r(\pi) = 0$ for all possible level $r < k$ in the tower. Note that by an obvious reason in this case

$\theta_1(\pi) = 0$ and $\theta_k(\pi)$ for $k > 2n$ must not be cuspidal by the nonsingularity of cuspidal representations [**Ral1**] and [**Li1**].

Of course, one can fix the orthogonal group and let the symplectic group vary, and then obtain another type of the tower of dual reductive pairs. The similar phenomenon occurs. The precise description of such general phenomenon is Rallis' theory of 'tower of theta liftings'. See [**ArGe**], [**Ral1**], [**KuRa2**], and [**KRS**] for more information on this aspect.

0.3. Motivations. The automorphic L-function we are concerned with is the degree 16 standard L-function of $GSp(2) \times GSp(2)$, where $GSp(2)$ is the reductive group of symplectic similitudes of rank 2. More precisely, the L-function is defined as follows: Let π_1 and π_2 be irreducible automorphic cuspidal representations of $GSp(2, \mathbb{A})$ with trivial central character and ρ_1 the standard representation of $GSp(2, \mathbb{C})$, the complex dual group of $GSp(2)$ [**Bor1**]. The degree 16 standard L-function is $L^S(s, \pi_1 \otimes \pi_2, \rho_1 \otimes \rho_1)$.

There are two fundamental reasons for us to study this degree 16 standard L-function $L^S(s, \pi_1 \otimes \pi_2, \rho_1 \otimes \rho_1)$:

(1) The L-function $L^S(s, \pi_1 \otimes \pi_2, \rho_1 \otimes \rho_1)$ has certain distinguished analytic properties, which can be predicted by means of Langlands principle of functoriality and Rallis' theory of 'tower of theta liftings'.

(2) A global integral of Rankin-Selberg type, which potentially represents the L-function $L^S(s, \pi_1 \otimes \pi_2, \rho_1 \otimes \rho_1)$, can be motivated via Kudla's seesaw identity for dual reductive pairs of similitude groups and Jacquet, Piatetski-Shapiro, and Shalika's theory of Rankin-Selberg convolution of $GL(n)$.

We shall explain (1) and (2) more precisely as follows:
Let us consider following two towers of dual reductive pairs:

$$
\begin{array}{ccc}
& \nearrow\ GO(4,4) & \\
GSp(2) & \rightarrow\ GO(3,3) \quad ; \\
& \searrow\ GO(2,2) &
\end{array}
\qquad
\begin{array}{ccc}
& \nearrow\ GSp(4) & \\
GO(2,2) & \rightarrow\ GSp(3) & \\
& \searrow\ GSp(2) & \\
& \ \ \ GSp(1) &
\end{array}
\ .
$$

According to Rallis' theory of 'tower of theta liftings', generic square integrable cusp forms on $GSp(2, \mathbb{A})$ may have a nonzero theta lifting to $L^2_{cusp}(O(2,2,\mathbb{A}))$ or to $L^2_{cusp}(O(3,3,\mathbb{A}))$, and generic square integrable cusp forms on $O(2,2,\mathbb{A})$ may have nonzero theta lifting to $L^2_{cusp}(Sp(2,\mathbb{A}))$. In particular, we obtain a chain of the theta liftings for generic cusp forms:

$$GO(2,2) \rightarrow GSp(2) \rightarrow GO(3,3). \tag{4}$$

Since the central characters of the cuspidal representations under consideration are assumed to be trivial, the above theta correspondences can be viewed as the theta correspondences among the projective groups, i.e.

$$PGO(2,2) \to PGSp(2) \to PGO(3,3). \tag{5}$$

By the accidental isomorphisms of small rank classical groups, one can 'identity' $PGO(2,2)$ with $PGL(2) \times PGL(2)$, $PGSp(2)$ with $PGO(3,2)$, and $PGO(3,3)$ with $PGL(4)$.

If we consider their corresponding complex dual groups, we obtain the following commutative diagram with L-homomorphisms: $\sigma_2 = \sigma_1 \circ \sigma$ and

$$
\begin{array}{ccc}
 & \sigma & \\
Sp(2,\mathbb{C}) & \to & SL(4,\mathbb{C}), \\
\sigma_1 \searrow & & \nearrow \sigma_2 \\
 & SL(2,\mathbb{C}) \times SL(2,\mathbb{C}) &
\end{array}
$$

where $Sp(2,\mathbb{C}) = {}^L[PGSp(2)]$, $SL(4,\mathbb{C}) = {}^L[PGL(4)]$, and $SL(2,\mathbb{C}) \times SL(2,\mathbb{C}) = {}^L[PGL(2) \times PGL(2)]$.

According to the Langlands principle of functoriality, there should exist functorial liftings among the L-packets of these three groups. The natural question must be if the theta liftings obtained from the Rallis' towers are functorial.

According to an early work of Rallis [**Ral3**] and a recent work of J. Adams [**Adm**], theta liftings are in general not functorial. In other words, theta liftings may not map L-packets of one group to L-packets of the other group when these two groups form a dual reductive pair and their L-groups have an L-homomorphism from one to another. In general, one hopes that the theta correspondeces should preserve the Arthur packets of automorphic representations of the relevant groups in the situation as in the diagram above.

The notion of Arthur packets of automorphic representations was suggested by the study of trace formula and its application to automorphic forms [**Art1**], [**Art2**]. Arthur packets should give another type of partition in the set of equivalence classes of irreducible unitary automorphic representations of $G(\mathbb{A})$ parameterized by Arthur parameters. Arthur packets are essentially different from L-packets. Only in the case where we consider tempered representations, the Arthur packets will coincide with L-packets. According to a series of conjectures of J. Arthur, suggested by the study of trace formula, Arthur packets should be the right object for study when we consider general square integrable automorphic representations.

Another important notion also suggested by the study of trace formula is that of endoscopic groups, the precise definition of which can be found for example in [**LaSh**], [**Art1**], and [**Art2**]. In general, to a reductive group G, there are families of endoscopic groups H associated. From the structure of groups, the group G and its endoscopic groups H are closely related in their dual group sides. For example, when $G = Sp(2n)$, its complex dual group is ${}^L G = SO(2n + 1, \mathbb{C})$. Let ${}^L H_k := SO(2n - 2k + 1, \mathbb{C}) \times SO(2k, \mathbb{C})$. Then $H_k = Sp(2n - 2k) \times SO(2k)$ is one of the families of the endoscopic groups of G.

Back to our special case above, one notice that the group $PGL(2) \times PGL(2)$ is the only nontrivial cuspidal (or elliptic) endoscopic group of $PGSp(2)$ [**Art1**] and also a noncuspidal endoscopic group of $PGL(4)$. Although it is unknown if the above theta liftings coincide with the functorial liftings and if the principle of functoriality holds for this example, these conjectured results can be used to foretell the analytic properties of the L-function $L^S(s, \pi_1 \otimes \pi_2, \rho_1 \otimes \rho_1)$.

Under the assumption of Langlands principle of functoriality for the examples we are concerned with, for a fixed Arthur parameter ψ of the endoscopic group $PGL(2) \times PGL(2)$, which gives rise to an Arthur parameter ψ for $PGSp(2)$ and an Arthur parameter ψ for $PGL(4)$, there should be the following commutative diagram among the corresponding Arthur packets:

$$
\begin{array}{ccc}
\Pi_\psi(PGSp(2)) & \overset{\sigma}{\longrightarrow} & \Pi_\psi(PGL(4)) \\
{\scriptstyle\sigma_1}\searrow & & \nearrow{\scriptstyle\sigma_2} \\
& \Pi_\psi(PGL(2) \times PGL(2)) &
\end{array}
,
$$

where the functorial lifting σ_2 from $\Pi_\psi(PGL(2) \times PGL(2))$ to $\Pi_\psi(PGL(4))$ should be the automorphic induction following [**Lan4**], [**JPSS**], and [**ArGe**], and the map σ_1 from $\Pi_\psi(PGL(2) \times PGL(2))$ to $\Pi_\psi(PGSp(2))$ is the endoscopic lifting. This means $\sigma \circ \sigma_1 = \sigma_2$.

According to the above arguments, the analytic properties of the L-function $L^S(s, \pi_1 \otimes \pi_2, \rho_1 \otimes \rho_1)$ for $Re(s) > 0$ should be determined case by case as follows:

(1) *If none of π_1 and π_2 is an endoscopic lifting via σ_1, then*

$$
L^S(s, \pi_1 \otimes \pi_2, (\rho_1 \otimes \rho_1) \circ \sigma) = L^S(s, \sigma(\pi_1) \otimes \sigma(\pi_2), \rho_1 \otimes \rho_1)
$$

is holomorphic for all s except at $s = 1$ where the L-function $L^S(s, \pi_1 \otimes \pi_2, \rho_1 \otimes \rho_1)$ has a simple pole if and only if $\sigma(\pi_2) = \sigma(\pi_1)^\vee$, the contragredient representation of $\sigma(\pi_1)$.

(2) *If only one of π_1 and π_2 is an endoscopic lifting via σ_1, then the L-function $L^S(s, \pi_1 \otimes \pi_2, \rho_1 \otimes \rho_1)$ is holomorphic for all s.*

In fact, if, say, $\pi_1 = \sigma_1(\pi_1^{(1)} \otimes \pi_2^{(1)})$, an endoscopic lifting via σ_1, then one has

$$
\sigma(\pi_1) = \sigma_2(\pi_1^{(1)} \otimes \pi_2^{(1)}) = \pi_1^{(1)} \oplus \pi_2^{(1)} \text{ (automorphic induction)}.
$$

Thus the L-function $L^S(s, \pi_1 \otimes \pi_2, (\rho_1 \otimes \rho_1) \circ \sigma)$ has following properties:

$$L^S(s, \pi_1 \otimes \pi_2, (\rho_1 \otimes \rho_1) \circ \sigma)$$
$$= L^S(s, \sigma(\pi_1) \otimes \sigma(\pi_2), \rho_1 \otimes \rho_1)$$
$$= L^S(s, (\pi_1^{(1)} \oplus \pi_2^{(1)}) \otimes \sigma(\pi_2), \rho_1 \otimes \rho_1)$$
$$= L^S(s, \pi_1^{(1)} \otimes \sigma(\pi_2), \rho_1 \otimes (\rho_1 \circ \sigma_2)) \cdot L^S(s, \pi_2^{(1)} \otimes \sigma(\pi_2), \rho_1 \otimes (\rho_1 \circ \sigma_2)).$$

Since $L^S(s, \pi_1^{(1)} \otimes \sigma(\pi_2), \rho_1 \otimes (\rho_1 \circ \sigma_2))$ and $L^S(s, \pi_2^{(1)} \otimes \sigma(\pi_2), \rho_1 \otimes (\rho_1 \circ \sigma_2))$ are L-functions of $PGL(2) \times PGL(4)$, they are holomorphic for all s. Thus the product of these two L-functions is holomorphic for all s.

(3) *If both of π_1 and π_2 are endoscopic liftings by means of σ_1, then the L-function $L^S(s, \pi_1 \otimes \pi_2, \rho_1 \otimes \rho_1)$ is holomorphic for all s except for $s = 1$ where the L-function may achieve a pole of degree at most two, according to the following discussion:*

Assume that $\pi_1 = \sigma_1(\pi_1^{(1)} \otimes \pi_2^{(1)})$ and $\pi_2 = \sigma_1(\pi_1^{(2)} \otimes \pi_2^{(2)})$ are endoscopic liftings via σ_1, then one has

$$L^S(s, \pi_1 \otimes \pi_2, (\rho_1 \otimes \rho_1) \circ \sigma)$$
$$= L^S(s, \sigma(\pi_1) \otimes \sigma(\pi_2))$$
$$= L^S(s, (\pi_1^{(1)} \oplus \pi_2^{(1)}) \otimes (\pi_1^{(2)} \oplus \pi_2^{(2)}))$$
$$= L^S(s, \pi_1^{(1)} \otimes \pi_1^{(2)}) \cdot L^S(s, \pi_1^{(1)} \otimes \pi_2^{(2)}) \cdot L^S(s, \pi_2^{(1)} \otimes \pi_1^{(2)}) \cdot L^S(s, \pi_2^{(1)} \otimes \pi_2^{(2)}).$$

Each of the L-functions $L^S(s, \pi_i^{(1)} \otimes \pi_j^{(2)})$ is a standard L-function of $PGL(2) \times PGL(2)$, which is holomorphic except for $s = 1$ where the L-function $L^S(s, \pi_i^{(1)} \otimes \pi_j^{(2)})$ has a simple pole if and only if $\pi_i^{(1)}$ is the contragredient representation of $\pi_j^{(2)}$). Note that automorphic cuspidal representations of $PGL(2)$ are self-dual. Thus one has following cases:

(3a) If $L^S(s, \pi_1 \otimes \pi_2, \rho_1 \otimes \rho_1)$ has a pole at $s = 1$ of fourth degree, then each of $L^S(s, \pi_i^{(1)} \otimes \pi_j^{(2)})$ has a simple pole at $s = 1$. Thus all these four representations are equivalent to each other. This implies that $\pi_1 = \pi_2 = \sigma_1(\pi \otimes \pi)$ for a cusp form π of $PGL(2)$. According to Rallis' theory of 'tower of theta liftings', the first occurrence of the theta lifting of the automorphic cuspidal representation $\pi \otimes \pi$ of $PGO(2, 2)$ $(PGL(2) \times PGL(2))$ is on the group $PGSp(1)$. Therefore $\sigma_1(\pi \otimes \pi)$ is no longer a cusp form on $PGSp(2)$. It follows that the degree 16 standard L-function $L^S(s, \pi_1 \otimes \pi_2, \rho_1 \otimes \rho_1)$ of $GSp(2) \times GSp(2)$ can not have a pole at $s = 1$ of degree greater than three. Note that the L-function $L^S(s, \pi_1 \otimes \pi_2, \rho_1 \otimes \rho_1)$ can not have a pole at $s = 1$ of degree three following the same argument.

(3b) If $L^S(s, \pi_1 \otimes \pi_2, \rho_1 \otimes \rho_1)$ has a pole at $s = 1$ of degree two, the only case that both $\pi_1 = \sigma_1(\pi_1^{(1)} \otimes \pi_2^{(1)})$ and $\pi_2 = \sigma_1(\pi_1^{(2)} \otimes \pi_2^{(2)})$ are cusp forms on $PGSp(2)$

is: $\pi_1^{(1)} = \pi_1^{(2)}$ and $\pi_2^{(1)} = \pi_2^{(2)}$ with $\pi_1^{(1)} \neq \pi_2^{(1)}$. This implies that π_2 is the contragredient representation of π_1 and $\pi_1 = \sigma_1(\pi_1^{(1)} \otimes \pi_2^{(1)})$ is a cusp form in the image of theta lifting from $PGO(2,2)$ to $PGSp(2)$.

(3c) If $L^S(s, \pi_1 \otimes \pi_2, \rho_1 \otimes \rho_1)$ has only a simple pole at $s = 1$, then one has $\pi_1^{(1)} \neq \pi_2^{(1)} \neq \pi_2^{(2)} \neq \pi_1^{(1)}$ and $\pi_1^{(1)} = \pi_1^{(2)}$. It follows that $\pi_1 = \sigma_1(\pi_1^{(1)} \otimes \pi_2^{(1)})$ and $\pi_2 = \sigma_1(\pi_1^{(1)} \otimes \pi_2^{(2)})$ are cusp forms on $PGSp(2)$, which are not contragredient to each other.

Based on the above analysis under the assumption of the functorial liftings and the endoscopic liftings for our example, we *might* make the following **conclusions:**

(*) *For a generic cusp form π on $GSp(2, \mathbb{A})$ with trivial central character, the automorphic L-function $L^S(s, \pi \otimes \pi^\vee, \rho_1 \otimes \rho_1)$ is holomorphic for all s except for $s = 1$ where the L-function $L^S(s, \pi \otimes \pi^\vee, \rho_1 \otimes \rho_1)$ has a pole of degree at most two.*

(**) *For a generic cusp form π on $GSp(2, \mathbb{A})$ with trivial central character, the automorphic L-function $L^S(s, \pi \otimes \pi^\vee, \rho_1 \otimes \rho_1)$ achieves the second degree pole at $s = 1$ if and only if the generic cusp form π is a nonzero endoscopic lifting of a generic cusp form on $PGL(2, \mathbb{A}) \times PGL(2, \mathbb{A})$.*

In order to prove these delicate properties of the L-function, one may either try to prove the functoriality for this example, or try to study the L-function directly, say, by Rankin-Selberg method. Note that the automorphic L-function under consideration is not on Shahidi's list.

A Rankin-Selberg integral for the L-function can be constructed by the following observation of Kudla's seesaw identity and Jacquet, Piatetski-Shapiro, and Shalika' theory of Rankin-Selberg convolution of $GL(n)$: The L-function $L^S(s, \pi_1 \otimes \pi_2, \rho_1 \otimes \rho_1)$ should look like the standard L-function of $GL(4) \times GL(4)$, which is represented by the following global integral integral by Jacquet and Shalika [**JaSh**]

$$\int_{GL(4,F) \backslash GL(4,\mathbb{A})} \phi_1(g)\phi_2(g) E_{3,1}^4(g, s) dg \tag{6}$$

where $E_{3,1}^4(g, s)$ is an Eisenstein series of $GL(4)$ attached to a section in the degenerate principal series representation of $GL(4)$ induced from the standard maximal parabolic (mirabolic) subgroup $P_{3,1}^4$ with its Levi part isomorphic to $GL(3) \times GL(1)$. The point is that one can view this integral as an integral over the group $GO(3,3)$, that is,

$$\int_{GO(3,3;F) \backslash GO(3,3;\mathbb{A})} \phi_1(g)\phi_2(g) E_3^3(g, s) dg \tag{7}$$

where $E_3^3(g, s)$ is an Eisenstein series of $GO(3,3)$ of Siegel type. This later integral can be viewed as one side of a seesaw identity of the following dual reductive pairs of similitude groups,

$$
\begin{array}{ccc}
GSp(4) & & [GO(3,3) \times GO(3,3)]^\circ \\
\uparrow & \times & \uparrow \\
[GSp(2) \times GSp(2)]^\circ & & GO(3,3)
\end{array} \ .
$$

Following the seesaw identity in the sense of Kudla [**Kud**], one may guess the other side of the seesaw identity, that is, if we set $H := [GSp(2) \times GSp(2)]^\circ$,

$$
\int_{C(\mathbb{A})H(F)\backslash H(\mathbb{A})} \theta(\phi_1)(g_1)\theta(\phi_2)(g_2)\mathcal{E}_3^4((g_1,g_2),s)dg_1dg_2
$$
$$
= \int_{C(\mathbb{A})GO(3,3;F)\backslash GO(3,3;\mathbb{A})} \phi_1(g)\phi_2(g)E_3^3(g,s)dg
$$
$$
= \int_{C(\mathbb{A})GL(4,F)\backslash GL(4,\mathbb{A})} \phi_1(g)\phi_2(g)E_{3,1}^4(g,s)dg
$$

where $\mathcal{E}_3^4((g_1,g_2),s)$ is the theta lifting of the Eisenstein series $E_3^3(g,s)$ from $GO(3,3)$ to $GSp(4)$, given by Kudla-Rallis' regularized theta integral, and is an Eisenstein series on $GSp(4)$ associated to a special section in the degenerate principal series of $GSp(4,\mathbb{A})$ induced from the standard maximal parabolic subgroup P_3^4 with its Levi factor isomorphic to $GL(3) \times GSp(1)$. According to Jacquet and Shalika [**JaSh**], the general version of the last integral represents the degree 16 standard L-function of $GL(4) \times GL(4)$. One naturally hope that the general version of the first integral should represent the degree 16 standard L-function of $GSp(2) \times GSp(2)$.

0.4. Summary. Let $G := GSp(4)$. Then $H := [GSp(2) \times GSp(2)]^\circ$ can be embedded into G via the doubling method. let π_1 and π_2 be irreducible automorphic cuspidal representations of $GSp(2,\mathbb{A})$ with trivial central characters in the sense of [**BoJa**]. Let $E_3^4(g,s;f_s)$ be an Eisenstein series associated to a section f_s in the degenerate principal series representation $I_3^4(s)$ of $G(\mathbb{A})$ induced from a standard maximal parabolic subgroup P_3^4 with its Levi part isomorphic to $GL(3) \times GSp(1)$. For any cusp forms $\phi_1 \in \pi_1$, $\phi_2 \in \pi_2$, we set our Rankin-Selberg global convolution (integral) as follows:

$$
\mathcal{Z}(s,\phi_1,\phi_2,f_s) = \int_{C(\mathbb{A})H(F)\backslash H(\mathbb{A})} E_3^4((g_1,g_2),s;f_s)\phi_1(g_1)\phi_2(g_2)dg_1dg_2 \tag{8}
$$

where $C = C_4$ is the center of $GSp(4)$.

The first main result proved here is the following **basic identity:**

$$
\mathcal{Z}(s,\phi_1,\phi_2,f) = \frac{L^S(s',\pi_1 \otimes \pi_2, \rho_1 \otimes \rho_1)}{d_G^S(s)} \cdot \prod_{v \in S} \mathcal{Z}_v(s, W_{\phi_{1v}}^{\psi_v}, W_{\phi_{2v}}^{\overline{\psi_v}}, f_v) \tag{9}
$$

where $L^S(s',\pi_1 \otimes \pi_2, \rho_1 \otimes \rho_1)$ is the Langlands L-function for the standard tensor product representation $\rho_1 \otimes \rho_1$ of $GSp(2) \times GSp(2)$, which is of degree 16, $s' = \frac{s+1}{2}$, $d_G^S(s)$ is the normalizing factor of our Eisenstein series, and $\mathcal{Z}_v(s, W_1, W_2, f_s)$ for $v \in S$

are ramified local integrals. Note that those integrals (both local and global) require that the representations π_1 and π_2 have nonzero Whittaker models. We assume from now on that the representations π_1 and π_2 are *generic*.

This basic identity is the starting point of my Ph.D. dissertation research project under the guidance of Professor Stephen Rallis. We predict that the degree 16 Langlands L-function $L^S(s', \pi_1 \otimes \pi_2, \rho_1 \otimes \rho_1)$ should enjoy following properties:

(1) (**Langlands Conjecture**). $L^S(s', \pi_1 \otimes \pi_2, \rho_1 \otimes \rho_1)$ converges absolutely for $Re(s)$ large and continues to a meromorphic function on the whole complex plane which has only finitely many poles and satisfies certain functional equation relating its value at s to its value at $1 - s$.

(2) $L^S(s', \pi_1 \otimes \pi_2, \rho_1 \otimes \rho_1)$ is holomorphic for $Re(s) > 1$ and has an at most double pole at $s = 1$. If $L^S(s', \pi_1 \otimes \pi_2, \rho_1 \otimes \rho_1)$ achieves the double pole at $s = 1$, then the irreducible automorphic representation π_2 should be equal to the contragredient representation $\pi_1^\vee$ of π_1.

(3) For a generic cusp form π over $GSp(2, \mathbb{A})$ with a trivial central character, the L-function $L^S(s', \pi \otimes \pi^\vee, \rho_1 \otimes \rho_1)$ achieves the double pole at $s = 1$ if and only if the image $\theta_2(\pi)$ under the theta lifting for the dual reductive pair $(GSp(2), GO(2, 2))$ is a nonzero cusp form over $GO(2, 2; \mathbb{A})$. (We believe that this theta lifting is functorial and so become an endoscopic lifting.)

The significance of these properties of the L-function $L^S(s', \pi_1 \otimes \pi_2, \rho_1 \otimes \rho_1)$ is clear. For instance, (2) is an analogue of Jacquet and Shalika's results for $GL(n)$ in [**JaSh**] and implies important applications in the theory of automorphic forms; (3) is an analogue of Waldspurger's description of Shimura liftings [**Wald**]and is predicted by Rallis' theory of 'towers of theta liftings' [**Ral1**]. Both (2) and (3) will imply that the degree 10 Langlands L-function $L^S(s', \pi, Ad)$ is holomorphic and nonvanishing at $s = 1$, following the simple observation: $\rho_1 \otimes \rho_1 = (Sym^2\rho_1) \oplus (\wedge^2\rho_1)$.

Based on this fundamental identity, our study of the degree 16 standard L-function $L^S(s', \pi_1 \otimes \pi_2, \rho_1 \otimes \rho_1)$ will break into the following four basic steps:

Step (I): Determine the poles of the Eisenstein series $E_3^4(g, s; f_s)$ and understand the residue representation of the Eisenstein series $E_3^4(g, s; f_s)$ at the poles.

Step (II): Determine the analytic properties of the global integral $\mathcal{Z}(s, \phi_1, \phi_2, f)$.

Step (III): Develop the local theory of Rankin-Selberg convolutions for the local zeta integral $\mathcal{Z}_v(s, W_1, W_2, f_s)$ so that the analytic properties (the location and possible degree of the poles) of the partial L-function $L^S(s', \pi_1 \otimes \pi_2, \rho_1 \otimes \rho_1)$ will follow from that of the global zeta integral $\mathcal{Z}(s, \phi_1, \phi_2, f)$.

Step (IV): Find the boundary of the absolute convergence of the Euler product $L^S(s', \pi_1 \otimes \pi_2, \rho_1 \otimes \rho_1)$ by refining the results on the nonarchimedean local zeta integral $\mathcal{Z}_v(s, W_1, W_2, f_s)$ and analyzing the local L-factor $L_v(s', \pi_{1,v} \otimes \pi_{2,v}, \rho_1 \otimes \rho_1)$.

The main results we have accomplished in this paper are the following four Theorems:

THEOREM 0.4.1 (Fundamental Identity). *Let π_1, π_2 be irreducible automorphic cuspidal generic representations of $GSp(2, \mathbb{A})$ with trivial central characters. Let $\phi_1 \in \pi_1$, $\phi_2 \in \pi_2$, and $f_s \in I_3^4(s)$, which are factorizable. Then we have*

$$\mathcal{Z}(s, \phi_1, \phi_2, f) = \frac{L^S(s', \pi_1 \otimes \pi_2, \rho_1 \otimes \rho_1)}{d_G^S(s)} \cdot \prod_{v \in S} \mathcal{Z}_v(s, W_{\phi_{1v}}^{\psi_v}, W_{\phi_{2v}}^{\overline{\psi_v}}, f_v) \qquad (10)$$

where S is the finite set of places of F, including all infinite places of F and determined by the data (π_1, π_2, f_s), and $d_G^S(s)$ is the normalizing factor of the Eisenstein series $E_3^4(g, s; f_s)$ attached to the section f_s in $I_3^4(s)$.

Let $I_r^n(s)$ be the (global) degenerate principal series representation of $Sp(n, \mathbb{A})$ (the symplectic group of rank n) induced from the one-dimensional representation of the standard maximal parabolic subgroup P_r^n with its Levi factor isomorphic to $GL(r) \times Sp(n - r)$. The precise definition of the representation $I_r^n(s)$ can be found in Chapter III. As usual, to a section $f_s \in I_r^n(s)$, one can define an Eisenstein series as follows:

$$E_r^n(g, s; f_s) = \sum_{\gamma \in P_r^n \backslash Sp(n)} f_s(\gamma g)$$

THEOREM 0.4.2 (Poles of Eisenstein Series). *For $r = n - 1$, we have*

(1) *The Eisenstein series $E_{n-1}^n(g, s; f_s)$ is holomorphic for $Re(s) \geq 0$ except for $s \in X_n^+ = \{\frac{e(n)}{2}, \cdots, \frac{n-2}{2}, \frac{n}{2}, \frac{n+2}{2}\}$, where $e(n) = 1$ if n is odd and 2 if n is even.*
(2) *At $s = \frac{n}{2}, \frac{n+2}{2}$, $E_{n-1}^n(g, s; f_s)$ achieves a simple pole, and at $s = \frac{n-2}{2}$, $E_{n-1}^n(g, s; f_s)$ achieves a pole of degree two.*
(3) *At $s \in X_n^+ \setminus \{\frac{n-2}{2}, \frac{n}{2}, \frac{n+2}{2}\}$, the Eisenstein series $E_{n-1}^n(g, s; f_s)$ may achieves a pole of degree at most two.*

The next main result is about the residue representations of the Eisenstein series at its poles. At this moment, we only have the result for $n = 4$. Consider the following Laurent expansions of three Eisenstein series:

$$E_3^4(g, s; f_s) = \frac{\Lambda_{-2}^{4,3}(s_0; f_s)}{(s - s_0)^2} + \frac{\Lambda_{-1}^{4,3}(s_0; f_s)}{(s - s_0)} + \cdots .$$

$$E_4^4(g, s; f_s') = \frac{\Lambda_{-1}^{4,4}(s_0'; f_s')}{(s - s_0')} + \Lambda_0^{4,4}(s_0'; f_s') + \cdots .$$

$$E_1^4(g, s; f_s'') = \Lambda_0^{4,1}(s_0''; f_s'') + \Lambda_1^{4,1}(s_0''; f_s'')(s - s_0'') + \cdots .$$

THEOREM 0.4.3 (First Term Identities). *With the above notations, we have*

$$\Lambda_{-2}^{4,3}(1; f_s) = c_1 \cdot \Lambda_{-1}^{4,4}(\frac{1}{2}; f_s') \qquad (11)$$

$$\Lambda_{-1}^{4,3}(2; f_s) = c_2 \cdot \Lambda_0^{4,1}(1; f_s''). \qquad (12)$$

The first identity is a special case of Kudla-Rallis' first term identity, but here the identity is valid for general sections. The second identity is new, which does not involved in the Siegel Eisenstein series. By means of these two identities, the analytic properties of the global integral $\mathcal{Z}(s, \phi_1, \phi_2, f)$ for $Re(s) \geq 0$ can be completely determined.

The last Theorem we are going to state is about the local theory of Rankin-Selberg convolution.

THEOREM 0.4.4 (Local Convolution).

(1)(**nonarchimedean case***)*

(a) $\mathcal{Z}_v(s, W_{1v}, W_{2v}, f_{sv})$ *converges absolutely for* $Re(s)$ *large and continues a meromorphic function to the whole s-plane.*

(b) *One can pick up data* (W_{1v}, W_{2v}, f_{sv}), *so that* $\mathcal{Z}_v(s, W_{1v}, W_{2v}, f_{sv}) = 1$.

(2)(**archimedean case***)*

(a) $\mathcal{Z}_v(s, W_{1v}, W_{2v}, f_{sv})$ *converges absolutely for* $Re(s)$ *large with any smooth data* (W_{1v}, W_{2v}, f_{sv}).

(b) *For a given* s_0, *we can choose* W_{1v} *and* W_{2v} *to be K-finite, and* f_{sv} *to be smooth, so that* $\mathcal{Z}_v(s_0, W_{1v}, W_{2v}, f_{sv}) \neq 0$.

It should be mentioned that our results in the archimedean case are not complete for our study of the degree 16 standard L-function $L^S(s', \pi_1 \otimes \pi_2, \rho_1 \otimes \rho_1)$. We hope to finish this point in the future.

0.5. Content. Let us describe the content of this manuscript briefly.

In Chapter II, we shall establish (in section 2.3) the global integral $\mathcal{Z}(s, \phi_1, \phi_2, f_s)$ of Rankin-Selberg type via the doubling method. After the standard unfolding, $\mathcal{Z}(s, \phi_1, \phi_2, f_s)$ becomes a global integral against two Whittaker functions $W_{\phi_1}^{\psi}$ and $W_{\phi_2}^{\overline{\psi}}$, and the section f_s, which is eulerian by the uniqueness of Whittaker models [**Shl**]. The unramified computation consists of two parts:

(1) the computation of the local zeta integral with the unramified integrating data (section 2.3), and

(2) the computation of the predicted Langlands local L-factor (section 2.4).

Following Casselman and Shalika's formula for the unramified Whittaker functions, the computation of the unramified local integral $\mathcal{Z}_v(s, W_1^\circ, W_2^\circ, f_s^\circ)$ is reduced to that of an local integral $I(a_1, a_2; f_s^\circ, \psi)$ (section 2.3), which can be eventually reformulated as certain 'Partial Fourier Coefficient' or a special case of degenerate Jacquet integrals [**Wal**]. The most technical part of this Chapter is the computation of the Langlands local L-factor $L_v(s', \pi_{1v} \otimes \pi_{2v}, \rho_1 \otimes \rho_1)$ (section 2.4). The computation is based on the algorithm of Kostant and Rallis [**KoRa**], which gives an explicit spectral decomposition of the space of harmonic polynomial functions over a symmetric space, which is in our case $(Sp(4), Sp(2) \times Sp(2))$. For technical reasons, we also need the multiplicity one branch formula for $(Sp(2), Sp(1) \times Sp(1))$ and the Borel-Weil-Bott theorem about the realization of finite dimensional irreducible complex representations of a complex algebraic group on the space of holomorphic (or anti-holomorphic) sections of a certain line bundle.

Chapter III goes generally for $Sp(n)$, symplectic group of rank n. We are concerned with a family of non-Siegel type Eisenstein series $E_{n-1}^n(g, s; f_s)$. We shall determine the location and the order of possible poles of $E_{n-1}^n(g, s; f_s)$ for general holomorphic sections f_s in $I_{n-1}^n(s)$, the (global) degenerate principal series representation of $Sp(n)$ (Theorem 4.0.4), which is an analogue of the weak form of Kudla and Rallis' Theorem of this type [**KuRa**]. The ideas and the methods used here were initially developed by Piatetski-Shapiro and Rallis [**PSRa**] and [**PSRa1**], and Kudla and Rallis [**KuRa**] and [**KuRa1**], for the case of Eisenstein series of Siegel type. For our non-Siegel type Eisenstein series, the situation becomes more comprehensive and complicated. For example, when n is even, we have to prove that at $s = \frac{1}{2}$ the Eisenstein series $E_{n-1}^n(g, s; f_s)$ is actually holomorphic although $s = \frac{1}{2}$ is a pole of one of the intertwining operators involved in the expansion of the constant term. This is proved in a representation theoretic way. With an inductive formula (Theorem 4.2.3), the information about the poles of the constant term $E_{n-1,P_1^n}^n(g, s; f_s)$ of the Eisenstein series $E_{n-1}^n(g, s; f_s)$ along the maximal parabolic subgroup P_1^n will be given by that of Eisenstein series of lower ranked group $Sp(n-1)$ and that of the relevant intertwining operators. Following the inductive formula, the most technical part in our proof of Theorem 4.0.4 will be reduced to the special case of $n = 3$ and $s = \frac{1}{2}$ (section 4.4). For general n, we have proved that the Eisenstein series $E_{n-1}^n(g, s; f_s)$ always achieves a double pole at $s = \frac{n-2}{2}$ and a simple poles at $s = \frac{n}{2}$ and $s = \frac{n+2}{2}$. The existence of the double pole at $s = \frac{n-2}{2}$ of the Eisenstein series $E_{n-1}^n(g, s; f_s)$ was predicted by means of Kudla and Rallis' first term identity [**KuRa**]. It should be mentioned that the useful Lemma of Rallis also plays a critical role in our study of the poles of relevant intertwining operators.

Then in Chapter IV, we are going to prove two first term identities as stated above. In order to do that, we have to know specifically the irreducible quotient representations of the degenerate principal series representation $I_3^4(s)$ ($n = 4$). In fact, at any local place v, the degenerate principal series representation $I_{3,v}^4(s)$ has a unique (up to isomorphism) irreducible quotient representation. When the place v is finite, this result can be proved as in the author's Thesis [**Jia**] by computation of certain Hecke operators associated to the parahori subgroup related to the maximal parabolic subgroup P_3^4, or follows from Jantzen's recent work [**Jan1**]. However, when the place v is infinite, i.e. $F_v = \mathbb{R}$ since we assume that the underlying number field is totally real, the uniqueness of the irreducible quotient representation of $I_{3,\infty}^4(s)$ is proved by Bruhat theory of intertwining distributions and an argument on the existence of certain quasi-invariant distributions suggested by Prof. Rallis. This will be done in section 4.2.

In the last Chapter, the local theory of the Rankin-Selberg convolution, i.e. the local integral $\mathcal{Z}_v(s, W_1, W_2, f_s)$ is developed. Following the standard estimates of Whittaker functions on the splitting torus, which is an analogue of the estimates made in [**JPSS**], [**JaSh**], and [**Sou**], the technical part is the estimates of the local integral $I(a_1, a_2; f_s, \psi)$ for a general section f_s and a additive character ψ (subsection 3.1.3). In nonarchimedean cases, the theory is completed. In other words, we have proved that the local integral $\mathcal{Z}_v(s, W_1, W_2, f_s)$ converges absolutely for $Re(s)$ large and continues a meromorphic function on the complex plane, and after choosing an appropriate data (W_1, W_2, f_s), the integral $\mathcal{Z}_v(s, W_1, W_2, f_s)$ can be made to be one. In archimedean cases, however, the theory is not yet completed. We have proved:

(i) the local integral $\mathcal{Z}_v(s, W_1, W_2, f_s)$ converges absolutely for $Re(s)$ large with W_1, W_2 being any K-finite Whittaker functions on $GSp(2, \mathbb{A})$, and f_s any smooth section in $I^4_{3,\infty}(s)$ (Theorem 3.3.5); and

(ii) nonvanishing of $\mathcal{Z}_v(s, W_1, W_2, f_s)$ for a suitable choice of K-finite Whittaker functions W_1 and W_2 and a smooth section f_s (Theorem 3.2.4).

One of the subtle points is that the general theory of Eisenstein series assumes the K-finiteness of the section f_s at archimedean local places. In order to apply the local theory to the L-function $L^S(s', \pi_1 \otimes \pi_2, \rho_1 \otimes \rho_1)$, we thus have to prove either one of following statements:

(1) The local integral $\mathcal{Z}_v(s, W_1, W_2, f_s)$ continues to a meromorphic function on the complex plane with K-finite data (W_1, W_2, f_s), and for a given value s_0 there exists a K-finite data (W_1, W_2, f_{s_0}) so that $\mathcal{Z}_v(s_0, W_1, W_2, f_{s_0})$ does not vanish.

(2) The local integral $\mathcal{Z}_v(s, W_1, W_2, f_s)$ continues to a meromorphic function on the complex plane with K-finite W_1, W_2 and smooth section f_s, and for fixed W_1, W_2 (which can be K-finite), the local integral $\mathcal{Z}_v(s, W_1, W_2, f_s)$ is a continuous functional over the space of smooth sections f_s (which is Frechet space in a natural topology).

I do not have a complete proof for Statement (1) or (2) as yet. It is sure that the well developed local theory of the Rankin-Selberg convolution $\mathcal{Z}_v(s, W_1, W_2, f_s)$ will lead very interesting applications of the degree 16 standard L-function $L^S(s, \pi_1 \otimes \pi_2, \rho_1 \otimes \rho_1)$ to both automorphic representation theory and number theory.

0.6. Acknowledgment. As mentioned above, this manuscript exposes the main parts of my Ph. D. dissertation at The Ohio State University 1994. I would like to take this opportunity to express my deepest gratitude to my thesis advisor, Prof. Stehpen Rallis, for his introducing me to the topic of automorphic L-functions and representations, for his sharing with me his wonderful ideas and insights to various problems of mathematics, and for his constant encouragement and support.

I will like to thank David Ginzburg, Cary Rader, David Soudry, and Robert Stanton for valued conversations during my working with the thesis problems. My thanks will also go to Daniel Barbasch, Daniel Bump, Chris Jantzen, Tomasz Przebinda, and Nolan Wallach for mathematical conversations and communications.

Finally, I would like to thank the referee for his generous suggestions implemented in rewriting parts of the original Introduction. This paper was revised when I was a Postdoctoral Fellow at Mathematical Sciences Research Institute for the academic year 1994-95. The research at MSRI was partly supported by NSF grant DMS 9022140. I would like to thank MSRI for the kind hospitality.

CHAPTER 2

Degree 16 Standard L-function of $GSp(2) \times GSp(2)$

In this Chapter, we are going to establish, via the doubling method, a global integral of Rankin-Selberg type,, which represents the degree 16 standard L-function of $GSp(2) \times GSp(2)$.

1. Preliminaries

We shall first recall some basic facts on representation theory of algebraic groups for our later use. For the general theory, we prefer T. A. Springer [**Spr**]. We only consider two algebraic groups H and G of symplectic similitudes, which relates to each other via the doubling method. We will study the H-orbit decomposition of some flag variety constructed from G and the negligibility of those H-orbits. General discussion of the doubling method can be found in S. Rallis' IMC paper [**Ral**].

1.1. The Group of Symplectic Similitudes. Let $(V, (\ ,\ ))$ be a 4-dimensional non-degenerate symplectic vector space over a field F with characteristic zero and $GSp(V)$ the group of similitudes of $(V, (\ ,\ ))$, i.e., $GSp(V) = \{g \in GL(4) : (gu, gv) = \nu(g)(u, v)$ for $u, v \in V$ and $s(g)$ is a scalar in $F^\times\}$. Let $(W, \langle\ ,\ \rangle)$ be the doubling symplectic space of $(V, (\ ,\ ))$, i.e. $W = V^+ \oplus V^-$, where $V^+ = \{(v, 0)\ v \in V\}$ and $V^- = \{(0, v)\ v \in V\}$, and the symplectic form is defined by

$$\langle (u_1, u_2), (v_1, v_2) \rangle = (u_1, v_1) - (u_2, v_2) \quad \text{for } u_1, u_2, v_1, v_2 \in V.$$

Then $(W, \langle\ ,\ \rangle)$ is an 8-dimensional non-degenerate symplectic vector space over the field F. We denote by $G = GSp(W)$ be the group of similitudes of $(W, \langle\ ,\ \rangle)$.

In $(V, (\ ,\ ))$, choose a symplectic basis $\{e_1, e_2, e_1', e_2'\}$ so that the underlying vector space V is identified with F^4 (row vectors) and the form $(\ ,\ )$ corresponds to the matrix $J_2 = \begin{pmatrix} 0 & I_2 \\ -I_2 & 0 \end{pmatrix}$. Then in $(W, \langle\ ,\ \rangle)$, we have a typical symplectic basis

$$\{(e_1, 0), (e_2, 0), (0, -e_1), (0, -e_2), (e_1', 0), (e_2', 0), (0, e_1'), (0, e_2')\} \tag{13}$$

under which the underlying vector space W is identified with F^8 (row vectors) and the form $\langle\ ,\ \rangle$ of W corresponds to the matrix $J_4 = \begin{pmatrix} 0 & I_4 \\ -I_4 & 0 \end{pmatrix}$. Under the chosen basis, the group of symplectic similitudes can be embedded into a general linear group, i.e., $GSp(V) = GSp(2) = \{g \in GL(4) : gJ_2{}^t g = s(g)J_2\}$ and $GSp(W) = GSp(4) = \{g \in$

$GL(8) : gJ_4{}^tg = s(g)J_4\}$, and the action of $GSp(V)$, $GSp(W)$ on V, W corresponds to that of $GSp(2)$, $GSp(4)$ on F^4, F^8 to the right, respectively.

Let $H = (GSp(2) \times GSp(2))^\circ = \{(g_1, g_2) \in GSp(2) \times GSp(2) : s(g_1) = s(g_2)\}$. Then the reductive group H can be embedding into the reductive group G in a canonical way: $i : H \hookrightarrow G$, $(v_1, v_2) \cdot i(g_1, g_2) = (v_1 g_1, v_2 g_2)$. In other words,

$$
\begin{pmatrix} A & B \\ C & D \end{pmatrix} \times \begin{pmatrix} A' & B' \\ C' & D' \end{pmatrix} \longmapsto \begin{pmatrix} A & & B & \\ & A' & & -B' \\ C & & D & \\ & -C' & & D' \end{pmatrix}. \tag{14}
$$

From now on we will identify the group H with its image $i(H)$ in G.

Let $L_\circ = F(e_1', 0) \oplus F(e_2', 0) \oplus F(0, e_1')$. Then $L_\circ$ is a three dimensional isotropic subspace in W. Let $P_3^4 = \text{Stab}_G(L_\circ)$. Then P_3^4 is a maximal parabolic subgroup of G, whose Levi decomposition is $P_3^4 = M_3^4 N_3^4$ with

$$
\begin{aligned}
M_3^4 &= (GL(3) \times GSp(1)) \\
&= \{ \begin{pmatrix} a & 0 & 0 & 0 \\ 0 & x_1 & 0 & x_2 \\ 0 & 0 & a' & 0 \\ 0 & x_3 & 0 & x_4 \end{pmatrix} \in GSp(4) : a' = (x_1 x_4 - x_2 x_3)^t a^{-1}, \ a \in GL(3) \}
\end{aligned}
$$

and

$$
N_3^4 = \{ \begin{pmatrix} I_3 & x & w & y \\ 0 & 1 & y' & 0 \\ 0 & 0 & I_3 & 0 \\ 0 & 0 & x' & 1 \end{pmatrix} \in GSp(4) : x, y, {}^t x', {}^t y' \in M(3, 1) \}.
$$

where N_3^4 is the unipotent radical of P_3^4. We take, as a maximal F-split torus $T_4 \subset P_3^4$,

$$
T_4(F) = \{t = h(t_1, t_2, t_3, t_4, t_5, t_6, t_7, t_8) : \ t_1 t_5 = t_2 t_6 = t_3 t_7 = t_4 t_8\},
$$

where $h(\cdots)$ indicates a diagonal matrix element in G under the basis. Let B_4 be the standard Borel subgroup of G with $T_4 \subset B_4 \subset P_3^4$ and N^4 be the unipotent radical of B_4.

1.2. Root Systems and Representations. Let $X^*(T_4)$ be the group of characters of T_4 and ε_i be such a character that $\varepsilon_i(t) = t_i$. Then

$$
X^*(T_4) = \{\sum_{i=1}^8 n_i \varepsilon_i : \ \varepsilon_1 + \varepsilon_5 = \varepsilon_2 + \varepsilon_6 = \varepsilon_3 + \varepsilon_7 = \varepsilon_4 + \varepsilon_8\}.
$$

Let $\Phi_G = \Phi(G, T_4)$ be the set of roots of T_4 in G, Φ_G^+ the set of positive roots of Φ_G determined by N^4, and $\triangle_G$ the set of simple roots in Φ_G^+. Then we have the root system for G

$$
\begin{aligned}
\Phi_G &= \{\pm(\varepsilon_i \pm \varepsilon_j), \pm 2\varepsilon_i \ : i < j, \ i, j = 1, 2, 3, 4\}, \\
\Phi_G^+ &= \{(\varepsilon_i \pm \varepsilon_j), 2\varepsilon_i \ : i < j, \ i, j = 1, 2, 3, 4\}, \\
\triangle_G &= \{\alpha_1 = \varepsilon_1 - \varepsilon_2, \alpha_2 = \varepsilon_2 - \varepsilon_3, \alpha_3 = \varepsilon_3 - \varepsilon_4, \alpha_4 = 2\varepsilon_4\}.
\end{aligned} \tag{15}
$$

PROPOSITION 1.2.1. *For the complex group $GSp(2, \mathbb{C})$, we have*

(a) *The fundamental dominant weights are $\lambda_1 = \varepsilon_1$ and $\lambda_2 = \varepsilon_1 + \varepsilon_2$,*
(b) *For any dominant weight $n_1\varepsilon_1 + n_2\varepsilon_2$, $n_1 \geq n_2 \geq 0$, there is an irreducible complex representation $\rho_{(n_1,n_2)}$ of $GSp(2)$ with highest weight $n_1\varepsilon_1 + n_2\varepsilon_2$ and any irreducible complex representation of $GSp(2)$ is equivalent to the product of some power of the character $s(g)$ (the factor of similitudes) and some $\rho_{(n_1,n_2)}$.*

Actually, $\rho_{(n_1,n_2)}$ is the representation of the derived group of $GSp(2)$, which is $Sp(2)$, with highest weight $n_1\varepsilon_1 + n_2\varepsilon_2, n_1 \geq n_2 \geq 0$, and is trivial at the 'similitude' part, i.e. $\{\begin{pmatrix} I_2 & 0 \\ 0 & dI_2 \end{pmatrix} \in GSp(4) : d$ is non-zero scalar$\}$.

PROPOSITION 1.2.2. [**Bor**] *The complex dual group of $GSp(2)$ is $GSp(2, \mathbb{C})$.*

1.3. The Negligibility of Orbits. In this subsection, we consider the H-orbital decomposition on the flag variety $P_3^4 \setminus G$ and describe the negligibility of these H-orbits. Those results will be used to construct our Rankin-Selberg global integral in the next section. The notion of negligibility was introduced in [**PSRa**].

Let $\mathcal{L} = \{$all 3-dimensional isotropic subspaces L of $W\}$ and $L_o = F(e_1', 0) \oplus F(e_2', 0) \oplus F(0, e_1')$ as chosen in §1.1. Fix once for all the following isomorphism from $P_3^4 \setminus G$ onto $\mathcal{L}$ via $g \longmapsto L_o g$.

Let $\pi^\pm$ be the projections from W onto V^+ or V^-, respectively. Let L be any 3-dimensional isotropic subspace of W. Denote $L^\pm = L \cap V^\pm$ and $L' = \pi^+(L)$, $L'' = \pi^-(L)$. Then it is easy to check that $\dim L^+ + \dim L^- \leq 3$, $\dim L^\pm \leq 2$, and $\dim L' + \dim L^- = \dim L'' + \dim L^+ = 3$.

THEOREM 1.3.1. *Let $\kappa^+(L) = \dim L^+$ and $\kappa^-(L) = \dim L^-$. Then $(\kappa^+(L), \kappa^-(L))$ completely determines H-orbits of $L \in \mathcal{L}$; that is, for $L, M \in \mathcal{L}$, $(\kappa^+(L), \kappa^-(L)) = (\kappa^+(M), \kappa^-(M))$ if and only if there is $g \in H$ such that $Lg = M$.*

PROOF. It is evident that $(\kappa^+(L), \kappa^-(L))$ is an invariant of the H-orbits on $\mathcal{L}$ since $[L(g_1, g_2)]^+ = L^+g_1$ and $[L(g_1, g_2)]^- = L^-g_2$. Moreover, $(\kappa^+(L), \kappa^-(L))$ completely determines H-orbits because, for $L, M \in \mathcal{L}$, we can use the same argument as in [**PSRa**] to prove that there are $g_1, g_2 \in Sp(2)$ such that $L(g_1, g_2) = M$. □

By straightforward calculation, the flag variety $\mathcal{L}$ has a decomposition of H-orbits as

$$\mathcal{L} = \mathcal{L}_{(2,1)} \cup \mathcal{L}_{(1,2)} \cup \mathcal{L}_{(1,1)} \cup \mathcal{L}_{(1,0)} \cup \mathcal{L}_{(0,1)} \cup \mathcal{L}_{(0,0)}, \tag{16}$$

where $\mathcal{L}_{(i,j)}$ is the H-orbit with invariants $(\kappa^+(L), \kappa^-(L)) = (i, j)$, and the only non-negligible H-orbit in the sense of Piatetski-Shapiro and Rallis [**PSRa**] is $\mathcal{L}_{(0,0)}$. The unique nonnegligible H-orbit is represented by a three-dimensional isotropic subspace

$L_{(0,0)} = F(e_1', e_1') \oplus F(e_2', e_2') \oplus F(e_2, e_2)$ or $L_{(0,0)} = L_\circ \gamma_\circ$ where

$$
\gamma_\circ = \begin{pmatrix}
1 & & & & 0 & & & \\
& 1 & & & & 0 & & \\
& & 0 & & & & 0 & 1 \\
& & & 0 & & & 1 & 0 \\
0 & & & & 1 & & 1 & \\
& 0 & & & & 1 & & 1 \\
& 1 & 0 & -1 & & & 0 & \\
1 & & -1 & 0 & & & & 0
\end{pmatrix}.
\tag{17}
$$

It is easy to see that $L_{(0,0)}$ is the diagonal embedding into W of a three-dimensional subspace $L^\star = Fe_1' \oplus Fe_2' \oplus Fe_2$ of V. Let $Q = \mathrm{Stab}_H(L_{(0,0)})$. Then we deduce that

$$
Q = \{(g_1, g_2) \in H : g_1|_{L^\star} = g_2|_{L^\star},\ L^\star g_i = L^\star,\ i = 1,\ 2\}.
\tag{18}
$$

Let $P_1^2 = \mathrm{Stab}_{GSp(2)}(L^\star)$. Then P_1^2 is a maximal parabolic subgroup of $GSp(2)$, which is of form, under the chosen basis $\{e_1, e_2, e_1', e_2'\}$,

$$
P_1^2 = \{ \begin{pmatrix}
a & x & z & s \\
0 & x_1 & s' & x_2 \\
0 & 0 & d & 0 \\
0 & x_3 & x' & x_4
\end{pmatrix} \in GSp(2)\ :\ ad = x_1 x_4 - x_2 x_3 \neq 0\}.
$$

If $g \in GSp(2)$ such that $g|_{L^\star} = 1$, then $e_1' g = e_1' + z e_1$, i.e. $g = \chi_{2\varepsilon_1}(z)$, where $\chi_{2\varepsilon_1}$ is the one-parameter subgroup of $GSp(2)$ associated to the root $2\varepsilon_1$. Let $Z_2 = \{\chi_{2\varepsilon_1}(t)\ :\ t \in F\}$. Then we obtain that

$$
Q = \mathrm{Stab}_H(L_\circ \gamma_\circ) = P_1^{2,\triangle}(Z_2 \times I_4)
\tag{19}
$$

where $P_1^{2,\triangle}$ is the diagonal embedding of P_1^2 into Q, and $P_3^4 \gamma_\circ H$ corresponds to $\mathcal{L}_{(0,0)} = L_{(0,0)} H = L_\circ \gamma_\circ H$.

2. Global Integral of Rankin-Selberg Type

We assume from now on that F be a number field and $\mathbb{A}$ its ring of adeles. Let F_v be the local field of F associated to the place v. When v is finite, we denote by $\mathcal{O}_v$ the ring of local integers in F_v. By automorphic representations of adelic groups we mean that in the sense of Borel and Jacquet's Corvallis paper [BoJa]. We shall establish a global zeta integral via the doubling method, which will be a Rankin-Selberg convolution of two cusp forms of $GSp(2)$ against an Eisenstein series of $GSp(4)$. The location and the order of poles of such a family of Eisenstein series are explicitly determined in Chapter III. Before doing so, we shall first study Whittaker functions on $GSp(2, \mathbb{A})$.

2.1. The Whittaker functions on $GSp(2)$. The properties of Whittaker functions studied here will play a critical role in our proof of the eulerian properties of our global zeta integral in the next subsection. For general description of Whittaker models, see, J. Shalika [Shl] or S. Gelbart and F. Shahidi [GeSh].

Let π be irreducible admissible automorphic cuspidal representation of $GSp(2, \mathbb{A})$ with trivial central character. Let ψ be a generic character of the standard maximal unipotent subgroup N^2 of $GSp(2, \mathbb{A})$. Without loss of generality, we may assume that ψ has form:

$$\psi\left(\begin{pmatrix} 1 & x & z & w \\ 0 & 1 & w' & y \\ 0 & 0 & 1 & 0 \\ 0 & 0 & -x & 1 \end{pmatrix}\right) = \psi_\circ(x + y), \tag{20}$$

where $\psi_\circ(t)$ is a nontrivial additive unitary character on $F\backslash\mathbb{A}$. We define, for $\phi \in \pi$, a function $\phi^\star$ by the following integral

$$\phi^\star(g) = \int_{Z_2(F)\backslash Z_2(\mathbb{A})} \phi(zg)dz \tag{21}$$

where Z_2 is the one-parameter subgroup attached to the root $2\varepsilon_1$, as defined in section 1. The ψ-Whittaker function associated to ϕ is defined as

$$W_\phi^\psi(g) = \int_{N^2(F)\backslash N^2(\mathbb{A})} \phi(ug)\psi^{-1}(u)du. \tag{22}$$

Then we call the representation π ψ-generic if there exists a function ϕ such that $W_\phi^\psi \neq 0$. In this case, we call ϕ ψ-generic. The idea to introduce the function $\phi^\star$ is from Gelbart and Piatetski-Shapiro [**GePS**]. Let C_2 be the center of $GSp(2)$. Let

$$R_1 = \{\begin{pmatrix} a & b & * \\ c & d & * \\ 0 & 0 & 1 \end{pmatrix} \in GL(3)\}. \tag{23}$$

Then we have

LEMMA 2.1.1.   (a) $C_2 Z_2$ is a normal subgroup of P_1^2;
(b) $C_2 Z_2 \backslash P_1^2 \cong R_1$;
(c) $\phi^\star(g)$ is left $[C_2 Z_2](\mathbb{A})$-invariant and left $P_1^2(F)$-invariant;
(d) There is a 'Fourier' development for $\phi^\star(g)$ i.e.

$$\phi^\star(g) = \sum_{\beta \in U_2(F)\backslash GL(2,F)} W_\phi^\psi(\beta g)$$

where the group $GL(2)$ is embedded into R_1 as in (20) and U_2 is the standard unipotent subgroup of $GL(2)$.

PROOF. We can write any $p \in P_1^2$ in form: $p = cmn$, with $c \in C_2$, $n \in N_1$, and

$$m = \begin{pmatrix} e & 0 & 0 & 0 \\ 0 & a & 0 & b \\ 0 & 0 & 1 & 0 \\ 0 & c & 0 & d \end{pmatrix}.$$

(a) is true since Z_2 is the center of the unipotent radical N_1 of P_1^2 and

$$
\begin{pmatrix} e & 0 & 0 & 0 \\ 0 & a & 0 & b \\ 0 & 0 & 1 & 0 \\ 0 & c & 0 & d \end{pmatrix}^{-1}
\begin{pmatrix} 1 & 0 & z & 0 \\ 0 & 0 & 0 & 0 \\ 0 & 0 & 1 & 0 \\ 0 & 0 & 0 & 1 \end{pmatrix}
\begin{pmatrix} e & 0 & 0 & 0 \\ 0 & a & 0 & b \\ 0 & 0 & 1 & 0 \\ 0 & c & 0 & d \end{pmatrix}
=
\begin{pmatrix} 1 & 0 & e^{-1}z & 0 \\ 0 & 0 & 0 & 0 \\ 0 & 0 & 1 & 0 \\ 0 & 0 & 0 & 1 \end{pmatrix}.
\tag{24}
$$

To prove (b), we consider the projection $\theta : P_1^2 \to R_1$

$$
\begin{pmatrix} e & 0 & 0 & 0 \\ 0 & a & 0 & b \\ 0 & 0 & 1 & 0 \\ 0 & c & 0 & d \end{pmatrix}
\mapsto
\begin{pmatrix} a & b & 0 \\ c & d & 0 \\ 0 & 0 & 1 \end{pmatrix}
\quad \text{and} \quad
\begin{pmatrix} 1 & x & 0 & s \\ 0 & 1 & s & 0 \\ 0 & 0 & 1 & 0 \\ 0 & 0 & -x & 1 \end{pmatrix}
\mapsto
\begin{pmatrix} 1 & 0 & -s \\ 0 & 1 & x \\ 0 & 0 & 1 \end{pmatrix},
$$

where $e = ad - bc$. It is not difficult to check that θ is surjective and $ker(\theta) = CZ_2$. Thus θ induces an isomorphism $C_2 Z_2 \backslash P_1^2 \cong R_1$. The left $[C_2 Z_2](\mathbb{A})$-invariance of $\phi^\star$ is easy to check since the central character of π is trivial, while the left $P_1^2(F)$-invariance of $\phi^\star$ can be proved in the following way: For any $p = cnm \in P_1^2(F)$, we have

$$
\phi^\star(pg) = \int_{Z_2(F)\backslash Z_2(\mathbb{A})} \phi(zpg)dz = \int_{Z_2(F)\backslash Z_2(\mathbb{A})} \phi(zmg)dz
$$

since Z_2 is the center of N_1. By means of identity (20) and the fact that ϕ is automorphic, we conjugate the integrating variable z by m change the variable: $e^{-1}z \mapsto z$, and then we have $\phi^\star(pg) = |e|_{\mathbb{A}}\phi^\star(g) = \phi^\star(g)$ since $e \in F$.

We are now going to prove (d). Let us denote $\varphi(r) = \phi^\star(\theta^{-1}(r)g)$, which is a cusp form on $R_1(\mathbb{A})$, and let $U_3 = \{ \begin{pmatrix} 1 & y & s \\ 0 & 1 & x \\ 0 & 0 & 1 \end{pmatrix} \in R_1 \}$ and $\psi \begin{pmatrix} 1 & y & s \\ 0 & 1 & x \\ 0 & 0 & 1 \end{pmatrix} = \psi_\circ(x + y)$. It is evident that φ is ψ-generic if ϕ is. Since the following isomorphisms of varieties $C_2 N^2 \backslash P_1^2 \cong U_3 \backslash R_1 \cong U_2 \backslash GL(2)$, it is well known that the generic cusp form φ can be recovered by its ψ-Whittaker functions, that is,

$$
\varphi(r) = \sum_{\beta \in U_2(F)\backslash GL(2,F)} \varphi_\psi(\beta r),
\tag{25}
$$

where

$$
\varphi_\psi(r) = \int_{U_3(F)\backslash U_3(\mathbb{A})} \varphi(nr)\overline{\psi(n)}dn.
$$

On the other hand, the restriction of θ to the unipotent subgroup N^2 gives us a central extension of unipotent groups

$$
1 \to Z_2 \to N^2 \to U_3 \to 1
$$

and as algebraic varieties $N^2 = Z_2 \times U_3$. We thus have $N^2(\mathbb{A}) = Z_2(\mathbb{A}) \times U_3(\mathbb{A})$ and $N^2(F) = Z_2(F) \times U_3(F)$, and for $\beta \in [U_3(F) \setminus R_1(F)] \ (\cong U_2(F) \setminus GL(2, F))$,

$$
\begin{aligned}
\varphi_\psi(\beta) &= \int_{U_3(F)\setminus U_3(\mathbb{A})} \varphi(n\beta)\overline{\psi(n)}dn \\
&= \int_{U_3(F)\setminus U_3(\mathbb{A})} \phi^\star(\theta^{-1}(n\beta)g)\overline{\psi(n)}dn \\
&= \int_{U_3(F)\setminus U_3(\mathbb{A})} \int_{Z_2(F)\setminus Z_2(\mathbb{A})} \phi(z\theta^{-1}(n\beta)g)\overline{\psi(z\theta^{-1}(n))}dzdn \\
&= \int_{N^2(F)\setminus N^2(\mathbb{A})} \phi(u\beta g)\overline{\psi(u)}du \\
&= W_\phi^\psi(\beta g).
\end{aligned}
$$

Now let $r = 1$, we obtain the 'Fourier' expansion for the function $\phi^\star$, i.e.

$$
\begin{aligned}
\phi^\star(g) = \varphi(1) &= \sum_{\beta \in U_2(F)\setminus GL(2,F)} \varphi_\psi(\beta) \\
&= \sum_{\beta \in U_2(F)\setminus GL(2,F)} W_\phi^\psi(\beta g).
\end{aligned}
$$

The proof is finished.   □

2.2. The global zeta-integral. The global integral we are going to establish is a Rankin-Selberg convolution of two cusp forms of $GSp(2)$ against the Eisenstein series of $GSp(4)$. We state the basic properties of such Eisenstein series, the proof of which will be given in Chapter IV.

Let P_3^4 be the maximal parabolic subgroup of $G = GSp(4)$ as in §1.1 and $a(p) = det(a)$. The modulus character $\delta = \delta_{P_3^4}$ will be

$$
\delta(p) = |a(p)|^6 |x_1x_4 - x_2x_3|^{-9}. \tag{26}
$$

Let $K_4 = \prod_v K_{4v}$ be the maximal compact subgroup of $G(\mathbb{A})$ with K_{4v} the standard maximal compact subgroup of $G(F_v)$ in the usual sense for each place v of F.

Let $\delta_s(p) := \delta^{\frac{s}{6}}(p)$. Then δ_s can be extended to be a character of $P_3^4(\mathbb{A})$ which is trivial over $P_3^4(F)N_3^4(\mathbb{A})$. We denote by $I_3^4(s) = Ind_{P_3^4(\mathbb{A})}^{G(\mathbb{A})}(\delta_s)$ the degenerate principal series representation of $G(\mathbb{A})$, which consists of smooth functions $f(\cdot, s) : G(\mathbb{A}) \to \mathbb{C}$ satisfying the following condition: $f(pg, s) = \delta_s(p)\delta^{\frac{1}{2}}(p)f(g, s)$ for $p \in P_3^4(\mathbb{A})$, $g \in G(\mathbb{A})$. The group action is given by right translation. By smoothness of the function f we mean that the function f is locally constant as a function over the non-archimedean local variables and smooth in the usual sense as a function over the archimedean local variables. A section $f(g, s)$ is called holomorphic or entire if the section $f(g, s)$, as a function of one complex variable s, is holomorphic or entire. We also assume that sections $f(g, s)$ is right K-finite. This implies that $I_3^4(s)$ is, in fact, a representation of $(\mathfrak{g}_\infty, K_\infty) \times G(\mathbb{A}_f)$, where $\mathfrak{g}_\infty$ is the Lie algebra of G_∞ and $\mathbb{A}_f$ indicates the finite adeles.

We define, for any section $f_s = f(\cdot, s) \in I_3^4(s)$, an Eisenstein series as follows:

$$E_3^4(g, s; f_s) = \sum_{\gamma \in P_3^4(F) \backslash G(F)} f_s(\gamma g). \tag{27}$$

The normalization $E_3^{4,*}(g, s; f_s)$ of the Eisenstein series $E_3^4(g, s; f_s)$ by the normalizing factor $d_G^S(s)$ is defined as

$$E_3^{4,*}(g, s; f_s) = d_G^S(s) E_3^4(g, s; f_s), \tag{28}$$

where $d_G^S(s) = \prod_{v \notin S} \zeta_v(2s + 2)\zeta_v(s + 1)\zeta_v(s + 2)\zeta_v(s + 3)$ and the finite set S is determined by the ramification of the section f_s, see §3. We state below the basic properties of $E_3^{4,*}(g, s; f_s)$, which will be proved in section 4, Chapter III.

THEOREM 2.2.1. *For any holomorphic section $f(\cdot, s) \in I_3^4(s)$, we have*

(a) *$E_3^4(g, s, ; f_s)$ converges absolutely for Re(s) large and has a meromorphic continuation to the whole complex plane with finitely many poles, and satisfies a functional equation which relates its value at s to its value at $-s$ [**Lan**] and [**Art**].*

(b) *The normalized Eisenstein series $E_3^{4,*}(g, s, ; f_s)$ is holomorphic except for s in $X_3 = \{\pm 1, \pm 2, \pm 3\}$, has at most a simple pole at $s = -2, -3$ and at most a double pole at $s = -1$, and achieves a simple pole at $s = 2, 3$ and a double pole at $s = 1$.*

Let π_1, π_2 be irreducible automorphic cuspidal representations of $GSp(2, \mathbb{A})$ with trivial central characters. Then an element in an irreducible automorphic cuspidal representation π will be a cusp form [**BoJa**]. Recall that $H = (GSp(2) \times GSp(2))^\circ$. We define, for $\phi_1 \in \pi_1; \phi_2 \in \pi_2$, our (unnormalized) global integral as follows:

$$\mathcal{Z}(s, \phi_1, \phi_2, f_s) = \int_{C(\mathbb{A})H(F) \backslash H(\mathbb{A})} E_3^4((g_1, g_2), s; f_s) \phi_1(g_1) \phi_2(g_2) dg_1 dg_2 \tag{29}$$

where $C = C_4$ is the center of $GSp(4)$. The (normalized) global integral is defined as

$$\mathcal{Z}^*(s, \phi_1, \phi_2, f_s) = d_G^S(s) \cdot \mathcal{Z}(s, \phi_1, \phi_2, f_s). \tag{30}$$

The Haar measure $dg_1 dg_2$ is canonically chosen as [**HaKu**] and [**Har**].

THEOREM 2.2.2.   (a) *If ϕ_1, ϕ_2 are cusp forms in π_1, π_2, resp., then the global integral $\mathcal{Z}^*(s, \phi_1, \phi_2, f_s)$ converges absolutely for all complex value s except possibly for those finitely many values of s at which the normalized Eisenstein series $E_3^{4,*}(g, s; f_s)$ may achieve a pole, and satisfies a functional equation which relates the value s to the value $-s$.*

(b) *If ϕ_1 and ϕ_2 are generic, then we have the 'basic identity' in the sense of* [**GeSh**], *i.e.*

$$\mathcal{Z}^*(s, \phi_1, \phi_2, f_s) = d_G^S(s) \int_{\mathcal{D}(\mathbb{A})} f_s(\gamma_\circ(g_1, g_2)) W_{\phi_1}^\psi(g_1) W_{\phi_2}^{\overline{\psi}}(g_2) dg_1 dg_2,$$

where $\mathcal{D}(\mathbb{A})$ denotes $C(\mathbb{A})N^{2,\triangle}(\mathbb{A})(Z_2 \times I_4)(\mathbb{A}) \backslash H(\mathbb{A})$ and $N^{2,\triangle}$ is the diagonal embedding of the standard maximal unipotent subgroup N^2 of $GSp(2)$ into H.

PROOF. The first part of (a) follows since cusp forms are rapidly decreasing and Eisenstein series are slowly increasing when they are restricted on an appropriate Siegel domain, and the second part of (a) follows from the corresponding functional equation of the Eisenstein series. Fix from now on such a value of s where the integral is absolutely convergent. It is enough to prove statement (b) for the unnormalized integral $\mathcal{Z}(s, \phi_1, \phi_2, f)$.

Unfolding the Eisenstein series, we have

$$\mathcal{Z}(s, \phi_1, \phi_2, f)$$
$$= \int_{C(\mathbb{A})H(F)\backslash H(\mathbb{A})} \sum_{\gamma_\circ \in P_3^4(F)\backslash G(F)} f_s(\gamma_\circ(g_1, g_2))\phi_1(g_1)\phi_2(g_2) dg_1 dg_2$$
$$= \sum_{\gamma_\circ \in P_3^4(F)\backslash G(F)/H(F)} \int_{C(\mathbb{A})H(F)_\circ^\gamma \backslash H(\mathbb{A})} f_s(\gamma_\circ(g_1, g_2))\phi_1(g_1)\phi_2(g_2) dg_1 dg_2,$$

where $H_\circ^\gamma = \gamma_\circ^{-1} P_3^4 \gamma_\circ \cap H$ is the stabilizer in H of $\gamma_\circ$ or $L_\circ \gamma_\circ$. As in Piatetski-Shapiro and Rallis [**PSRa**], the integral will vanish over the negligible H-orbits $P_3^4 \gamma_\circ H$. Therefore our global integral become:

$$\mathcal{Z}(s, \phi_1, \phi_2, f) = \int_{C(\mathbb{A})H(F)^{\gamma_\circ} \backslash H(\mathbb{A})} f(\gamma_\circ(g_1, g_2))\phi_1(g_1)\phi_2(g_2) dg_1 dg_2 \qquad (31)$$

where $P_3^4 \gamma_\circ H$ is the unique nonnegligible H-orbit, whose stabilizer $H^{\gamma_\circ}$ is the subgroup $P_1^{2,\triangle}(F)(Z_2 \times I_4)$. Applying the function $\phi_1^\star$ defined in (18), we deduce that the right hand side of (28) equals

$$= \int_{C(\mathbb{A})P_1^d(F)(Z_2 \times I_4)(F)\backslash H(\mathbb{A})} f(\gamma_\circ(g_1, g_2))\phi_1(g_1)\phi_2(g_2) dg_1 dg_2$$
$$= \int_{C(\mathbb{A})P_1^{2,\triangle}(F)(Z_2 \times I_4)(\mathbb{A})\backslash H(\mathbb{A})} f(\gamma_\circ(g_1, g_2)) \int_{Z_2(F)\backslash Z_2(\mathbb{A})} \phi_1(zg_1) dz \phi_2(g_2) dg_1 dg_2$$
$$= \int_{C(\mathbb{A})P_1^{2,\triangle}(F)(Z_2 \times I_4)(\mathbb{A})\backslash H(\mathbb{A})} f(\gamma_\circ(g_1, g_2))\phi_1^\star(g_1)\phi_2(g_2) dg_1 dg_2$$

since $\gamma_\circ(Z_2 \times I_4)\gamma_\circ^{-1}$ is included in P_3^4 and is unipotent. Recall from Lemma 1 in subsection 2.1 that $\phi_1^\star(g_1)$ has a Fourier expansion $\phi_1^\star(g_1) = \sum_{\beta \in U_2 \backslash GL(2)} W_{\phi_1}^\psi(\beta g_1)$, and $U_2 \backslash GL(2) \cong N^2 \backslash P_1^2$. So our last integral becomes

$$\int_{C(\mathbb{A})P_1^{2,\triangle}(F)(Z_2 \times I_4)(\mathbb{A}) \backslash H(\mathbb{A})} f(\gamma_\circ(g_1, g_2)) \sum_{\beta \in U_2 \backslash GL(2)} W_{\phi_1}^{\psi}(\beta g_1)\phi_2(g_2)dg_1 dg_2$$

$$= \int_{C(\mathbb{A})P_1^{2,\triangle}(F)(Z_2 \times I_4)(\mathbb{A}) \backslash H(\mathbb{A})} f(\gamma_\circ(g_1, g_2)) \sum_{\beta \in N^2 \backslash P_1^2} W_{\phi_1}^{\psi}(\beta g_1)\phi_2(\beta g_2)dg_1 dg_2$$

$$= \int_{C(\mathbb{A})N^{2,\triangle}(F)(Z_2 \times I_4)(\mathbb{A}) \backslash H(\mathbb{A})} f(\gamma_\circ(g_1, g_2))W_{\phi_1}^{\psi}(g_1)\phi_2(g_2)dg_1 dg_2,$$

here we have used in the first step of the computation the condition that ϕ_2 is automorphic and in the last step the fact that P_1^d is in the stabilizer of $\gamma_\circ$. Continuing our computation, the last integral equals

$$= \int_{\mathcal{D}(\mathbb{A})} f(\gamma_\circ(g_1, g_2)) \int_{N^{2,\triangle}(F) \backslash N^{2,\triangle}(\mathbb{A})} W_{\phi_1}^{\psi}(ug_1)\phi_2(ug_2)du dg_1 dg_2$$

$$= \int_{\mathcal{D}(\mathbb{A})} f(\gamma_\circ(g_1, g_2))W_{\phi_1}^{\psi}(g_1) \int_{N^2(F) \backslash N^2(\mathbb{A})} \phi_2(ug_2)\psi(u)du dg_1 dg_2$$

$$= \int_{\mathcal{D}(\mathbb{A})} f(\gamma_\circ(g_1, g_2))W_{\phi_1}^{\psi}(g_1)W_{\phi_2}^{\overline{\psi}}(g_2)dg_1 dg_2.$$

The $\overline{\psi}$-Whittaker function of ϕ_2 is defined by

$$W_{\phi_2}^{\overline{\psi}}(g_2) = \int_{N^2(F) \backslash N^2(\mathbb{A})} \phi_2(ug_2)\psi(u)du dg_1 dg_2.$$

Thus we obtain that

$$\mathcal{Z}(s, \phi_1, \phi_2, f) = \int_{\mathcal{D}(\mathbb{A})} f_s(\gamma_\circ(g_1, g_2))W_{\phi_1}^{\psi}(g_1)W_{\phi_2}^{\overline{\psi}}(g_2)dg_1 dg_2. \qquad (32)$$

Multiplying both sides by the normalizing factor $d_G^S(s)$, we achieve what we want to prove. $\qquad \square$

Now, if we choose the additive characters ψ and $\overline{\psi}$, and $\phi_1 \in \pi_1, \phi_2 \in \pi_2$, and $f \in I_3^4(s)$ all decomposable, that is, $\psi = \otimes_v \psi_v$, and $\overline{\psi} = \otimes_v \overline{\psi}_v$, and $f = \otimes'_v f_v$, $\phi_1 = \otimes'_v \phi_{1v}$, and $\phi_2 = \otimes'_v \phi_{2v}$, where $\otimes'$ denotes the restricted tensor product, then, by the uniqueness of the local and global Whittaker models, and the uniqueness of the restricted tensor decomposition of $I_3^4(s)$, see [**Shl**] and [**GeSh**], we obtain the eulerian property of our Rankin-Selberg global integral.

COROLLARY 2.2.1. *The global integral can be decomposed into an Euler product:*

$$\mathcal{Z}(s, \phi_1, \phi_2, f) = \prod_v \int_{\mathcal{D}(F_v)} f_v(\gamma_\circ(g_1, g_2), s)W_{\phi_{1v}}^{\psi_v}(g_1)W_{\phi_{2v}}^{\overline{\psi}_v}(g_2)dg_1 dg_2,$$

where $\psi = \otimes_v \psi_v$, *and* $\overline{\psi} = \otimes_v \overline{\psi}_v$ *are generic characters of the standard maximal unipotent subgroup* N^2 *of* $GSp(2, \mathbb{A})$.

Here we have assumed that for each place v, the local integral in the infinite product converges absolutely for $Re(s) >> 0$. The identity in the Corollary holds for $Re(s)$ large and then for all value of s by meromorphic continuation. The absolute convergence and the meromorphic continuation of such local integrals will be given in Chapter V.

Let us introduce a notation for the local zeta integral. For $W_{1v} \in \mathcal{W}(\pi_\infty, \psi)$, $W_{2v} \in \mathcal{W}(\pi_\in, \overline{\psi})$, and $f_{s,v} \in I^4_{3v}(s)$, we define

$$\mathcal{Z}_v(s, W_{1v}, W_{2v}, f_v) = \int_{\mathcal{D}(F_v)} f_v(\gamma_\circ(g_1, g_2), s) W_{1v}(g_1) W_{2v}(g_2) dg_1 dg_2. \qquad (33)$$

Then we have $\mathcal{Z}(s, \phi_1, \phi_2, f) = \prod_v \mathcal{Z}_v(s, W^{\psi_v}_{\phi_{1v}}, W^{\overline{\psi_v}}_{\phi_{2v}}, f_v)$. We call $\mathcal{Z}_v(s, W_{1v}, W_{2v}, f_v)$ the (unnormalized) local zeta integral. The (normalized) local zeta integral is defined as

$$\mathcal{Z}^*_v(s, W_{1v}, W_{2v}, f_v) := d_{v,G}(s) \cdot \mathcal{Z}_v(s, W_{1v}, W_{2v}, f_v), \qquad (34)$$

where $d_{v,G}(s)$ is the v-factor of the normalizing factor $d^S_G(s)$.

3. Unramified Local Zeta Integrals

The 'unramified computation' in the Rankin-Selberg method is to identify the local zeta integral with the local Langlands factor at any 'unramified' finite place v of F. In this section, we shall evaluate the normalized local zeta integral $\mathcal{Z}^*_v(s, W_{1v}, W_{2v}, f_v)$ with unramified data $(W_{1v}, W_{2v}, f_v, \psi_v)$. This will be done by computations of certain partial 'Fourier transform' of the unramified f°_v and application of Casselman and Shalika's formula for unramified Whittaker functions [**CaSh**].

Let F_v be the local completion of F at the local finite place v and $\mathcal{O}_v$ the ring of integers in F_v. Let $K_{4,v} = GSp(4, \mathcal{O}_v)$ and $K_{2,v} = GSp(2, \mathcal{O}_v)$ be the chosen maximal compact subgroup in $GSp(4, F_v)$ and $GSp(2, F_v)$, respectively. Then $(K_{2,v} \times K_{2,v})^\circ$ is the standard maximal compact subgroup in $H(F_v) = (GSp(2) \times GSp(2))^\circ(F_v)$, and $(K_{2,v} \times K_{2,v})^\circ \hookrightarrow K_{4,v}$ under the canonical embedding of $H(F_v)$ into $G(F_v)$. Let π_{1v} and π_{2v} be irreducible admissible representations of $GSp(2, F_v)$ (v-component of irreducible automorphic representations π_1 and π_2, resp.) and ψ_v and $\overline{\psi_v}$ are generic character of the standard maximal unipotent subgroup N^2 of $GSp(2, F_v)$ induced canonically by ψ and $\overline{\psi}$.

The integral we are going to study is $\mathcal{Z}_v(s, W_{1v}, W_{2v}, f_v)$ as in (30). We will define an unramified local finite place of F for our local zeta integral as follows:

DEFINITION 3.0.1. *A local finite place v of F is called unramified if the following hold*

(a) *The local component $\pi_{1,v}$ and $\pi_{2,v}$ of the automorphic representations π_1 and π_2 are unramified, or of class one, or right $K_{2,v}$-spherical;*

(b) *The local component $\psi_{\circ,v}$ of the nontrivial additive unitary character $\psi_\circ$ on $F \backslash \mathbb{A}$ is unramified, that is, the conductor of $\psi_{\circ,v}$ is $\mathcal{O}_v$.*

It is well known that almost all local finite places are unramified. At any unramified place v, we can pick up unramified integration data $(W_{1,v}^\circ, W_{2,v}^\circ, f_v^\circ)$, where f_v° is the unique normalized right $K_{4,v}$-spherical section in $I_{3,v}^4(s)$ so that $f_v^\circ(1) = 1$, and Whittaker functions W_{1v}° and W_{2v}° are right $K_{2,v}$-spherical and normalized so that $W_{1v}^\circ(1) = 1$ and $W_{2v}^\circ(1) = 1$. The local zeta integral with the unramified integration data is called the unramified local zeta integral.

For simplification of notations, we will drop v from the subscripts of our notations if no serious confusion will occur.

3.1. Reduction. We will reduce our computation of the unramified local zeta integral to that of certain 'Fourier coefficient' of the unramified section f°. To this end, we consider the Iwasawa decomposition of the reductive group $H(F)$, that is,

$$H(F) = (N^2 \times N^2)(T_2 \times T_2)^\circ (K_2 \times K_2)^\circ,$$

where T_2 is the standard maximal split torus in $GSp(2, F)$, so that $N^2 T_2 K_2$ is the Iwasawa decomposition of the reductive group $GSp(2, F)$. Then the domain of our local integration has the following decomposition:

$$\mathcal{D} = [CN^{2,\triangle}(Z_2 \times I_4)]\backslash H = [Z_2 \backslash N^2][C \backslash (T_2 \times T_2)^\circ](K_2 \times K_2)^\circ. \tag{35}$$

For our chosen unramified data $(\phi_1^\circ, \phi_2^\circ, f_s^\circ)$, we have the following

LEMMA 3.1.1. *The (unnormalized) unramified local zeta integral*

$$\mathcal{Z}(s, \phi_1^\circ, \phi_2^\circ, f_s^\circ) \tag{36}$$
$$= \int_{C\backslash(T_2\times T_2)^\circ} W_1^\circ(t_1)W_2^\circ(t_2)[\int_{Z_2\backslash N^2} f_s^\circ(\gamma_\circ(ut_1, t_2))\psi(u)du]\delta_H^{-1}(t_1, t_2)dt_1 dt_2$$

where δ_H is the modulus character of the standard Borel subgroup of H (restricted to $C \backslash (T_2 \times T_2)^\circ$).

PROOF. By the Iwasawa decomposition, the element $(g_1, g_2) \in [CN^{2,\triangle}(Z_2 \times I_4)]\backslash H$ can be written as $(g_1, g_2) = (ut_1 k_1, t_2 k_2)$. Then we have that

$$f_s^\circ(\gamma_\circ(ut_1 k_1, t_2 k_2)) = f_s^\circ(\gamma_\circ(ut_1, t_2)),$$

and $W_1^\circ(ut_1 k_1) = \psi(u)W_1^\circ(t_1)$ and $W_2^\circ(t_2 k_2) = W_2^\circ(t_2)$. Thus the unnormalized, unramified local zeta integral $\mathcal{Z}(s, W_1^\circ, W_2^\circ, f_s^\circ)$ equals

$$= \int_{[CN^{2,\triangle}(Z_2\times I_4)\backslash H](F)} f_s^\circ(\gamma_\circ(g_1, g_2))W_1^\circ(g_1)W_2^\circ(g_2)dg_1 dg_2$$
$$= \int_{C\backslash(T_2\times T_2)^\circ} W_1^\circ(t_1)W_2^\circ(t_2)[\int_{Z_2\backslash N^2} f_s^\circ(\gamma_\circ(ut_1, t_2))\psi(u)du]\delta_H^{-1}(t_1, t_2)dt_1 dt_2.$$

The lemma is proved.   □

From this lemma, it suffices to compute the integral

$$I(t_1, t_2; f_s^\circ, \psi) = \int_{Z_2 \backslash N^2} f_s^\circ(\gamma_\circ(ut_1, t_2))\psi(u)du. \tag{37}$$

This will be carried out in the next subsection. We conclude this subsection with results on $\delta_H(t_1, t_2)$.

LEMMA 3.1.2. *The modulus character*

$$\delta_H(t_1, t_2) = \delta_{GSp(2)}(t_1) \cdot \delta_{GSp(2)}(t_2) = |a^3 a'^{-1} b'^{-2}| \cdot |c^3 c'^{-1} d'^{-2}|,$$

if $(t_1, t_2) = h(a, b, a', b') \times h(c, d, c', d') \in T_2 \times T_2.$

PROOF. By the definition of the modulus character, we have

$$\begin{aligned}
\delta_H(t_1, t_2) &= |det[Ad(t_1, t_2)|_{\mathfrak{n}}]| \\
&= |det[Ad(t_1)|_{\mathfrak{n}_2}]| \cdot |det[Ad(t_2)|_{\mathfrak{n}_2}]| \\
&= \delta_{GSp(2)}(t_1) \cdot \delta_{GSp(2)}(t_2),
\end{aligned}$$

where $\mathfrak{n}$ is the standard maximal nilpotent subalgebra of the Lie algebra of H and $\mathfrak{n}_2$ is that of the Lie algebra of $GSp(2)$. Then the result follows from some standard computations. $\quad\square$

Note that $(T_2 \times T_2)^\circ = \{h(a, b, a', b') \times h(c, d, c', d') : aa' = bb' = cc' = dd' \neq 0\}$. we can choose parameters so that

$$C \backslash (T_2 \times T_2)^\circ = \{h(ab, a, b^{-1}, 1) \times h(cd, c, c^{-1}d^{-1}a, c^{-1}a) : abcd \neq 0\}. \tag{38}$$

From now on, we will fix this parameter system on $C \backslash (T_2 \times T_2)^\circ$ and have

COROLLARY 3.1.1. *If* $(t_1, t_2) \in C \backslash (T_2 \times T_2)^\circ$, *then* $\delta_H(t_1, t_2) = |b^4 c^6 d^4|.$

3.2. Computation of $I(t_1, t_2; f_s^\circ, \psi)$**.** The computation of $I(t_1, t_2; f_s^\circ, \psi)$ will be reduced to that of a partial 'Fourier transform' of certain Schwartz-Bruhat function over F^5. To this end, we need several lemmas. Let $\chi_\alpha(t)$ denote a one-parameter subgroup of G corresponding to a root α and

$$\chi(w, y, x, u, v) = \chi_{-\varepsilon_3-\varepsilon_4}(w)\chi_{-2\varepsilon_3}(y)\chi_{-\varepsilon_2-\varepsilon_4}(x)\chi_{-\varepsilon_1-\varepsilon_4}(u)\chi_{-\varepsilon_2-\varepsilon_3}(v). \tag{39}$$

LEMMA 3.2.1. *Let* f° *be the spherical section chosen above. Then for* $u \in Z_2 \backslash N^2$ *and* $(t_1, t_2) \in C \backslash (T_2 \times T_2)^\circ$, *we have*

$$f_s^\circ(\gamma_\circ(ut_1, t_2)) = f_s^\circ(\chi(-w, -y, x, 1, 1)h(ab, a, c^{-1}a, c^{-1}d^{-1}a, b^{-1}, 1, c, cd)).$$

PROOF. As mentioned before, for $(t_1, t_2) \in C \backslash (T_2 \times T_2)^\circ$, we can rewrite it as, under the embedding $H \hookrightarrow G$, $(t_1, t_2) = h(ab, a, cd, c, b^{-1}, 1, c^{-1}d^{-1}a, c^{-1}a)$. Then its conjugation $\gamma_\circ(t_1, t_2)\gamma_\circ^{-1}$ by $\gamma_\circ$ is

$$\chi_{-\varepsilon_1-\varepsilon_4}(1)\chi_{-\varepsilon_2-\varepsilon_3}(1)h_1(a, b, c, d)\chi_{-\varepsilon_2-\varepsilon_3}(-1)\chi_{-\varepsilon_1-\varepsilon_4}(-1)$$

and

$$\chi_{-\varepsilon_2-\varepsilon_3}(-1)\chi_{-\varepsilon_1-\varepsilon_4}(-1)\gamma_\circ = w_0.$$

Similarly, for $(u, 1) \in N^2 \times I_4$, its conjugation $\gamma_\circ(u, 1)\gamma_\circ^{-1}$ by $\gamma_\circ$ is equal to

$$\begin{pmatrix} 1 & x & -w & -z-wx & z & w \\ & 1 & -y & & -w & w' & y \\ & & 1 & & & \\ & & & 1 & & \\ & & & & -x & 1 \\ & & & & w' & y & 1 \\ & & & -z-wx & z & w & 1 \end{pmatrix} \chi(-w, -y, x, 0, 0) = p(w, x, y, z)\chi(-w, -y, x, 0, 0) \tag{40}$$

with $p(w, x, y, z) \in P_3$. Hence we have that

$$\begin{aligned} f_s(\gamma_\circ(ut_1, t_2)) &= f_s(\gamma_\circ(u, 1)\gamma_\circ^{-1}\gamma_\circ(t_1, t_2)\gamma_\circ^{-1}\gamma) \\ &= f_s(\chi(-w, -y, x, 1, 1)h_1(a, b, c, d)w_0). \end{aligned}$$

When $f_s = f_s^\circ$, we obtain what we need. □

LEMMA 3.2.2. *Set* $\chi(w) = \chi(w, 0, 0, 0, 0)$, $\chi(w, y) = \chi(w, y, 0, 0, 0)$, *and so on. Then we have following identities:*

(a) $h(t_1, t_2, t_3, t_4, t_5, t_6, t_7, t_8)^{-1}\chi(w, y, x, u, v)h(t_1, t_2, t_3, t_4, t_5, t_6, t_7, t_8)$
$= \chi(t_3t_8^{-1}w, t_3t_7^{-1}y, t_2t_8^{-1}x, t_1t_8^{-1}u, t_2t_7^{-1}v);$

(b) $\chi_{-2\varepsilon_3}(y) = h(1, 1, -y^{-1}, 1, 1, 1, -y, 1)\chi_{2\varepsilon_3}(y)k_y$, *for* $|y| > 1;$

(c) $\chi_{-\varepsilon_2-\varepsilon_4}(x) = h(1, -x^{-1}, 1, -x^{-1}, 1, -x, 1, -x, 1)\chi_{\varepsilon_2+\varepsilon_4}(x)k_x$, *for* $|x| > 1;$

(d) $\chi_{-\varepsilon_1-\varepsilon_4}(u) = h(-u^{-1}, 1, 1, -u^{-1}, -u, 1, 1, -u)\chi_{\varepsilon_1+\varepsilon_4}(u)k_u$, *for* $|u| > 1;$

(e) $\chi_{-\varepsilon_2-\varepsilon_3}(v) = h(1, -v^{-1}, -v^{-1}, 1, 1, -v, -v, 1)\chi_{\varepsilon_2+\varepsilon_3}(v)k_v$, *for* $|v| > 1;$

(f) $\chi(w, y)\chi_{\varepsilon_2+\varepsilon_4}(x) = \chi_{\varepsilon_2-\varepsilon_3}(-wx)\chi_{\varepsilon_2+\varepsilon_4}(x)\chi(w, y);$

(g) $\chi(w)\chi_{2\varepsilon_3}(y) = \chi_{\varepsilon_3-\varepsilon_4}(-wy)\chi_{2\varepsilon_3}(y)\chi_{-2\varepsilon_4}(-w^2y)\chi(w);$

(h) $\chi(w, y, x)\chi_{\varepsilon_1+\varepsilon_4}(u) = \chi_{\varepsilon_1-\varepsilon_2}(-xu)\chi_{\varepsilon_1-\varepsilon_3}(-wu)\chi_{\varepsilon_1+\varepsilon_4}(u)\chi(w, y, x);$

(i) $\chi(w, y, x)\chi_{\varepsilon_2+\varepsilon_3}(v) = p_1(w, y, x, v)\chi(w, y, x);$

(j) $\chi(w, y, x, u)\chi_{\varepsilon_2+\varepsilon_3}(v) = p_2(w, y, x, v)\chi(w - xyv, y, x, u),$

where $k_y, k_x, k_u, k_v \in K_4$ *and* $p_1(w, y, x, v), p_2(w, y, x, v) \in P_3^4(F)$ *and also*

$$\delta_{P_3^4}(P_1^2(w, y, x, v)) = \delta_{P_3^4}(p_2(w, y, x, v)) = 1.$$

The proof of the lemma is straightforward. The integral $I(t_1, t_2; f_s^\circ, \psi)$ as defined in (34) can be reduced as follows.

PROPOSITION 3.2.1. *Let* $\delta = \delta_H(h(ab, a, b^{-1}, 1) \times h(cd, c, c^{-1}d^{-1}a, c^{-1}a)) = |b^4c^6d^4|$. *Then the integral* $I(t_1, t_2; f_s^\circ, \psi)$ *equals*

$$\delta^{\frac{1}{2}}(|a|^{\frac{3}{2}}|b||c|^{-1})^{s+1}\int_{F^3} f^\circ(\chi(w, y, x, abc^{-1}d^{-1}, ac^{-1}))\psi_\circ(a^{-1}cdx + a^{-1}c^2y)dwdxdy.$$

PROOF. Let $h_1(a, b, c, d) = h(ab, a, c^{-1}a, c^{-1}d^{-1}a, b^{-1}, 1, c, cd)$. Since the element $\gamma_\circ(ut_1, t_2)\gamma_\circ^{-1}$ is equal to $\chi(w, y, x, 1, 1)h_1(a, b, c, d)\chi_{-\varepsilon_1-\varepsilon_4}(-1)\chi_{-\varepsilon_2-\varepsilon_3}(-1)$ and the element $\chi(w, y, x, 1, 1) \cdot h_1(a, b, c, d)$ is equal to

$$h_1(a, b, c, d)\chi(ac^{-2}d^{-1}w, ac^{-2}y, ac^{-1}d^{-1}x, abc^{-1}d^{-1}, ac^{-1}),$$

we have

$$f_s(\gamma_\circ(ut_1, t_2)) = |a^3bc^{-1}|^{s+3}|a|^{-\frac{3}{2}(s+3)}$$
$$\cdot f_s(\chi(ac^{-2}d^{-1}w, ac^{-2}y, ac^{-1}d^{-1}x, abc^{-1}d^{-1}, ac^{-1})w_\circ).$$

Thus we can deduce our integral $I(t_1, t_2; f_s, \psi)$ as follows:

$$\int_{Z_2\backslash N^2} f_s(\gamma_\circ(ut_1, t_2))\psi(u)du$$
$$= |a^3bc^{-1}|^{s+3}|a|^{-\frac{3}{2}(s+3)}$$
$$\cdot \int_{F^3} f_s(\chi(ac^{-2}d^{-1}w, ac^{-2}y, ac^{-1}d^{-1}x, abc^{-1}d^{-1}, ac^{-1})w_\circ)\psi_\circ(x+y)dwdxdy$$
$$= \delta^{\frac{1}{2}}(|a|^{\frac{3}{2}}|b||c|^{-1})^{s+1}$$
$$\cdot \int_{F^3} f_s(\chi(w, y, x, abc^{-1}d^{-1}, ac^{-1})w_\circ)\psi_\circ(a^{-1}cdx + a^{-1}c^2y)dwdxdy,$$

here we change the variables by $ac^{-2}d^{-1}w \mapsto w$; $ac^{-2}y \mapsto y$; and $ac^{-1}d^{-1}x \mapsto x$. Set $f_s = f_s^\circ$, we get the formula. $\qquad\square$

Now we have to compute an integral of following type:

$$I(u, v, \alpha, \beta; f_s^\circ, \psi) = \int_{F^3} f_s^\circ(\chi(w, y, x, u, v))\psi_\circ(\alpha x + \beta y)dwdxdy. \qquad (41)$$

LEMMA 3.2.3. *For the normalized unramified section f_s° in $I_{3,v}^4(s)$ and the un-ramified additive unitary character ψ, the integral $I(u, v, \alpha, \beta; f^\circ, \psi_\circ)$ equals*

$$= \begin{cases} I(0, 0, \alpha, \beta; f_s^\circ, \psi_\circ) & if\ |u|, |v| \le 1; \\ |v|^{-2s-2}I(0, 0, -v\alpha, v^2\beta; f_s^\circ, \psi_\circ) & if\ |u| \le 1, |v| > 1; \\ |u|^{-s-1}I(0, 0, -u\alpha, \beta; f_s^\circ, \psi_\circ) & if\ |u| > 1, |v| \le 1; \\ |v|^{-2s-2}|u|^{-s-1}I(0, 0, uv\alpha, v^2\beta; f_s^\circ, \psi_\circ) & if\ |u| > 1, |v| > 1. \end{cases}$$

PROOF. We will evaluate the integral $I(u, v, \alpha, \beta; f_s, \psi)$ case by case.

(1) If $|u| \le 1$ and $|v| \le 1$, then there is nothing to do with because by definition $\chi(w, y, x, u, v) = \chi(w, y, x)\chi(u, v)$ and $\chi(u, v) \in K_4$.

(2) If $|u| \le 1$ and $|v| > 1$, then by Lemma 3.2.2 (e) and (i),

$$f_s^\circ(\chi(w, y, x, u, v))$$
$$= f_s^\circ(\chi(w, y, x, v))$$
$$= f_s^\circ(h(1, -v^{-1}, -v^{-1}, 1, 1, -v, -v, 1)\chi(-v^{-1}w, v^{-2}y, -v^{-1}x)\chi_{\varepsilon_2+\varepsilon_3}(v)k(v))$$
$$= |v|^{-2s-6}f_s^\circ(\chi(-v^{-1}w, v^{-2}y, -v^{-1}x)\chi_{\varepsilon_2+\varepsilon_3}(v))$$
$$= |v|^{-2s-6}f_s^\circ(\chi(-v^{-1}w, v^{-2}y, -v^{-1}x)).$$

Hence the integral can be deduced as

$$I(u, v, \alpha, \beta; f_s^\circ, \psi) = |v|^{-2s-2}\int_{F^3} f_s^\circ(\chi(w, y, x))\psi_\circ(-v\alpha x + v^2\beta y)dwdxdy,$$

here the variables are changed by setting $-v^{-1}w \mapsto w, v^{-2}y \mapsto y$, and $-v^{-1}x \mapsto x$.

(3) If $|u| > 1$ and $|v| \leq 1$, then by Lemma 3.2.2 (d) and (h),

$$
\begin{aligned}
& f_s^{\circ}(\chi(w,y,x,u,v)) \\
= \ & f_s^{\circ}(\chi(w,y,x,u)) \\
= \ & f_s^{\circ}(h(-u^{-1},1,1,-u^{-1},-u,1,1,-u)\chi(-u^{-1}w,y,-u^{-1}x)\chi_{\varepsilon_2+\varepsilon_3}(u)k(u)) \\
= \ & |u|^{-s-3}f_s^{\circ}(\chi(-u^{-1}w,y,-u^{-1}x)\chi_{\varepsilon_2+\varepsilon_3}(u)) \\
= \ & |u|^{-s-3}f_s^{\circ}(\chi(-u^{-1}w,y,-u^{-1}x)).
\end{aligned}
$$

Hence the integral has form:

$$
I(u,v,\alpha,\beta;f_s^{\circ},\psi) = |u|^{-s-1}\int_{F^3}f_s^{\circ}(\chi(w,y,x))\psi_{\circ}(-u\alpha x + \beta y)dwdxdy,
$$

here the variables are changed by setting $-u^{-1}w \mapsto w, y \mapsto y$, and $-u^{-1}x \mapsto x$

(4) If $|u| > 1$ and $|v| > 1$, then by Lemma 3.2.2 (d), (e), (h), (i), and (j), we deal with the variable v first and then with u and obtain that

$$
\begin{aligned}
& f_s^{\circ}(\chi(w,y,x,u,v)) \\
= \ & f_s^{\circ}(h(1,-v^{-1},-v^{-1},1,1,-v,-v,1)\chi(-v^{-1}w,v^{-2}y,-v^{-1}x,u)\chi_{\varepsilon_2+\varepsilon_3}(v)k(v)) \\
= \ & |v|^{-2s-6}f_s^{\circ}(\chi(-v^{-1}w,v^{-2}y,-v^{-1}x,u)\chi_{\varepsilon_2+\varepsilon_3}(v)) \\
= \ & |v|^{-2s-6}f_s^{\circ}(\chi(-v^{-1}w+v^{-2}yx,v^{-2}y,-v^{-1}x,u)) \\
= \ & |u|^{-s-3}|v|^{-s-6}f_s^{\circ}(\chi((vu)^{-1}w,v^{-2}y,(vu)^{-1}x)).
\end{aligned}
$$

Hence the integral is equal to

$$
I(u,v,\alpha,\beta;f_s^{\circ},\psi) = |v|^{-2s-2}|u|^{-s-1}\int_{F^3}f_s^{\circ}(\chi(w,y,x))\psi_{\circ}(uv\alpha x + v^2\beta y)dwdxdy,
$$

here the variables are changed by setting $(uv)^{-1}w \mapsto w, v^{-2}y \mapsto y$, and $(uv)^{-1}x \mapsto x$. $\square$

According to the above reductions, we have to compute the 'Fourier transform' of function $f_s^{\circ}(\chi(w,y,x))$ with respect to additive characters $\psi_{\circ}(\alpha x + \beta y)$ for $\alpha, \beta \in F^{\times}$, that is, the integral as follows:

$$
I(\alpha,\beta) := \int_{F^3} f^{\circ}(\chi(w,y,x))\psi_{\circ}(\alpha x + \beta y)dwdxdy. \tag{42}
$$

This is done in the following lemmas.

LEMMA 3.2.4. *The integral $I(\alpha, \beta)$ can reduced to a product of several one-variable integrations as follows:*

$$
\begin{aligned}
I(\alpha, \beta) &= \int_{F^3} f^\circ(\chi(w, y, x))\psi_\circ(\alpha x + \beta y)dwdxdy \\
&= [1 + \int_{|w|>1} |w|^{-s-3}dw] \cdot [\int_{|x|\leq 1} \psi_\circ(\alpha x)dx + \int_{|x|>1} |x|^{-s-2}\psi_\circ(\alpha x)dx] \\
&\quad \cdot [\int_{|y|\leq 1} \psi_\circ(\beta y)dy + \int_{|y|>1} |y|^{-s-2}\psi_\circ(\beta y)dy].
\end{aligned}
$$

PROOF. If $|x| \leq 1$, then

$$
\begin{aligned}
I(\alpha, \beta) &= \int_{F^3, |x|\leq 1} f^\circ(\chi(w, y, x))\psi_\circ(\alpha x + \beta y)dwdxdy \\
&= \int_{F^3, |x|\leq 1} f^\circ(\chi(w, y))\psi_\circ(\alpha x + \beta y)dwdxdy \\
&= \int_{|x|\leq 1} \psi_\circ(\alpha x)dx \int_{F^2} f^\circ(\chi(w, y))\psi_\circ(\beta y)dwdxdy.
\end{aligned}
$$

If $|x| > 1$, then by Lemma 3.2.2 (f),

$$
\begin{aligned}
\chi(w, y, x) &= h(1, -x^{-1}, 1, -x^{-1}, 1, -x, 1, -x)\chi(-x^{-1}w, y)\chi_{\varepsilon_2+\varepsilon_4}(x) \\
&= h(1, -x^{-1}, 1, -x^{-1}, 1, -x, 1, -x)p(w, x)\chi(-x^{-1}w, y),
\end{aligned}
$$

where $p(w, x) = \chi_{\varepsilon_2-\varepsilon_3}(-wx)\chi_{\varepsilon_2+\varepsilon_4}(x)$ belongs to P_3^4. Hence we have $f^\circ(\chi(w, y, x)) = |x|^{-s-3}f^\circ(\chi(-x^{-1}w, y))$ and the integral $I(\alpha, \beta)$ equals

$$
\begin{aligned}
&\int_{F^3, |x|>1} f^\circ(\chi(w, y, x))\psi_\circ(\alpha x + \beta y)dwdxdy \\
&= \int_{F^3, |x|>1} |x|^{-s-3}f^\circ(\chi(-x^{-1}w, y))\psi_\circ(\alpha x + \beta y)dwdxdy \\
&= \int_{|x|>1} |x|^{-s-2}\psi_\circ(\alpha x)dx \int_{F^2} f^\circ(\chi(w, y))\psi_\circ(\beta y)dwdy,
\end{aligned}
$$

where we change the variables by $-x^{-1}w \mapsto w, x \mapsto x$, and $y \mapsto y$. Therefore we deduce that the integral $I(\alpha, \beta)$ is equal to

$$
\begin{aligned}
&\int_{F^3} f^\circ(\chi(w, y, x))\psi_\circ(\alpha x + \beta y)dwdxdy \\
&= [\int_{|x|\leq 1} \psi_\circ(\alpha x)dx + \int_{|x|>1} |x|^{-s-2}\psi_\circ(\alpha x)dx] \int_{F^2} f^\circ(\chi(w, y))\psi_\circ(\beta y)dwdy.
\end{aligned}
$$

In a similar way, we can integrate the other two variable w, y

$$\int_{F^2} f^\circ(\chi(w,y))\psi_\circ(\beta y)dwdy$$

$$= [\int_{|y|\leq 1} \psi_\circ(\beta y)dy + \int_{|y|>1} |y|^{-s-2}\psi_\circ(\beta y)dy] \int_F f^\circ(\chi(w))dw$$

$$= [\int_{|y|\leq 1} \psi_\circ(\beta y)dy + \int_{|y|>1} |y|^{-s-2}\psi_\circ(\beta y)dy][\int_{|w|\leq 1} dw + \int_{|w|>1} |w|^{-s-3}dw].$$

The proof is finished. $\qquad\qquad\square$

LEMMA 3.2.5. *Let q be the cardinality of the residue class field of F. The one-variable integrals in the last lemma can be evaluated as follows:*

(a) $1 + \int_{|w|>1} |w|^{-s-3}dw = (1 - q^{-s-3})(1 - q^{-s-2})^{-1};$

(b) $\int_{|x|\leq 1} \psi_\circ(\alpha x)dx + \int_{|x|>1} |x|^{-s-2}\psi_\circ(\alpha x)dx =$

$$= \begin{cases} 0 & if\ |\alpha| > 1, \\ (1 - q^{-s-2})(1 - q^{-(ord\alpha+1)(s+1)})(1 - q^{-s-1})^{-1} & if\ |\alpha| \leq 1 \end{cases}$$

PROOF. The computation to establish those formulas is standard and is omitted here. $\qquad\qquad\square$

We are finally able to complete our computation of the integral $I(t_1, t_2; f_s^\circ, \psi)$ as in (34) and state it in the following theorem.

THEOREM 3.2.1. *Let $m_1 = ord_v(a)$, $m_2 = ord_v(b)$, $n_1 = ord_v(c)$, and $n_2 = ord_v(d)$. Let $\delta^{\frac{1}{2}} = \delta_H^{\frac{1}{2}}(h(ab, a, b^{-1}, 1) \times h(cd, c, c^{-1}d^{-1}a, c^{-1}a)) = |b^2c^3d^2|$ and $t = -(s+1)$. The integral $I(t_1, t_2; f_s^\circ, \psi)$ equals a product of $\delta_H^{\frac{1}{2}}(t_1, t_2)\frac{(1-q^{t-2})(1-q^{t-1})}{(1-q^t)^2}$ and*

$$\begin{cases} q^{(\frac{3}{2}m_1+m_2-n_1)t}(1-q^{(n_1+n_2-m_1+1)t})(1-q^{(2n_1-m_1+1)t}) & if\ \begin{smallmatrix} m_1+m_2\geq n_1+n_2\geq m_1\\ 2n_1\geq m_1\geq n_1 \end{smallmatrix}; \\ q^{(m_2-\frac{1}{2}m_1+n_1)t}(1-q^{(n_2+1)t})(1-q^{(m_1+1)t}) & if\ \begin{smallmatrix} m_1+m_2\geq n_1+n_2\\ n_1>m_1\geq 0, n_2\geq 0 \end{smallmatrix}; \\ q^{(\frac{1}{2}m_1+n_2)t}(1-q^{(m_2+1)t})(1-q^{(2n_1-m_1+1)t}) & if\ \begin{smallmatrix} m_1+m_2<n_1+n_2, m_2\geq 0\\ 2n_1\geq m_1\geq n_1 \end{smallmatrix}; \\ q^{(2n_1+n_2-\frac{3}{2}m_1)t}(1-q^{(m_1+m_2-n_1+1)t})(1-q^{(m_1+1)t}) & if\ \begin{smallmatrix} n_1\leq m_1+m_2<n_1+n_2\\ n_1>m_1\geq 0 \end{smallmatrix}; \\ 0 & otherwise \end{cases}$$

PROOF. By Proposition 3.2.1, one has that

$$I(t_1, t_2; f_s^\circ, \psi) = \delta^{\frac{1}{2}}(|a|^{\frac{3}{2}}|b||c|^{-1})^{s+1}I(abc^{-1}d^{-1}, ac^{-1}, a^{-1}cd, a^{-1}c^2; f_s^\circ, \psi_\circ)$$

$$= \delta^{\frac{1}{2}}q^{(\frac{3}{2}m_1+m_2-n_1)t}I(abc^{-1}d^{-1}, ac^{-1}, a^{-1}cd, a^{-1}c^2; f_s^\circ, \psi_\circ).$$

Then, by Lemma 3.2.3, the integral $I(abc^{-1}d^{-1}, ac^{-1}, a^{-1}cd, a^{-1}c^2; f_s^\circ, \psi_\circ)$ can be reduced as follows:

(1) If $m_1 + m_2 \geq n_1 + n_2$ and $m_1 \geq n_2$, then

$$I(abc^{-1}d^{-1}, ac^{-1}, a^{-1}cd, a^{-1}c^2; f_s^\circ, \psi_\circ) = I(a^{-1}cd, a^{-1}c^2).$$

(2) If $m_1 + m_2 \geq n_1 + n_2$ and $m_1 < n_2$, then

$$I(abc^{-1}d^{-1}, ac^{-1}, a^{-1}cd, a^{-1}c^2; f_s^\circ, \psi_\circ) = q^{2(n_1-m_1)t}I(-d, a).$$

(3) If $m_1 + m_2 < n_1 + n_2$ and $m_1 \geq n_2$, then

$$I(abc^{-1}d^{-1}, ac^{-1}, a^{-1}cd, a^{-1}c^2; f_s^\circ, \psi_\circ) = q^{(n_1+n_2-m_1-m_2)t}I(-b, a^{-1}c^2).$$

(4) If $m_1 + m_2 < n_1 + n_2$ and $m_1 < n_2$, then

$$I(abc^{-1}d^{-1}, ac^{-1}, a^{-1}cd, a^{-1}c^2; f_s^\circ, \psi_\circ) = q^{(3n_1+n_2-3m_1-m_2)t}I(abc^{-1}, a).$$

Now the theorem follows from Lemma 3.2.4 and Lemma 3.2.5. $\square$

3.3. The Unramified Local Zeta Integral. With the preparation of last two subsections, we can complete our computation of the local zeta integral. To this end, we first recall the formula of Casselman-Shalika for the unramified Whittaker functions, which works for any quasi-split reductive groups over a p-adic field [**CaSh**]. For our special case, such a formula can also be found in Bump's survey [**Bum**].

THEOREM 3.3.1 (Casselman-Shalika[**CaSh**]). *For the unramified Whittaker function* $W^\circ \in W(\pi_v)$, $W^\circ(h(ab, a, b^{-1}, 1))$ *vanishes if* $ord_v(a) < 0$ *or* $ord_v(b) < 0$ *and also*

$$W^\circ(h(ab, a, b^{-1}, 1)) = \delta_{GSp(2)}^{1/2}(h(ab, a, b^{-1}, 1))tr(ord_v(a), ord_v(b)), \qquad (43)$$

where $tr(m, n)$ *is the trace of the representation* $\rho_{(m,n)}$ *evaluating at the conjugacy class of semisimple elements in* $GSp(2, \mathbb{C})$, *the L-group of* $GSp(2)$[**Bor**], *which is determined by* π_v *via Satake isomorphism and* $\rho_{(m,n)}$ *is the irreducible representation of* $GSp(2, \mathbb{C})$ *with highest weight* $(m, n) = m\varepsilon_1 + n(\varepsilon_1 + \varepsilon_2)$, $m, n \geq 0$.

THEOREM 3.3.2. *Let* $X = q^{-(s+1)/2}$, $k_1 = ord_v(a)$, $k_2 = ord_v(b)$, $l_1 = ord_v(c)$, *and* $l_2 = ord_v(d)$, *with* $k_1, k_2, l_1, l_2 \geq 0$. *Then the (normalized) unramified local zeta integral* $\mathcal{Z}^*_v(s, W_{1v}^\circ, W_{2v}^\circ, f_v^\circ) = d_{v,G}(s)\mathcal{Z}_v(s, W_{1v}^\circ, W_{2v}^\circ, f_v^\circ)$ *equals*

$$\frac{1}{(1-X^2)(1-X^4)} \cdot$$

$$\Big\{ \sum_{\substack{k_1+k_2\geq l_1+l_2\geq k_1 \\ 2l_1\geq k_1\geq l_1}} (k_1, k_2|2l_1 - k_1, l_2)P_1(k_1, k_2, l_1, l_2; X)$$

$$+ \sum_{\substack{k_1+k_2\geq l_1+l_2 \\ l_1>k_1}} (k_1, k_2|2l_1 - k_1, l_2)P_2(k_1, k_2, l_1, l_2; X)$$

$$+ \sum_{\substack{k_1+k_2<l_1+l_2 \\ 2l_1\geq k_1\geq l_1}} (k_1, k_2|2l_1 - k_1, l_2)P_3(k_1, k_2, l_1, l_2; X)$$

$$+ \sum_{\substack{l_1\leq k_1+k_2<l_1+l_2 \\ k_1<l_1}} (k_1, k_2|2l_1 - k_1, l_2)P_4(k_1, k_2, l_1, l_2; X)\Big\},$$

where $d_{v,G}(s) = \zeta_v(2s+2)\zeta_v(s+1)\zeta_v(s+2)\zeta_v(s+3)$ *is the local v-component of the normalizing factor,* $(m_1, m_2|n_1, n_2)$ *indicates the trace of the tensor product of representations* $\rho_{(m_1,m_2)}$, *and* $\rho_{(n_1,n_2)}$ *of* $GSp(2,\mathbb{C})$ *evaluated at the semi-simple conjugacy class associated to* $\pi_1 \times \pi_2$ *in* $(GSp(2,\mathbb{C}) \times GSp(2,\mathbb{C}))^\circ$, *and*

$$P_1(k_1, k_2, l_1, l_2; X) = X^{3k_1+2k_2-2l_1}\frac{(1 - X^{2(l_1+l_2-k_1+1)})(1 - X^{2(2l_1-k_1+1)})}{(1-X^2)^2},$$

$$P_2(k_1, k_2, l_1, l_2; X) = X^{2k_2-k_1+2l_1}\frac{(1 - X^{2(l_2+1)})(1 - X^{2(k_1+1)})}{(1-X^2)^2},$$

$$P_3(k_1, k_2, l_1, l_2; X) = X^{k_1+2l_2}\frac{(1 - X^{2(k_2+1)})(1 - X^{2(2l_1-k_1+1)})}{(1-X^2)^2},$$

and

$$P_4(k_1, k_2, l_1, l_2; X) = X^{4l_1+2l_2-3k_1}\frac{(1 - X^{2(k_1+k_2-l_1+1)})(1 - X^{2(k_1+1)})}{(1-X^2)^2}.$$

PROOF. The unnormalized local zeta-integral is

$$\mathcal{Z}_v(s, W_{1v}^\circ, W_{2v}^\circ, f_v^\circ) = \int_{C\backslash(T_2\times T_2)^\circ} W_{1v}^\circ(t_1)W_{2v}^\circ(t_2)I(t_1, t_2; f_s^\circ, \psi)\delta_H^{-1}(t_1, t_2)dt_1 dt_2.$$

Applying the formula of Casselman-Shalika for the case that $t_1 = h(ab, a, b^{-1}, 1)$ and $t_2 = h(cd, c, c^{-1}d^{-1}a, c^{-1}a)$, we obtain that the nonvanishing of

$$W_{2v}^\circ(t_2) = W_{2v}^\circ(h(cd, c, c^{-1}d^{-1}a, c^{-1}a))$$

implies that $2l_1 - k_1, l_2 \geq 0$ and the nonvanishing of $W_{1v}^\circ(t_1)$ implies that $k_1, k_2 \geq 0$, and also

$$W_{1v}^\circ(t_1) = |a^3b^4|^{1/2}tr(k_1, k_2), \quad W_{2v}^\circ(t_2) = |c^6d^4a^{-3}|^{1/2}tr(2l_1 - k_1, l_2).$$

Thus we obtain that

$$\begin{aligned} W_{1v}^\circ(t_1)W_{2v}^\circ(t_2) &= |b^2c^3d^2|tr(k_1, k_2)tr(2l_1 - k_1, l_2) \\ &= \delta_H^{\frac{1}{2}}(k_1, k_2|2l_1 - k_1, l_2). \end{aligned}$$

Note that $\zeta_v(s) = (1 - q^{-s})^{-1}$. Applying these data and Theorem 3.2.1 to our (unnormalized) unramified local zeta integral, we will get what we needed. □

We are now going to state our main theorem on the unramified computation, which equates our local unramified zeta integral to the local Langlands L-factor.

THEOREM 3.3.3. *Let* $s' = (s+1)/2$. *Let* ρ_1 *be the irreducible representation of* $GSp(2, \mathbb{C})$ *with highest weight* ε_1. *Then*

$$\mathcal{Z}_v^*(s, W_{1v}^\circ, W_{2v}^\circ, f^\circ) = L_v(s', \pi_{1v} \otimes \pi_{2v}, (\rho_1 \otimes \rho_1)_v)$$

Note that the local Langlands L-factor $L_v(s', \pi_{1v} \otimes \pi_{2v}, (\rho_1 \otimes \rho_1)_v)$ is of degree 16. By the definition of local Langlands L-factor,

$$L_v(s', \pi_{1v} \otimes \pi_{2v}, (\rho_1 \otimes \rho_1)_v) = \frac{1}{\det[1 - (\rho_1 \otimes \rho_1)_v(\tau_{1v}, \tau_{2v})X]}$$

where $X = q^{-s'}$ and τ_{1v}, τ_{2v} are representatives in semisimple conjugacy class in $^LGSp(2, F_v)$ determined by π_{1v}, π_{2v}, respectively. Since an element in the maximal split torus T_2 can be written as $h(t_1, t_2, t_3, t_4) = h(t_1, t_2, t_1^{-1}, t_2^{-1})h(1, 1, t_1 t_3, t_2 t_4)$ and $\rho_1(h(t_1, t_2, t_3, t_4)) = \rho_1(h(t_1, t_2, t_1^{-1}, t_2^{-1}))$, see Proposition 1.2.1, we can assume that ρ_1 is the irreducible representation of $Sp(2, \mathbb{C})$ with highest weight ε_1 and $\tau_{1,v}$, $\tau_{2,v}$ are the associated semisimple conjugacy classes in $Sp(2, F_v)$. This theorem will be proved in the next section.

4. Local Langlands Factor of Degree 16 for $GSp(2) \times GSp(2)$

We are going to study the 16 degree local Langlands L-factor $L_v(s', \pi_{1v} \otimes \pi_{2v}, (\rho_1 \otimes \rho_1)_v)$ by means of the representation theory of the complex dual group $(GSp(2, \mathbb{C}) \times GSp(2, \mathbb{C}))^\circ$. As explained in Proposition 1.2.1, it is actually reduced to the representation theory of $Sp(2, \mathbb{C}) \times Sp(2, \mathbb{C})$. The method we will apply is the Kostant-Rallis' algorithm about the explicit spectral decomposition of the space of harmonic polynomial functions over a symmetric space described in [**KoRa**]. We will recall first the algorithm of Kostant-Rallis in general, and then apply it to the computation of our local Langlands L-factor. Let us mention that the normality of some nilpotent orbits is a critical sufficient condition to make the Kostant-Rallis algorithm work. However, it is usually not an easy job to check the normality of a nilpotent orbit in the case of symmetric spaces. Instead of checking the normality of the relevant nilpotent orbit, we will use branching formula to make the Kostant-Rallis algorithm work in our special case.

4.1. Kostant-Rallis Algorithm.
We will recall some notions from Kostant-Rallis [**KoRa**] and J. Sekiguchi [**Sek**], and restate their main results as a theorem.

Let $\mathfrak{g}$ be a complex Lie algebra and θ a complex linear involution of $\mathfrak{g}$. Then we have as usual the following decomposition as linear spaces: $\mathfrak{g} = \mathfrak{t} + \mathfrak{p}$, where $\mathfrak{t} = \{x \in \mathfrak{g} : \theta(x) = x\}$ and $\mathfrak{p} = \{x \in \mathfrak{g} : \theta(x) = -x\}$. In this way, we obtain a (complex) symmetric pair $(\mathfrak{g}, \mathfrak{t})$ in terminology of Sekiguchi [**Sek**]. The subspace $\mathfrak{p}$ is called the vector space associated to the symmetric pair. Let G be the adjoint group of $\mathfrak{g}$ and K the analytic subgroup of G corresponding to $\mathfrak{t}$. Let $K_\theta = \{g \in G : \theta(g) = g\}$ (the involution θ of a Lie algebra will induce an involution of the adjoint group, which is denoted by the same notation θ). According to Lemma 1.1 in [**Sek**], one has that K_θ coincides with the normalizer of K in G.

It follows from the decomposition $\mathfrak{g} = \mathfrak{t} + \mathfrak{p}$ that $\mathfrak{p}$ has a structure of K_θ-module by the adjoint action. Naturally, we have the coadjoint representation of K_θ on $\mathcal{S}$ the ring of all polynomial functions in $\mathfrak{p}$, which is defined as follows: for any $g \in K_\theta$

and $f \in \mathcal{S}$, and any $x \in \mathfrak{p}$, $(g \cdot f)(x) = f(g^{-1} \cdot x)$. Now we restate the main results from Kostant-Rallis [**KoRa**], which gives an algorithm to obtain the implicit spectral decomposition of the space of all harmonic polynomial functions on $\mathfrak{p}$.

THEOREM 4.1.1 (Kostant-Rallis). *Let z be a regular semisimple element in $\mathfrak{p}$ and e a regular nilpotent element in $\mathfrak{p}$. Let $x_{\circ}$ be any element in $\mathfrak{t}$ such that $[x_{\circ}, e] = e$. Let Γ be the set of all equivalence classes of rational irreducible representations of K_{θ} and V_{γ} the space of the representation γ for $\gamma \in \Gamma$. Let O_z be the K_{θ}-orbit of z in $\mathfrak{p}$ and $R(O_z)$ the algebra of all everywhere defined rational functions on O_z. Let $\mathcal{S}$ be the ring of all polynomial functions on $\mathfrak{p}$. Then*

(a) *'Separation of Variables'*

$$\mathcal{S} = \mathcal{S}^{K_{\theta}} \otimes \mathcal{H}$$

where $\mathcal{S}^{K_{\theta}}$ is the subalgebra of K_{θ}-invariants in $\mathcal{S}$ and $\mathcal{H}$ is the space of all K_{θ}-harmonic polynomials on $\mathfrak{p}$.

(b) *the restriction map $r_{O_z} : f \mapsto f|_{O_z}$ gives a K_{θ}-isomorphism*

$$\mathcal{H} \to R(O_z).$$

(c) *'Multiplicity':*

$$\mathcal{H} = \bigoplus_{\gamma \in \Gamma} \mathcal{H}_{\gamma}, \quad \text{with } mult_{\mathcal{H}}(\gamma) = \dim V_{\gamma}^{K_{\theta}^z} = l(\gamma)$$

where $K_{\theta}^z = Stab_{K_{\theta}}(z)$, $V_{\gamma}^{K_{\theta}^z}$ is the subspace of all K_{θ}^z-fixed vectors in V_{γ}, and $\mathcal{H}_{\gamma}$ is the isotropic subspace of γ, that is, the set of all functions in H which transform under K_{θ} according to the representation γ.

(d) *'Homogeneity': The isotropic subspace can be described as*

$$\mathcal{H}_{\gamma} = \sum_{i=1}^{l(\gamma)} \mathcal{H}_{\gamma, i}$$

where $\mathcal{H}_{\gamma, i}$ is, for $i = 1, 2, \cdots, l(\gamma)$, an irreducible K_{θ}-module in $\mathcal{H}_{\gamma}$ consisting of homogeneous functions of degree $d_i(\gamma)$, corresponding to a unique monotonic sequence of increasing nonnegative integers $d_i(\gamma)$, $i = 1, 2, \cdots, l(\gamma)$.

(e) *For $\gamma \in \Gamma$, there is a $x_{\circ}$-stable $l(\gamma)$-dimensional subspace $V_{\gamma}(e) \subset V_{\gamma}^{K_{\theta}^e} \subset V_{\gamma}$ such that $x_{\circ}$ is diagonalizable on $V_{\gamma}(e)$ and the eigenvalues are exactly the nonnegative integers $d_i(\gamma), i = 1, 2, \cdots, l(\gamma)$, defined above.*

In general, for a regular element $e \in \mathfrak{p}$, the subspace $V_{\gamma}(e)$ is contained in the subspace $V_{\gamma}^{K_{\theta}^e}$, and those two subspaces may not be equal. It is not difficult to check that $V_{\gamma}(e) = V_{\gamma}^{K_{\theta}^e}$ if the K_{θ}-orbit of e is a normal algebraic variety [**KoRa**].

4.2. Symmetric Space $(Sp(4), Sp(2) \times Sp(2))$. In this subsection, we consider the symplectic group $G = Sp(4, \mathbb{C})$ and an involution θ defined by conjugation of the element $\eta = \begin{pmatrix} I_2 & 0 & 0 & 0 \\ 0 & -I_2 & 0 & 0 \\ 0 & 0 & I_2 & 0 \\ 0 & 0 & 0 & -I_2 \end{pmatrix}$ in G, that is, $\theta(g) = \eta g \eta^{-1}$ for $g \in G$. Then the subgroup $K_\theta = \{g \in Sp(4) : \theta(g) = g\}$ is in form: $\begin{pmatrix} X_1 & 0 & X_2 & 0 \\ 0 & Y_1 & 0 & Y_2 \\ X_3 & 0 & X_4 & 0 \\ 0 & Y_3 & 0 & Y_4 \end{pmatrix}$ satisfying the conditions below: for $X_i, i = 1, 2, 3, 4$,

$$-{}^tX_3X_1 + {}^tX_1X_3 = 0; \quad -{}^tX_4X_2 + {}^tX - 2X_4 = 0; \quad and \quad -{}^tX_3X_2 + {}^tX_1X_4 = 1,$$

and similarly for $Y_i, i = 1, 2, 3, 4$.

PROPOSITION 4.2.1. *There is an isomorphism between K_θ and $Sp(2) \times Sp(2)$ given by*

$$\begin{pmatrix} X_1 & X_2 \\ X_3 & X_4 \end{pmatrix} \begin{pmatrix} Y_1 & Y_2 \\ Y_3 & Y_4 \end{pmatrix} \mapsto \begin{pmatrix} X_1 & 0 & X_2 & 0 \\ 0 & Y_1 & 0 & -Y_2 \\ X_3 & 0 & X_4 & 0 \\ 0 & -Y_3 & 0 & Y_4 \end{pmatrix}.$$

On the level of Lie algebras, correspondingly, the involution θ gives following decomposition of $\mathfrak{sp}(4)$ the complex Lie algebra of $Sp(4)$

$$\mathfrak{sp}(4) = \mathfrak{t} + \mathfrak{p}, \tag{44}$$

with $[\mathfrak{t}, \mathfrak{p}] \subset \mathfrak{p}$. Under the adjoint action, the vector space $\mathfrak{p}$ has a K_θ-module structure, which can be described as follows.

PROPOSITION 4.2.2. *Let ρ_1 be the 4-dimensional standard complex representation of $Sp(2)$. Then one has the following isomorphism of K_θ-modules:*

$$\mathfrak{p} \cong \rho_1 \otimes \rho_1. \tag{45}$$

Let

$$\mathfrak{a} = \left\{z = \begin{pmatrix} 0 & A_1 & 0 & 0 \\ A_1 & 0 & 0 & 0 \\ 0 & 0 & 0 & -A_1 \\ 0 & 0 & -A_1 & 0 \end{pmatrix} : A_1 = diag(a, b)\right\}. \tag{46}$$

Then $\mathfrak{a}$ is a maximal abelian subalgebra of $\mathfrak{sp}(4)$ in $\mathfrak{p}$. This implies that our symmetric space $(Sp(4), Sp(2) \times Sp(2))$ is of rank two. Since K_θ is a semisimple algebraic group, according to the Invariant Theory, $\mathcal{S}^{K_\theta}$ is a polynomial algebra generated by two homogeneous polynomials u_1 and u_2 with degree 2 and 4, respectively. Then we can construct a fibration

$$u : \mathfrak{p} \to \mathbb{C}^2, \quad u(x) = (u_1(x), u_2(x))$$

According to Kostant and Rallis, u is equidimensional, each fiber $u^{-1}(\xi)$, $\xi \in \mathbb{C}^2$ is K_θ-stable and there is a unique maximal-dimensional K_θ-orbit in each fiber which consists of regular elements of $\mathfrak{sp}(4)$ in $\mathfrak{p}$. Note that the nilpotent variety $u^{-1}(0)$ may not be normal as mentioned in [**KoRa**].

Let us recall the identity for Poincaré series

$$\frac{1}{\det(I - AX)} = \sum_{l=0}^{\infty} trace(Sym^l(A))X^l, \tag{47}$$

where A is an arbitrary square complex matrix, X is a sufficiently small complex variable, and $Sym^l(A)$ is the l-th symmetric power of A. Let $X = q^{-(s+1)/2}$. Applying the identity to the matrix $A = (\rho_1 \otimes \rho_1)_v(\tau_{1v}, \tau_{2v})$, we have

$$L(s', \pi_{1v} \otimes \pi_{2v}, (\rho_1 \otimes \rho_1)_v) = \sum_{l=0}^{\infty} trace(Sym^l((\rho_1 \otimes \rho_1)_v(\tau_{1v}, \tau_{2v})))X^l. \tag{48}$$

By the separation of variables in Kostant-Rallis Theorem and the K_θ- module structure on $\mathfrak{p}$, one can easily see that the right hand side of (44) is equal to

$$\sum_{l=0}^{\infty} dim(Sym^l_{\mathcal{S}^{K_\theta}})X^l \cdot \sum_{l=0}^{\infty} trace(Sym^l_{\mathcal{H}}((\rho_1 \otimes \rho_1)_v(\tau_{1v}, \tau_{2v})))X^l,$$

where $Sym^l_{\mathcal{S}^{K_\theta}}$ denotes the l-th symmetric power representation on the polynomial algebra $\mathcal{S}^{K_\theta}$ and $Sym^l_{\mathcal{H}}$ denotes the restriction of the l-th symmetric power representation Sym^l to the algebra $\mathcal{H}$ of all K_θ-harmonic polynomials over $\mathfrak{p}$. According to the Invariant Theory, say [**Pop**] or [**gSch**], one has

$$\sum_{l=0}^{\infty} dim(Sym^l_{\mathcal{S}^{K_\theta}})X^l = \frac{1}{(1 - X^2)(1 - X^4)}. \tag{49}$$

Hence we have

PROPOSITION 4.2.3. *The local Langlands factor can be expressed as*

$$L(s', \pi_{1v} \otimes \pi_{2v}, (\rho_1 \otimes \rho_1)_v) = \frac{\sum_{l=0}^{\infty} trace(Sym^l_{\mathcal{H}}((\rho_1 \otimes \rho_1)_v(\tau_{1v}, \tau_{2v})))X^l}{(1 - X^2)(1 - X^4)}.$$

It is crucial to understand $\sum_{l=0}^{\infty} trace(Sym^l_{\mathcal{H}}((\rho_1 \otimes \rho_1)_v(\tau_{1v}, \tau_{2v})))X^l$. Let (m_1, m_2), (n_1, n_2) be the complex representations of $Sp(2)$ with highest weights $m_1\varepsilon_1 + m_2\varepsilon_2$, $n_1\varepsilon_1 + n_2\varepsilon_2$, $m_1 \geq m_2 \geq 0$ and $n_1 \geq n_2 \geq 0$ integers. Then we denote by $(m_1, m_2; n_1, n_2)$ the tensor product representation of (m_1, m_2) and (n_1, n_2), which is a representation of K_θ. Applying the Kostant-Rallis' algorithm to our special case, we obtain

PROPOSITION 4.2.4. *The following identity holds*

$$\sum_{l=0}^{\infty} trace(Sym^l_{\mathcal{H}}((\rho_1 \otimes \rho_1)_v(\tau_{1v}, \tau_{2v})))X^l = \sum_{\substack{m_1 \geq m_2 \geq 0 \\ n_1 \geq n_2 \geq 0}} (m_1, m_2|n_1, n_2)P(m_1, m_2, n_1, n_2; X),$$

where $(m_1, m_2 | n_1 n_2) = trace((m_1, m_2; n_1, n_2)(\tau_{1v}, \tau_{2v}))$ and $P(m_1, m_2, n_1, n_2; X)$ is a polynomial with non-negative integral coefficients and is in form:

$$P(m_1, m_2, n_1, n_2; X) = \sum_{l=0}^{q} p_l X^l, \tag{50}$$

and the term $p_l X^l$ indicates that the irreducible representation $(m_1, m_2; n_1, n_2)$ of K_θ occurs in the K_θ-submodule of $\mathcal{H}$ consisting of all homogeneous harmonic polynomial functions of degree l with multiplicity p_l.

Following Kostant-Rallis' algorithm, the polynomial $P(m_1, m_2, n_1, n_2; X)$ is determined by the subspace $(m_1, m_2; n_1, n_2)(e)$ for any regular nilpotent element $e \in \mathfrak{p}$ in following way: If $x_\circ$ is an element in $\mathfrak{t}$ so that $[x_\circ, e] = e$, then the l's are eigenvalues of $x_\circ$ in the subspace $(m_1, m_2; n_1, n_2)(e)$ and the p_l's are the multiplicity of l, and also $\sum_{l=0}^{q} p_l = l((m_1, m_2; n_1, n_2))$ is the multiplicity of the representation $(m_1, m_2; n_1, n_2)$ occurring in $\mathcal{H}$. However, the subspace $(m_1, m_2; n_1, n_2)(e)$ is abstractly defined in [**KoRa**]. In contrast, the subspace $(m_1, m_2; n_1, n_2)^{K_\theta^e}$ is easier to deal with by means of Borel-Weil-Bott Theorem and the Invariant Theory. Our scheme to compute the polynomials $P(m_1, m_2, n_1, n_2; X)$ is:

(1) Compute the multiplicity $l((m_1, m_2; n_1, n_2))$ by means of the known Branching Formula;
(2) Determine the eigenvalues and eigenspaces in $(m_1, m_2; n_1, n_2)^{K_\theta^e}$ for an element $x_\circ \in \mathfrak{t}$ so that $[x_\circ, e] = e$, by applying Borel-Weil-Bott Theorem and the Invariant Theory;
(3) Prove that $(m_1, m_2; n_1, n_2)(e) = (m_1, m_2; n_1, n_2)^{K_\theta^e}$ and determine the polynomial $P(m_1, m_2, n_1, n_2; X)$ in terms of results in (1) and (2).

4.3. Branching formula for $Sp(2) \downarrow SL(2) \times SL(2)$. We restate first Branching Rule for $\mathfrak{so}(2n + 1) \downarrow \mathfrak{so}(2n)$ from [**Zhe**] Let V_λ be the irreducible representation of the complex Lie algebra $\mathfrak{so}(2n + 1)$ with highest weight $\lambda = (\lambda_1, \lambda_2, \cdots, \lambda_n)$ (with respect to the standard Cartan subalgebra), which are simultaneously integers or half-integers satisfying $\lambda_1 \geq \lambda_2 \geq \cdots \geq \lambda_n \geq 0$, and V_μ the irreducible representation of the complex Lie algebra $\mathfrak{so}(2n)$ with highest weight $\mu = (\mu_1, \mu_2, \cdots, \mu_n)$ (with respect to the standard Cartan subalgebra), which are simultaneously integers or half-integers satisfying $\mu_1 \geq \mu_2 \geq \cdots \geq |\mu_n|$.

THEOREM 4.3.1. [**Zhe**] *The following Multiplicity One Branching Rule holds*

$$V_\lambda = \bigoplus_\mu V_\mu$$

as $\mathfrak{so}(2n)$-modules, where $\mu = (\mu_1, \mu_2, \cdots, \mu_n)$ are all dominant weights of $\mathfrak{so}(2n)$ such that

$$\lambda_1 \geq \mu_1 \geq \lambda_2 \geq \mu_2 \geq \cdots \geq \lambda_{n-1} \geq \mu_{n-1} \geq \lambda_n \geq |\mu_n|$$

and all numbers μ_i are integers or half-integers together with λ_i, $i = 1, 2, \cdots, n$.

Now ,we shall use Zhelobenko's Branching Rule to prove

THEOREM 4.3.2 (Branching Rule for $Sp(2) \downarrow (SL(2) \times SL(2))$). Let $V_{(n_1,n_2)}$ be an irreducible representation of $Sp(2,\mathbb{C})$ with highest weight (n_1, n_2), n_1, n_2 are integers satisfying $n_1 \geq n_2 \geq 0$. Let $V_{m_1} \otimes V_{m_2}$ be an irreducible representation of $SL(2,\mathbb{C}) \times SL(2,\mathbb{C})$ with highest weight (m_1, m_2), m_1, m_2 are integers satisfying $m_1 \geq 0, m_2 \geq 0$. Then the following multiplicity one Branching Rule holds

$$V_{(n_1,n_2)} = \bigoplus_{l_1,l_2} V_{l_1+l_2} \otimes V_{l_1-l_2},$$

where all l_1, l_2 are simultaneously integers or half-integers together with $(n_1+n_2)/2$, $(n_1-n_2)/2$, and satisfy the following inequalities:

$$|l_2| \leq (n_1 - n_2)/2 \leq l_1 \leq (n_1 + n_2)/2.$$

PROOF. It suffices to consider the same branching rule at level of Lie algebras since the groups involved are simply connected. By accidental isomorphisms of low ranks: $\mathfrak{sp}(2,\mathbb{C}) \cong \mathfrak{so}(5,\mathbb{C})$ and $\mathfrak{sl}(2,\mathbb{C}) \times \mathfrak{sl}(2,\mathbb{C}) \cong \mathfrak{so}(4,\mathbb{C})$, the theorem follows. $\square$

4.4. Multiplicity $l((m_1, m_2; n_1, n_2))$. Following Kostant-Rallis' algorithm, the multiplicity $l((m_1, m_2; n_1, n_2))$ equals the dimension of $(m_1, m_2; n_1, n_2)(z)$ for any regular element $z \in \mathfrak{p}$ and also equals $\dim(m_1, m_2; n_1, n_2)^{K_\theta^z}$ for any regular semisimple element z in $\mathfrak{p}$.

PROPOSITION 4.4.1. Let z be a semisimple element in $\mathfrak{a}$ with $a \neq b$ as (42). Then the stabilizer of z in K_θ is $K_\theta^z = Stab_{K_\theta}(z) = SL(2,\mathbb{C}) \times SL(2,\mathbb{C})$.

PROOF. Without loss of generality, we assume that $z = \begin{pmatrix} A & \\ & -A \end{pmatrix}$ with

$$A = \begin{pmatrix} & & 1 & 0 \\ & & 0 & -1 \\ 1 & 0 & & \\ 0 & -1 & & \end{pmatrix}.$$

Then the stabilizer $Stab_{K_\theta}(z) = \{k \in K_\theta : kzk^{-1} = z\}$. The isomorphism follows from easy computations. $\square$

COROLLARY 4.4.1. The element z with $a \neq b$ is a regular semisimple element in $\mathfrak{p}$.

PROOF. For such a semisimple element z, we have $\dim O_z = \dim u^{-1}(u(z))$. By definition in [**KoRa**], z is regular. $\square$

Let V and W be irreducible representations of $Sp(2,\mathbb{C})$. As we know, $SL(2,\mathbb{C}) \times SL(2,\mathbb{C}) \hookrightarrow Sp(2,\mathbb{C})$ is determined by

$$\begin{pmatrix} a_1 & a_2 \\ a_3 & a_4 \end{pmatrix} \times \begin{pmatrix} b_1 & b_2 \\ b_3 & b_4 \end{pmatrix} \mapsto \begin{pmatrix} a_1 & 0 & a_2 & 0 \\ 0 & b_1 & 0 & -b_2 \\ a_3 & 0 & a_4 & 0 \\ 0 & -b_3 & 0 & b_4 \end{pmatrix}$$

and $SL(2, \mathbb{C}) \times SL(2, \mathbb{C}) \cong K_\theta^z$ is determined by the diagonal embedding

$$(SL(2, \mathbb{C}) \times SL(2, \mathbb{C})) \overset{\triangle}{\hookrightarrow} Sp(2, \mathbb{C}) \times Sp(2, \mathbb{C}).$$

Then by the multiplicity one Branching rule for $Sp(2) \downarrow SL(2) \times SL(2)$, we have

$$V|_{SL(2) \times SL(2)} = \bigoplus_i V_i \text{ and } W|_{SL(2) \times SL(2)} = \bigoplus_j W_j$$

where V_i and W_j are irreducible representations of $SL(2) \times SL(2)$.

THEOREM 4.4.1. *Let V and W be irreducible representations of $Sp(2, \mathbb{C})$. Then*

$$l(V \otimes W) = \dim(V \otimes W)^{K_\theta^z}$$
$$= card\{(V_i, W_j) : V_i \cong W_j \text{ as } SL(2) \times SL(2) - modules\} \quad (51)$$

where (V_i, W_j) are such that $V_i \otimes W_j$ are summands of $V \otimes W$ for all possible i and j.

PROOF. By using the Branching rule and the fact that $V_i = V_{i1} \otimes V_{i2}$ and $W_j = W_{j1} \otimes W_{j2}$, where V_{il}, W_{jk} are irreducible representations of $SL(2)$, we have

$$V \otimes W = \bigoplus_{i,j} V_{i1} \otimes V_{i2} \otimes W_{j1} \otimes W_{j2}.$$

Thus the $SL(2) \times SL(2)$-invariant subspace of $V \otimes W$ is

$$(V \otimes W)^{SL(2) \times SL(2)} = (\bigoplus_{i,j} V_{i1} \otimes V_{i2} \otimes W_{j1} \otimes W_{j2})^{SL(2) \times SL(2)}$$
$$= \bigoplus_{i,j} (V_{i1} \otimes W_{j1})^{SL(2)} \otimes (V_{i2} \otimes W_{j2})^{SL(2)}.$$

On other hand, for $k = 1, 2$,

$$(V_{ik} \otimes W_{jk})^{SL(2)} = \mathrm{Hom}_{SL(2)}(V_{ik}^*, W_{jk}) = \begin{cases} \mathbb{C} & if \ V_{ik} \cong W_{jk}, \\ 0 & otherwise \end{cases}$$

Therefore we prove that

$$\dim(V \otimes W)^{K_\theta^z} = card\{(V_i, W_j) : V_i \cong W_j \text{ as } SL(2) \times SL(2) - modules\}.$$

$\square$

COROLLARY 4.4.2. *Let $\rho_{(m_1,m_2)}$ be the complex irreducible representation of $Sp(2)$ with highest weight (m_1, m_2), $m_1 \geq m_2 \geq 0$ and $(m_1, m_2; n_1, n_2) = \rho_{(m_1,m_2)} \otimes \rho_{(n_1,n_2)}$. Then $l((m_1, m_2; n_1, n_2)$ equals the number of the pairs (l_1, l_2) of integers satisfying the following conditions:*

$$\frac{(m_1 + m_2)}{2} \geq l_1 \geq \frac{(m_1 - m_2)}{2} \geq |l_2|; \quad \frac{(n_1 + n_2)}{2} \geq l_1 \geq \frac{(n_1 - n_2)}{2} \geq |l_2| \quad (52)$$

where l_1, l_2 are simultaneously integers or half-integers together with $(m_1 + m_2)/2$, $m_1 - m_2)/2$, $(n_1 + n_2)/2$, and $(n_1 - n_2)/2$.

PROOF. Using the corresponding between irreducible representations and dominant weights, we will achieve the result above.   □

COROLLARY 4.4.3. *Let* $m_1 \geq m_2 \geq 0$; $n_1 \geq n_2 \geq 0$ *be integers. We have*

(a) *If* $(m_1 + m_2) \geq (n_1 + n_2) \geq (m_1 - m_2) \geq (n_1 - n_2)$, *then*

$$l((m_1, m_2; n_1, n_2)) = \frac{1}{2}(n_1 + n_2 - m_1 + m_2 + 2)(n_1 - n_2 + 1);$$

(b) *If* $(m_1 + m_2) \geq (n_1 + n_2) \geq (n_1 - n_2) > (m_1 - m_2)$, *then*

$$l((m_1, m_2; n_1, n_2)) = (n_2 + 1)(m_1 - m_2 + 1);$$

(c) *If* $(n_1 + n_2) > (m_1 + m_2) \geq (m_1 - m_2) \geq (n_1 - n_2)$, *then*

$$l((m_1, m_2; n_1, n_2)) = (m_2 + 1)(n_1 - n_2 + 1);$$

(d) *If* $(n_1 + n_2) > (m_1 + m_2) \geq (n_1 - n_2) > (m_1 - m_2)$, *then*

$$l((m_1, m_2; n_1, n_2)) = \frac{1}{2}(m_1 + m_2 - n_1 + n_2 + 2)(m_1 - m_2 + 1);$$

(e) *If* $(m_1 - m_2) > (n_1 + n_2)$ *or* $(n_1 - n_2) > (m_1 + m_2)$, *then*

$$l((m_1, m_2; n_1, n_2)) = 0.$$

PROOF. It follows from solving the inequalities:

$$\frac{(m_1 + m_2)}{2} \geq l_1 \geq \frac{(m_1 - m_2)}{2} \geq |l_2|; \quad \frac{(n_1 + n_2)}{2} \geq l_1 \geq \frac{(n_1 - n_2)}{2} \geq |l_2|.$$

□

4.5. Orbit Decomposition on $(B \times B) \backslash K_\theta$. Let B be the standard Borel subgroup of $Sp(2, \mathbb{C})$. We will introduce a special subgroup R and study the orbit decomposition of $(B \times B) \backslash K_\theta$ under the action of R via the right translation. In particular, we will describe the all R-orbits in $(B \times B) \backslash K_\theta$ with codimension less than two.

PROPOSITION 4.5.1. *Let* $e = \begin{pmatrix} A & 0 \\ B & -{}^tA \end{pmatrix}$ *in* $\mathfrak{p}$ *and* $x_o = \begin{pmatrix} X & \\ & -{}^tX \end{pmatrix}$ *in* $\mathfrak{t}$ *with*

$$A = \begin{pmatrix} & & 0 & 0 \\ & & 1 & 0 \\ 0 & 0 & & \\ -1 & 0 & & \end{pmatrix}, B = \begin{pmatrix} & & 0 & 0 \\ & & 0 & 1 \\ 0 & 0 & & \\ 0 & 1 & & \end{pmatrix}, \text{ and } X = \begin{pmatrix} & & -1 & \\ & -1 & & \\ & & & -2 \\ & & & & 0 \end{pmatrix}.$$

Then

(a) e *is a regular nilpotent element in* $\mathfrak{p}$ *and* x_o *is in* $\mathfrak{t}$ *such that* $[x_o, e] = e$.

(b) *The stabilizer* $Stab_{K_\theta}(e) = K_\theta^e$ *of* e *in* K_θ *is*

$$\left\{ \begin{pmatrix} t & 0 & 0 & 0 \\ w & t^{-1} & 0 & 0 \\ z_1 & x' & t^{-1} & -w \\ x & y & 0 & t \end{pmatrix} \times \begin{pmatrix} t^{-1} & 0 & 0 & 0 \\ y & t & 0 & 0 \\ z_2 & -x & t & -y \\ -x' & w & 0 & t^{-1} \end{pmatrix} : x't + wy = xt^{-1} \right\}.$$

PROOF. The stabilizer $K_\theta^e = \{k \in K_\theta : kek^{-1} = e\}$. Part (b) follows the standard computation of matrices and Proposition 4.2.1. Since K_θ^e is a 6-dimensional subgroup of K_θ, the dimension of the K_θ-orbit of e is equal to the dimension of the zero fiber $u^{-1}(0)$, both of which are 14. Therefore e is regular. $\square$

Let $S = \{exp(tx_\circ) : t \in \mathbb{C}\}$, a one-parameter subgroup in K_θ, which is in form: $S = \{h_2(e^{-t}, e^{-t}, e^t, e^t) \times h_2(e^{-2t}, 1, e^{2t}, 1)\}$ where $h_2(\cdots)$ denote the diagonal matrix in $Sp(2)$. By setting $s = e^t$, we have

$$S = \{\underline{s} = h_2(s^{-1}, s^{-1}, s, s) \times h_2(s^{-2}, 1, s^2, 1) : s \in \mathbb{C}^\times\}. \tag{53}$$

and $\nu_p(\overline{s}) = s^p$ is a rational character of S where p is an integer.

PROPOSITION 4.5.2. *(a)* $S \subset N_{K_\theta}(K_\theta^e)$ *and (b)* SK_θ^e *is a 7-dimensional subgroup of* K_θ, *i.e.*

$$SK_\theta^e = \left\{ \begin{pmatrix} s^{-1}t & 0 & 0 & 0 \\ s^{-1}w & (st)^{-1} & 0 & 0 \\ sz_1 & sx' & st^{-1} & -sw \\ sx & sy & 0 & st \end{pmatrix} \begin{pmatrix} s^{-2}t^{-1} & 0 & 0 & 0 \\ y & t & 0 & 0 \\ s^2 z_2 & -s^2 x & s^2 t & -s^2 y \\ -x' & w & 0 & t^{-1} \end{pmatrix} : x't + wy = xt^{-1} \right\}.$$

PROPOSITION 4.5.3. *Let* $\overline{X} = \chi_{-\alpha_1-\alpha_2}(x)$ *be the one-parameter subgroup associated to the root* $-\alpha_1 - \alpha_2$ *and* $R = \overline{X}SK_\theta^e$ *as algebraic variety. Then* R *can be written as*

$$\left\{ \begin{pmatrix} s^{-1}t & 0 & 0 & 0 \\ s^{-1}w & (st)^{-1} & 0 & 0 \\ z_1 & x_1' & st^{-1} & -sw \\ x_1 & sy & 0 & st \end{pmatrix} \begin{pmatrix} s^{-2}t^{-1} & 0 & 0 & 0 \\ y & t & 0 & 0 \\ z_2 & x_2' & s^2 t & -s^2 y \\ x_2 & w & 0 & t^{-1} \end{pmatrix} : \begin{matrix} x_1't + syw = t^{-1}x_1 \\ x_2's^{-2}t^{-1} + yw = tx_2 \end{matrix} \right\}.$$

and enjoys the following properties:

(a) $\overline{X}S$ *is two dimension subgroup of* K_θ *and* $\overline{X}S$ *normalizes* K_θ^e,

(b) R *is an 8-dimensional subgroup of* K_θ *and including* K_θ^e *as a normal subgroup,*

PROOF. Since S normalizes $\overline{X}$ as subgroups in K_θ, $\overline{X}S$ is a group. By the previous proposition, S normalizes K_θ^e. It is easy to check that $\overline{X}$ commutes with the torus part of K_θ^e and normalizes the unipotent part of K_θ^e. Part (a) follows. Part (b) follows from part (a). $\square$

Since R is 8-dimensional subgroup of the complex group $K_\theta = Sp(2) \times Sp(2)$ and the dimension of the flag variety $X = (B \times B) \backslash K_\theta$ is also 8, the restriction to R of the action of K_θ on the flag variety X has a (Zariski) open dense R-orbit. Hence R is a spherical subgroup of of K_θ in the terminology of T. Matsuki [**Mat**]. Such a open dense R-orbit is unique as the flag variety is irreducible and the number of all R-orbits on the flag is finite [**Mat**]. By Invariant Theory, all of those codimension-one R-orbits are on the boundary of the unique open R-orbit.

We are going to find out the unique open R-orbit and all codimension-one R-orbits by means of the decomposition of Bruhat cells. Let B^- be the Borel subgroup opposite to the standard one. Then $(B \times B)(B^- \times B^-)$ is the open cell in the Bruhat decomposition of $Sp(2, \mathbb{C}) \times Sp(2, \mathbb{C})$ with respect to the standard Borel subgroup $B \times B$, and $(B \times B)(I_4 \times \omega_{\alpha_1})(B^- \times B^-)$, $(B \times B)(I_4 \times \omega_{\alpha_2})(B^- \times B^-)$, $(B \times B)(\omega_{\alpha_1} \times I_4)(B^- \times B^-)$, and $(B \times B)(\omega_{\alpha_2} \times I_4)(B^- \times B^-)$ are all codimension-one Bruhat cells, where ω_{α_i}, $i = 1, 2$, are the Weyl group elements corresponding to the simple reflections with respect to the simple roots α_i, $i = 1, 2$, respectively. Since

$$B^- \times B^- = (\begin{pmatrix} 1 & 0 & 0 & 0 \\ ab & b & 0 & 0 \\ 0 & 0 & 1 & -a \\ 0 & 0 & 0 & b^{-1} \end{pmatrix} \times \begin{pmatrix} d & 0 & 0 & 0 \\ c & 1 & 0 & 0 \\ 0 & 0 & d^{-1} & -cd^{-1} \\ 0 & 0 & 0 & 1 \end{pmatrix})R, \tag{54}$$

we know that the codimension of the Bruhat cell $(B \times B)(g \times h)(B^- \times B^-)$ must be less than or equal to 1 if the codimension of the double coset $(B \times B)(g \times h)R$ is lees than or equal to 1.

LEMMA 4.5.1. *In the Bruhat cells of codimension ≤ 1, R-orbits can be described as follows:*

(1) *In $(B \times B)(B^- \times B^-)$, R-orbits are of form $(B \times B)(\chi_{-\alpha_1}(c) \times \chi_{-\alpha_1}(d))R$ where $c, d \in \mathbb{C}$ and*

$$codim(B \times B)(\chi_{-\alpha_1}(c) \times \chi_{-\alpha_1}(d))R = \begin{cases} 0 & if\ c \neq 0\ and d \neq 0, \\ 2 & if\ c = d = 0, \\ 1 & otherwise \end{cases}$$

(2) *In $(B \times B)(I_4 \times \omega_{\alpha_1})(B^- \times B^-)$, R-orbits are $(B \times B)(\chi_{-\alpha_1}(a) \times \omega_{\alpha_1})R$ for $a \in \mathbb{C}$ and*

$$codim(B \times B)(\chi_{-\alpha_1}(a) \times \omega_{\alpha_1})R = \begin{cases} 2 & if\ a = 0, \\ 1 & otherwise \end{cases}$$

(3) *In $(B \times B)(I_4 \times \omega_{\alpha_2})(B^- \times B^-)$, R-orbits are $(B \times B)(I_4 \times \omega_{\alpha_2}\chi_{-\alpha_1}(a))R$ for $a \in \mathbb{C}$ and*

$$codim(B \times B)(I_4 \times \omega_{\alpha_2}\chi_{-\alpha_1}(a))R = \begin{cases} 2 & if\ a = 0, \\ 1 & otherwise \end{cases}$$

(4) *In $(B \times B)(\omega_{\alpha_1} \times I_4)(B^- \times B^-)$, R-orbits are $(B \times B)(\omega_{\alpha_1} \times \chi_{-\alpha_1}(a))R$ for $a \in \mathbb{C}$ and*

$$codim(B \times B)(\omega_{\alpha_1} \times \chi_{-\alpha_1}(a))R = \begin{cases} 2 & if\ a = 0, \\ 1 & otherwise \end{cases}$$

(5) In $(B \times B)(\omega_{\alpha_2} \times I_4)(B^- \times B^-)$, R-orbits are $(B \times B)(\omega_{\alpha_2}\chi_{-\alpha_1}(a) \times I_4)R$ for $a \in \mathbb{C}$ and

$$codim(B \times B)(\omega_{\alpha_2}\chi_{-\alpha_1}(a) \times I_4)R = \begin{cases} 2 & \text{if } a = 0, \\ 1 & \text{otherwise} \end{cases}$$

PROOF. (1) Since the 2-dimensional torus in section of R in $(B^- \times B^-)$ goes to $(B \times B)$, we have

$$(B \times B)(B^- \times B^-) = \cup_{c,d \in \mathbb{C}}(B \times B)(\chi_{-\alpha_1}(c) \times \chi_{-\alpha_1}(d))R.$$

It is easy to check that

$$codim[(B \times B)(\chi_{-\alpha_1}(c) \times \chi_{-\alpha_1}(d))R] = \dim[\mathrm{Stab}_R((B \times B)(\chi_{-\alpha_1}(c) \times \chi_{-\alpha_1}(d)))]$$

since $\dim R = \dim X$. On the other hand,

$$\mathrm{Stab}_R((B \times B)(\chi_{-\alpha_1}(c) \times \chi_{-\alpha_1}(d))) =$$

$$= \begin{cases} h(s^{-1}t, (st)^{-1}, st^{-1}, st) \times h(s^{-2}t^{-1}, t, s^2t, t^{-1}) & \text{if } c = d = 0, \\ h(s^{-1}, s^{-1}, s, s) \times h(s^{-2}, 1, s^2, 1) & \text{if } c \neq 0, d = 0, \\ (\pm h(t^2, 1, t^{-2}, 1)) \times h(t, t, t^{-1}, t^{-1}) & \text{if } c = 0, d \neq 0, \\ (\pm I_4) \times (\pm I_4) & \text{if } cd \neq 0 \end{cases}$$

(2) $(B \times B)(I_4 \times \omega_{\alpha_1})(B^- \times B^-) = (B \times B)(I_4 \times \omega_{\alpha_1})(\chi_{-\alpha_1}(a) \times \chi_{-\alpha_1}(c))R = (B \times B)(\chi_{-\alpha_1}(a) \times \omega_{\alpha_1})R$ since $\omega_{\alpha_1}\chi_{-\alpha_1}(c)\omega_{\alpha_1}^{-1} \in B$. The stabilizer

$$\mathrm{Stab}_R((B \times B)(\chi_{-\alpha_1}(a) \times \omega_{\alpha_1})) =$$

$$= \begin{cases} h(s^{-1}, s^{-1}, s,) \times h(s^{-2}, 1, s^2, 1) & \text{if } a \neq 0, \\ h(s^{-1}t, (st)^{-1}, st^{-1}, st) \times h(s^{-2}t^{-1}, t, s^2t, t^{-1}) & \text{if } a = 0 \end{cases}$$

(3) $(B \times B)(I_4 \times \omega_{\alpha_2})(B^- \times B^-) = (B \times B)(I_4 \times \omega_{\alpha_2})(\chi_{-\alpha_1}(a) \times \chi_{-\alpha_1}(c))R = (B \times B)(I_4 \times \omega_{\alpha_2}\chi_{-\alpha_1}(c)\chi_{-\alpha_2}(-a))R = (B \times B)(I_4 \times \omega_{\alpha_2}\chi_{-\alpha_2}(-a)\chi_{-\alpha_1}(c))R = (B \times B)(I_4 \times \omega_{\alpha_2}\chi_{-\alpha_1}(c))R$ since $\chi_{-\alpha_1}(a) \times \chi_{-\alpha_2}(a) \in R$ and $\omega_{\alpha_2}\chi_{-\alpha_2}(-a)\omega_{\alpha_2}^{-1} \in B$. The stabilizer

$$\mathrm{Stab}_R((B \times B)(I_4 \times \omega_{\alpha_2}\chi_{-\alpha_1}(c))) =$$

$$= \begin{cases} [\pm h(t^2, 1, t^{-2}, 1)] \times h(t, t, t^{-1}, t^{-1}) & \text{if } c = 0, \\ h(s^{-1}t, (st)^{-1}, st^{-1}, st) \times h(s^{-2}t^{-1}, t, s^2t, t^{-1}) & \text{if } c = 0 \end{cases}$$

(4) and (5) cad be proved in the way similar to (2) and (3), respectively. $\square$

LEMMA 4.5.2. *For the R-orbits of codimension ≤ 1, we have*

(a) $(B \times B)(\chi_{-\alpha_1}(c) \times \chi_{-\alpha_1}(d))R = (B \times B)(\chi_{-\alpha_1}(1) \times \chi_{-\alpha_1}(1))R,$

(b) $(B \times B)(\chi_{-\alpha_1}(c) \times I_4)R = (B \times B)(\chi_{-\alpha_1}(1) \times I_4)R,$

(c) $(B \times B)(I_4 \times \chi_{-\alpha_1}(d))R = (B \times B)(I_4 \times \chi_{-\alpha_1}(1))R,$

(d) $(B \times B)(\chi_{-\alpha_1}(a) \times \omega_{\alpha_1})R = (B \times B)(\chi_{-\alpha_1}(1) \times \omega_{\alpha_1})R,$

(e) $(B \times B)(I_4 \times \omega_{\alpha_2}\chi_{-\alpha_1}(a))R = (B \times B)(I_4 \times \omega_{\alpha_2}\chi_{-\alpha_1}(1))R,$

(f) $(B \times B)(\omega_{\alpha_1} \times \chi_{-\alpha_1}(a))R = (B \times B)(\omega_{\alpha_1} \times \chi_{-\alpha_1}(1))R,$

(g) $(B \times B)(\omega_{\alpha_2}\chi_{-\alpha_1}(a) \times I_4)R = (B \times B)(\omega_{\alpha_2}\chi_{-\alpha_1}(1) \times I_4)R.$

PROOF. Statements (a), (d), (e), (f), and (g) follow from the general fact: Let V be a complex connected quasi-projective variety , G a complex algebraic group, and G act on V regularly. Then there is at most one Zariski open G-orbit on V. However, statements (b) and (c) follow from easy computations of matrices:

$$(\chi_{-\alpha_1}(c) \times I_4) = [h(c^{\frac{-1}{2}}, c^{\frac{1}{2}}) \times h(c^{\frac{1}{2}}, c^{\frac{-1}{2}})](\chi_{-\alpha_1}(1) \times I_4)[h(c^{\frac{1}{2}}, c^{\frac{-1}{2}}) \times h(c^{\frac{-1}{2}}, c^{\frac{1}{2}})]$$

$$(I_4 \times \chi_{-\alpha_1}(d)) = [h(d^{\frac{1}{2}}, d^{\frac{-1}{2}}) \times h(d^{\frac{-1}{2}}, d^{\frac{1}{2}})](I_4 \times \chi_{-\alpha_1}(1))[h(d^{\frac{-1}{2}}, d^{\frac{1}{2}}) \times h(d^{\frac{1}{2}}, d^{\frac{-1}{2}})],$$

where $h(a,b) = h(a,b,a^{-1},b^{-1})$. $\square$

COROLLARY 4.5.1. *The Bruhat cells of codimension ≤ 1 can be decomposed into R-orbits as follows:*

(a) $(B \times B)(B^- \times B^-) = (B \times B)R \cup (B \times B)(\chi_{-\alpha_1}(1) \times I_4)R \cup (B \times B)(I_4 \times \chi_{-\alpha_1}(1))R \cup (B \times B)(\chi_{-\alpha_1}(1) \times \chi_{-\alpha_1}(1))R.$

(b) $(B \times B)(I_4 \times \omega_{\alpha_1})(B^- \times B^-) = (B \times B)(I_4 \times \omega_{\alpha_1})R \cup (B \times B)(\chi_{-\alpha_1}(1) \times \omega_{\alpha_1})R.$

(c) $(B \times B)(I_4 \times \omega_{\alpha_2})(B^- \times B^-) = (B \times B)(I_4 \times \omega_{\alpha_2})R \cup (B \times B)(I_4 \times \omega_{\alpha_2}\chi_{-\alpha_1}(1))R.$

(d) $(B \times B)(\omega_{\alpha_1} \times I_4)(B^- \times B^-) = (B \times B)(\omega_{\alpha_1} \times I_4)R \cup (B \times B)(\omega_{\alpha_1} \times \chi_{-\alpha_1}(1))R.$

(e) $(B \times B)(\omega_{\alpha_2} \times I_4)(B^- \times B^-) = (B \times B)(\omega_{\alpha_2} \times I_4)R \cup (B \times B)(\omega_{\alpha_2}\chi_{-\alpha_1}(1) \times I_4)R.$

THEOREM 4.5.1. *Let $\alpha_1 = \varepsilon_1 - \varepsilon_2$ be the short root of $Sp(2)$ and $\alpha_2 = 2\varepsilon_2$ the long root of $Sp(2)$. Let $\xi_0 = \chi_{-\alpha_1}(1) \times \chi_{-\alpha_1}(1)$. Define following two embeddings of $SL(2)$ into $Sp(2)$ corresponding to the roots α_1 and α_2, respectively:*

$$\imath_{\alpha_1} : \begin{pmatrix} a & b \\ c & d \end{pmatrix} \mapsto \begin{pmatrix} a & b & 0 & 0 \\ c & d & 0 & 0 \\ 0 & 0 & d & -c \\ 0 & 0 & -b & a \end{pmatrix} \text{ and } \imath_{\alpha_2} : \begin{pmatrix} e & -f \\ -g & h \end{pmatrix} \mapsto \begin{pmatrix} 1 & 0 & 0 & 0 \\ 0 & e & 0 & f \\ 0 & 0 & 1 & 0 \\ 0 & g & 0 & h \end{pmatrix}$$

and set $G_{\alpha_1} = im(\imath_{\alpha_1})$ and $G_{\alpha_2} = im(\imath_{\alpha_2})$. Then the open R-orbit and its codimension one boundaries can be described as follows:

(a) $(B \times B)\xi_0 R$ *is a Zariski open subset in $K_\theta = Sp(2, \mathbb{C}) \times Sp(2, \mathbb{C})$.*

(b) $(B \times B)(\chi_{-\alpha_1}(1) \times I_4)R$, $(B \times B)(I_4 \times \chi_{-\alpha_1}(1))R$, $(B \times B)(\chi_{-\alpha_1}(1) \times \omega_{\alpha_1})R$, $(B \times B)(\omega_{\alpha_1} \times \chi_{-\alpha_1}(1))R$, $(B \times B)(I_4 \times \omega_{\alpha_2}\chi_{-\alpha_1}(1))R$, *and* $(B \times B)(\omega_{\alpha_2}\chi_{-\alpha_1}(1) \times I_4)R$ *are all codimension-one boundaries of $(B \times B)\xi_0 R$ in K_θ.*

(c) *Those codimension-one boundaries can be distinguished in the following way: the disjoint union of the boundaries $(B \times B)(\chi_{-\alpha_1}(1) \times I_4)R$, $(B \times B)(\chi_{-\alpha_1}(1) \times \omega_{\alpha_1})R$, and $(B \times B)(\omega_{\alpha_2}\chi_{-\alpha_1}(1) \times I_4)R$ is contained in $(B \times B)(G_{\alpha_2} \times G_{\alpha_1})\xi_0 R$ and the disjoint union of the boundaries $(B \times B)(I_4 \times \chi_{-\alpha_1}(1))R$, $(B \times B)(\omega_{\alpha_1} \times \chi_{-\alpha_1}(1))R$, and $(B \times B)(I_4 \times \omega_{\alpha_2}\chi_{-\alpha_1}(1))R$ is included in $(B \times B)(G_{\alpha_1} \times G_{\alpha_2})\xi_0 R$.*

PROOF. The proof follows directly from Lemma 4.5.1 and 4.5.2 and the general fact about the orbital decomposition on a group variety, see [?]. $\square$

COROLLARY 4.5.2. *Let $\tilde{X} = (B \times B)[(G_{\alpha_1} \times G_{\alpha_2}) \cup (G_{\alpha_2} \times G_{\alpha_1})]\xi_0 R$. Then the codimension of $[(Sp(2, \mathbb{C}) \times Sp(2, \mathbb{C})) - \tilde{X}]$ is greater than 2.*

Finally we describe the intersections of the open dense double coset $(B \times B)\xi_0 R$ with $(B \times B)(G_{\alpha_1} \times G_{\alpha_2})\xi_0 R$ and $(B \times B)(G_{\alpha_2} \times G_{\alpha_1})\xi_0 R$, respectively, and choose six special paths that live in $(B \times B)\xi_0 R$ and go to each of those six codimension-one double cosets living in the boundary of $(B \times B)\xi_0 R$.

In $(B \times B)(G_{\alpha_1} \times G_{\alpha_2})\xi_0 R$, we set $Z_{1,2} := [(G_{\alpha_1} \times G_{\alpha_2})\xi_0 \overline{X}] \cap [(B \times B)\xi_0 R]$. Then $[(B \times B)\xi_0 R] \cap [(B \times B)(G_{\alpha_1} \times G_{\alpha_2})\xi_0 R] = (B \times B)\xi_0 Z_{1,2} SK_\theta^e$, which is a Zariski open subset in $(B \times B)(G_{\alpha_1} \times G_{\alpha_2})\xi_0 R$. Any $\underline{g}$ in $Z_{1,2}$ can be written as:

$$
\begin{aligned}
\underline{g} &= (\begin{pmatrix} a & b & 0 & 0 \\ c & d & 0 & 0 \\ 0 & 0 & d & -c \\ 0 & 0 & -b & a \end{pmatrix} \times \begin{pmatrix} 1 & 0 & 0 & 0 \\ 0 & e & 0 & f \\ 0 & 0 & 1 & 0 \\ 0 & g & 0 & h \end{pmatrix}) \xi_0 (I_4 \times \chi_{-\alpha_1-\alpha_2}(x)) \\
&= (\begin{pmatrix} d^{-1}t^{-2} & b & 0 & 0 \\ 0 & d & 0 & 0 \\ 0 & 0 & dt^2 & 0 \\ 0 & 0 & -bt^2 & d^{-1} \end{pmatrix} \times \begin{pmatrix} t^{-1} & 0 & 0 & 0 \\ 0 & (ht)^{-1} & 0 & tf \\ 0 & 0 & t & 0 \\ 0 & 0 & 0 & ht \end{pmatrix}) \xi_0 \\
&\quad \cdot (I_4 \times \chi_{-\alpha_1-\alpha_2}(x^*))(h_2(t,t,t^{-1},t^{-1}) \times h_2(t^2,1,t^{-2},1))k(w,t,g,h)
\end{aligned}
$$

where $(ts)^2 = 1$, $w = (t - t^{-1}) - t^{-1}cd^{-1} = -t^{-1}h^{-1}g$, $t^2 = gh^{-1} + cd^{-1} + 1 \neq 0$, $x^* = t^{-2}(x + gh^{-1})$, and $k(w,t,g,h)$ is in K_θ^e.

Let l be the parameter that will goes to zero. Then three special paths are defined by

$$
\begin{aligned}
\gamma_1(l) &= \chi_{-\alpha_1}(l-1) \times I_4 \mapsto \chi_{-\alpha_1}(-1) \times I_4. \\
\gamma_2(l) &= \begin{pmatrix} -1 & 1 & 0 & 0 \\ l-1 & -l & 0 & 0 \\ 0 & 0 & -l & 1-l \\ 0 & 0 & -1 & -1 \end{pmatrix} \times I_4 \mapsto \begin{pmatrix} -1 & 1 & 0 & 0 \\ -1 & 0 & 0 & 0 \\ 0 & 0 & 0 & 1 \\ 0 & 0 & -1 & -1 \end{pmatrix} \times I_4. \qquad (55) \\
\gamma_3(l) &= \chi_{-\alpha_1}(l-1) \times \begin{pmatrix} 1 & 0 & 0 & 0 \\ 0 & 0 & 0 & 1 \\ 0 & 0 & 1 & 0 \\ 0 & -1 & 0 & -l \end{pmatrix} \mapsto \chi_{-\alpha_1}(-1) \times \omega_{\alpha_2}.
\end{aligned}
$$

It is clear that those three paths are in the unique open subset $Z_{1,2}$ and go to the three codimension-one boundaries: $Y_1 := (B \times B)(\chi_{-\alpha_1}(-1) \times I_4)\xi_0 R$, $Y_2 := (B \times B)(\omega_{\alpha_1}\chi_{-\alpha_1}(-1) \times I_4)\xi_0 R$, and $Y_3 := (B \times B)(\chi_{-\alpha_1}(-1) \times \omega_{\alpha_2})\xi_0 R$.

Similarly, in $(B \times B)(G_{\alpha_2} \times G_{\alpha_1})\xi_0 R$, set $Z_{2,1} := [(G_{\alpha_2} \times G_{\alpha_1})\xi_0 \overline{X}] \cap [(B \times B)\xi_0 R]$. Then $[(B \times B)\xi_0 R] \cap [(B \times B)(G_{\alpha_2} \times G_{\alpha_1})\xi_0 R] = (B \times B)\xi_0 Z_{2,1} SK_\theta^e$, which is a Zariski open subset in $(B \times B)(G_{\alpha_2} \times G_{\alpha_1})\xi_0 R$. Any g in $Z_{2,1}$ is in form:

$$
\underline{g} = (\begin{pmatrix} 1 & 0 & 0 & 0 \\ 0 & e & 0 & f \\ 0 & 0 & 1 & 0 \\ 0 & g & 0 & h \end{pmatrix} \times \begin{pmatrix} a & b & 0 & 0 \\ c & d & 0 & 0 \\ 0 & 0 & d & -c \\ 0 & 0 & -b & a \end{pmatrix})\xi_0(I_4 \times \chi_{-\alpha_1-\alpha_2}(x))
$$

$$
= (\begin{pmatrix} s^{-1} & 0 & 0 & 0 \\ 0 & (sh)^{-1} & 0 & fs \\ 0 & 0 & s & 0 \\ 0 & 0 & 0 & hs \end{pmatrix} \times \begin{pmatrix} (s^2 d)^{-1} & b & 0 & 0 \\ 0 & d & 0 & 0 \\ 0 & 0 & ds^2 & 0 \\ 0 & 0 & -bs^2 & d^{-1} \end{pmatrix})\xi_0
$$

$$
\cdot (I_4 \times \chi_{-\alpha_1-\alpha_2}(x^*))(h_2(s,s,s^{-1},s^{-1}) \times_2 (s^2,1,s^{-2},1))k(z_1,z_2),
$$

where $t = 1$, $z_1 = gh^{-1}$, $z_2 = (x+z_1)(s^2+z_1)-(cd^{-1}+1)x-z_1 s^2$, $s^2 = gh^{-1}+cd^{-1}+1 \neq 0$, $x^* = s^{-2}(x+gh^{-1})$, and $k(z_1,z_2)$ is in K_θ^e. Three special paths γ_4, γ_5, and γ_6 from the open dense double coset to those three boundaries:

$$
\begin{aligned}
Y_4 &:= (B \times B)(I - 4 \times \chi_{-\alpha_1}(-1))\xi_0 R, \\
Y_5 &:= (B \times B)(I_4 \times \omega_{\alpha_1}\chi_{-\alpha_1}(-1))\xi_0 R, \\
Y_6 &:= (B \times B)(\omega_{\alpha_2} \times \chi_{-\alpha_1}(-1))\xi_0 R,
\end{aligned}
$$

respectively, are defined as follows:

$$
\gamma_4(l) = I_4 \times \chi_{-\alpha_1}(l-1) \mapsto I_4 \times \chi_{-\alpha_1}(-1).
$$

$$
\gamma_5(l) = I_4 \times \begin{pmatrix} -1 & 1 & 0 & 0 \\ l-1 & -l & 0 & 0 \\ 0 & 0 & -l & 1-l \\ 0 & 0 & -1 & -1 \end{pmatrix} \mapsto I_4 \times \begin{pmatrix} -1 & 1 & 0 & 0 \\ -1 & 0 & 0 & 0 \\ 0 & 0 & 0 & 1 \\ 0 & 0 & -1 & -1 \end{pmatrix}. \tag{56}
$$

$$
\gamma_6(l) = \begin{pmatrix} 1 & 0 & 0 & 0 \\ 0 & 0 & 0 & 1 \\ 0 & 0 & 1 & 0 \\ 0 & -1 & 0 & -l \end{pmatrix} \times \chi_{-\alpha_1}(l-1) \mapsto \omega_{\alpha_2} \times \chi_{-\alpha_1}(-1).
$$

4.6. Homogeneity $d_i((m_1,m_2;n_1,n_2))$. By Kostant-Rallis algorithm, the integers $d_i((m_1,m_2;n_1,n_2))$ are the eigenvalues of $x_\circ$ in $(m_1,m_2;n_1,n_2)(e)$, where $x_\circ \in \mathfrak{t}$ such that $[x_\circ, e] = e$ and is diagonalizable in the representation space $(m_1,m_2;n_1,n_2)$. Our proof is organized as follows:

(a) We realize $(m_1,m_2;n_1,n_2)^{K_\theta^e}$ as the space of right K_θ^e-invariant holomorphic sections of the line bundle attached to the highest weight $(m_1,m_2) \oplus (n_1,n_2)$ by Borel-Weil-Bott Theorem. In other words, $(m_1,m_2;n_1,n_2)^{K_\theta^e}$ can be realized as a space consisting of all regular functions $f : K_\theta \to \mathbb{C}$ satisfying the following property of quasi-invariance: for $(b_1,b_2) \in (B \times B)$, $k \in K_\theta^e$, and $g \in K_\theta$,

$$
f((b_1,b_2)gk) = (m_1,m_2)(b_1^{-1})(n_1,n_2)(b_2^{-1})f(g).
$$

(b) The restriction to the unique open dense R-orbit $(B \times B)\xi_0 R$ gives a canonical embedding:

$$
\imath^* : (m_1,m_2;n_1,n_2)^{K_\theta^e} \hookrightarrow \mathbb{C}[(B \times B)\xi_0 R]^{K_\theta^e} \tag{57}
$$

where $\mathbb{C}[(B \times B)\xi_0 R]^{K_\theta^e}$ is the ring of all right K_θ^e-invariant regular functions on $(B \times B)\xi_0 R$. Since K_θ^e is a normal subgroup of R by Proposition 4.5.3 in subsection 4.5, the embedding $\imath^*$ is actually a homomorphism of R-modules. Then the image of $\imath^*$ consists of all right K_θ^e-invariant regular functions f on $(B \times B)\xi_0 R$ enjoying following two properties:

(1) For $(b_1, b_2) \in (B \times B)$, $k \in K_\theta^e$, $\overline{x} \in \overline{X}$, and $\underline{s} \in S$,

$$f((b_1, b_2)\xi_0 \overline{x} \underline{s} k) = (m_1, m_2)(b_1^{-1})(n_1, n_2)(b_2^{-1})f(\xi_0 \overline{x} \underline{s}),$$

(2) f can be extended to be a regular function on the whole group K_θ.

(c) The implement of these extensions. Since any affine algebraic group is normal as an algebraic variety, see [**Hos**], it follows from [**Kos**] that any regular function f on the open dense double coset $(B \times B)\xi_0 R$ is regular on the whole group K_θ or the set of points where f is not defined has codimension one in K_θ. By Theorem 4.5.1 (c) and Corollary 4.5.2, the subset $\tilde{X}$ includes the unique open dense double coset and all six codimension-one double cosets with respect to the pair of subgroups $(B \times B, R)$, and the subset $K_\theta - \tilde{X}$ has codimension two. Hence if a regular function f on the open double coset $(B \times B)\xi_0 R$ can be extended to be regular on $\tilde{X}$, then f will be extended automatically to be regular on the whole group K_θ. One can find a similar argument in Bernstein-Gelfand-Gelfand [**BGG**]. First we consider the extensions along those special paths $\{\gamma_i\}$ of regular functions from the unique open dense double coset to its codimension-one boundaries. In other words we consider a subspace $(m_1, m_2; n_1, n_2)_{\{\gamma_i\}}^{K_\theta^e}$ of $\mathbb{C}[(B \times B)\xi_0 R]^{K_\theta^e}$ consisting of all functions f satisfying following two conditions:

(1) For $(b_1, b_2) \in (B \times B)$, $k \in K_\theta^e$, $\overline{x} \in \overline{X}$, and $\underline{s} \in S$,

$$f((b_1, b_2)\xi_0 \overline{x} \underline{s} k) = (m_1, m_2)(b_1^{-1})(n_1, n_2)(b_2^{-1})f(\xi_0 \overline{x} \underline{s}),$$

(2) $\lim_{l \to 0} f(\gamma_i(l))$ exist for $i = 1, 2, \cdots, 6$. (Note that this condition is weaker than the condition that f can be extended to be a regular function on $\tilde{X}$ and then to be regular on the whole group K_θ).

It is evident that $(m_1, m_2; n_1, n_2)^{K_\theta^e} \subset (m_1, m_2; n_1, n_2)_{\{\gamma_i\}}^{K_\theta^e}$. Further we have a filtration of subspaces:

$$(m_1, m_2; n_1, n_2)(e) \subset (m_1, m_2; n_1, n_2)^{K_\theta^e} \subset (m_1, m_2; n_1, n_2)_{\{\gamma_i\}}^{K_\theta^e}. \tag{58}$$

After computing the dimension of the subspace $(m_1, m_2; n_1, n_2)_{\{\gamma_i\}}^{K_\theta^e}$, we will show that

$$\dim(m_1, m_2; n_1, n_2)(e) = \dim(m_1, m_2; n_1, n_2)_{\{\gamma_i\}}^{K_\theta^e}. \tag{59}$$

In other words, we will obtain that $(m_1, m_2; n_1, n_2)(e) = (m_1, m_2; n_1, n_2)^{K_\theta^e}$.

(d) We will determine explicitly the eigenvalues and their multiplicities of the differential operator $x_\circ$ in the subspace $(m_1, m_2; n_1, n_2)(e)$.

Before going to compute the dimension of the subspace $(m_1, m_2; n_1, n_2)_{\{\gamma_i\}}^{K_\theta^e}$, we give first some necessary conditions about the existence of the K_θ^e-invariant functions $(m_1, m_2; n_1, n_2)^{K_\theta^e}$ and the occurance of eigenvalues ν_p of the subgroup S in $(m_1, m_2; n_1, n_2)^{K_\theta^e}$.

LEMMA 4.6.1. *Let $(m_1, m_2; n_1, n_2)^{K_\theta^e}(\nu_p)$ be the set of all S-eigenfunctions in the space $(m_1, m_2; n_1, n_2)^{K_\theta^e}$ with the eigenvalue ν_p $(f(g\underline{s}) = \nu_p(\underline{s})f(g) = s^p f(g))$. Then*

(a) *$(m_1, m_2; n_1, n_2)^{K_\theta^e} \neq 0$ implies that $(m_1 + m_2 + n_1 + n_2) \equiv 0 \pmod 2$.*

(b) *The space $(m_1, m_2; n_1, n_2)^{K_\theta^e}(\nu_p) \neq 0$ (or $(m_1, m_2; n_1, n_2)_{\{\gamma_i\}}^{K_\theta^e}(\nu_p) \neq 0$) implies that $(m_1 + m_2 + n_1 + n_2) \equiv 0 \pmod 2$ and $m_1 + m_2 \equiv p \pmod 2$.*

PROOF. (a) Since $(B \times B) \cap \xi_0 K_\theta^e \xi_0^{-1} = \{\pm(I_4 \times I_4)\}$, for any f in the subspace $(m_1, m_2; n_1, n_2)^{K_\theta^e}$, we have that $f(\xi_0 \bar{x}) = f(\xi_0 \bar{x}(-(I_4 \times I_4))) = (-1)^{m_1+m_2+n_1+n_2} f(\xi_0 \bar{x})$. Thus $f \neq 0$ implies that $m_1 + m_2 + n_1 + n_2 \equiv 0 \pmod 2$.

(b) Since $(B \times B) \cap \xi_0 S x i_0^{-1} = \{\pm I_4 \times I_4\}$, for any $f \in ((m_1, m_2; n_1, n_2))^{K_\theta^e}(\nu_p)$, we have that

$$f(\xi_0 \bar{x}((-I_4) \times I_4)) = (-1)^p f(\xi_0 \bar{x}) = f(((-I_4) \times I_4)\xi_0 \bar{x}) = (-1)^{m_1+m_2} f(\xi_0 \bar{x}).$$

Therefore $f \neq 0$ implies that $m_1 + m_2 \equiv p \pmod 2$.   □

THEOREM 4.6.1. *Let $\mathbb{C}[\overline{X}]$ be the affine algebra of the one-dimensional section $\overline{X}$ as defined in proposition 11. Let $\varphi(x) = \sum_{i=1}^q a_i x^i$, $\tilde{\varphi}(x) = \sum_{j=1}^q b_j x^j$ in $\mathbb{C}[\overline{X}]$ such that $\tilde{\varphi}(x+1) = \varphi(x)$, i.e. $b_j = \sum_{i=j}^q (-1)^{i-j} \binom{i}{j} a_i$. Then we have*

(a) *any S-eigenfunction f with eigenvalue s^p in $\mathbb{C}[(B \times B)\xi_0 R]^{K_\theta^e}$ can be expressed as*

$$f((b_1, b_2)\xi_0 \bar{x}\bar{s}k) = (m_1, m_2)(b_1^{-1})(n_1, n_2)(b_2^{-1})s^p f(\xi_0 \bar{x})$$
$$= (m_1, m_2)(b_1^{-1})(n_1, n_2)(b_2^{-1})s^p \varphi(x)$$

for $(b_1, b_2) \in (B \times B)$, $k \in K_\theta^e$, $\bar{x} \in \overline{X}$, and $\bar{s} \in S$. We call $\varphi(x)$ the polynomial of f.

(b) *the S-eigenfunction f with eigenvalue s^p belongs to $(m_1, m_2; n_1, n_2)_{\{\gamma_i\}}^{K_\theta^e}(\nu_p)$ if and only if the polynomial $\varphi(x)$ (or $\tilde{\varphi}(x)$) of f satisfies following four conditions:*

(c1) *$2q \leq m_1 + m_2 + 2n_1 - p$, and $a_i = 0$ if $2i < m_1 + m_2 + 2n_2 - p$,*

(c2) *$2q \leq n_1 + n_2 + 2m_1 - p$, and $a_i = 0$ if $2i < n_1 + n_2 + 2m_2 - p$,*

(c3) *if $2j < m_1 - m_2 + 2n_1 - p$, $b_j = 0$, and*

(c4) *if $2j < n_1 + n_2 + 2m_1 - p$, $b_j = 0$.*

PROOF. Part (a) follows straight from the quasi-invariant properties on both sides and the regularity of the function f on the unique open dense double coset. Part (b) will be more complicated and its proof will be separated into cases (b1) and (b2).

(b1) We consider the conditions on the existence of the extensions of the functions f to $(B \times B)(G_{\alpha_1} \times G_{\alpha_2})\xi_0 R$ along those three special paths γ_1, γ_2, and γ_3 as chosen in the previous subsection. In other words, we consider the conditions under which those three limits $\lim_{l \to 0} f(\gamma_i(l))$ exist for $i = 1, 2, 3$. When $\underline{g}$ in $Z_{1,2} := [(G_{\alpha_1} \times G_{\alpha_2})\xi_0 \overline{X}] \cap [(B \times B)\xi_0 R]$, it can be written as in (51). Thus for any S-eigenfunction f with eigenvalue s^p in $\mathbb{C}[(B \times B)\xi_0 R]^{K_\theta^e}$, one has

$$f(\underline{g}) = (dt^2)^{m_1} d^{-m_2} t^{n_1}(ht)^{n_2} t^{-p} \varphi(t^{-2}(x + gh^{-1})). \tag{60}$$

We are going to consider the existence of the limit $\lim f(\underline{g})$ when $\underline{g}$ goes to the codimension-one boundaries along those three paths $\gamma_1(l)$, $\gamma_2(l)$, and $\gamma_3(l)$ as in (51), where l is the parameter going to 0.

Along the path $\gamma_1(l)$, $t^2 = l$ goes to 0 and

$$f(\gamma_1(l)) = t^{2m_1 + n_1 + n_2 - p} \varphi(t^{-2}x).$$

When t goes to 0, the limit of $f(\gamma_1(l))$ exists if and only if

$$2q \le 2m_1 + n_1 + n_2 - p. \tag{61}$$

Along the path $\gamma_2(l)$, $t^2 = l^{-1}$ goes to ∞ and

$$f(\gamma_2(l)) = t^{2m_2 + n_1 + n_2 - p} \varphi(t^{-2}x).$$

If t goes to ∞, then the limit of $f(\gamma_2(l))$ exists if and only if

$$a_i = 0 \text{ when } 2i < 2m_2 + n_1 + n_2 - p. \tag{62}$$

Finally, along the path $\gamma_3(l)$, $t^2 = l - l^{-1}$ goes to ∞ (as l goes to 0) and

$$f(\gamma_3(l)) = t^{2m_1 + n_1 - n_2 - p} \varphi(t^{-2}x - 1) = t^{2m_1 + n_1 - n_2 - p} \tilde{\varphi}(t^{-2}x).$$

As t goes to ∞, the limit of $f(\gamma_3(l))$ exists if and only if

$$b_j = 0 \text{ when } 2j < 2m_1 + n_1 + n_2 - p. \tag{63}$$

In other words, the conditions on the existence of the extension of f to $(B \times B)(G_{\alpha_1} \times G_{\alpha_2})\xi_0 R$ along those three special paths γ_1, γ_2, and γ_3 are (c2) and (c4).

(b2) We consider the conditions on the existence of the extensions of the function f to $(B \times B)(G_{\alpha_2} \times G_{\alpha_1})\xi_0 R$ along those three special paths γ_4, γ_5, and γ_6. Again the element $\underline{g}$ in the open dense subset $Z_{2,1} = [(G_{\alpha_2} \times G_{\alpha_1})\xi_0 \overline{X}] \cap [(B \times B)\xi_0 R]$ can be written as (52). Hence any S-eigenfunction f with eigenvalue s^p in $\mathbb{C}[(B \times B)\xi_0 R]^{K_\theta^e}$ can be expressed as

$$f(\underline{g}) = s^{m_1}(hs)^{m_2}(ds^2)^{n_1} d^{-n_2} s^{-p} \varphi(s^{-2}(x + gh^{-1})). \tag{64}$$

By the same argument as that in case (b1), we obtain that

$$\begin{aligned} f(\gamma_4(l)) &= s^{m_1 + m_2 + 2n_1 - p} \varphi(s^{-2}x), (s^2 = l \to 0) \\ f(\gamma_5(l)) &= s^{-(m_1 + m_2 + 2n_2 - p)} \varphi(s^2 x), (s^2 = l - l^{-1} \to \infty) \\ f(\gamma_6(l)) &= s^{-(m_1 - m_2 + 2n_1 - p)} \tilde{\varphi}(s^2 x), (s^2 = l^{-1} \to \infty) \end{aligned} \tag{65}$$

where those three paths $\gamma_4(l)$, $\gamma_5(l)$, and $\gamma_6(l)$ are defined in (52). As l goes to 0, the existence of those limits $\lim_{l \to 0} f(\gamma_i(l))$ exist for $i = 4, 5, 6$, implies exactly the conditions (c1) and (c3). The theorem is proved. $\square$

The following lemma gives a way to count the dimension of the eigensubspace $(m_1, m_2; n_1, n_2)_{\{\gamma_i\}}^{K_\theta^e}(\nu_p)$ for each possible integer p.

LEMMA 4.6.2. Let $a_{i,j} = (-1)^{i-j} \binom{i}{j} = (-1)^{i-j} \frac{i!}{j!(i-j)!}$ and

$$
M(l,m,n) = \begin{pmatrix}
a_{l,0} & \cdots & a_{m,0} & \cdots & a_{n,0} \\
a_{l,1} & \cdots & a_{m,1} & \cdots & a_{n,1} \\
\cdots & \cdots & & & \\
a_{l,l} & \cdots & a_{m,l} & \cdots & a_{n,l} \\
& \ddots & \vdots & & \vdots \\
0 & & a_{m,m} & \cdots & a_{n,m}
\end{pmatrix}, \quad 0 \le l \le m;\ 0 \le l \le n,
$$

$$
N(l,m,n) = \begin{pmatrix}
a_{l,0} & \cdots & a_{n,0} \\
a_{l,1} & \cdots & a_{n,1} \\
\cdots & \cdots & \cdots \\
a_{l,m} & \cdots & a_{n,m}
\end{pmatrix}, \quad 0 \le l \le n;\ 0 \le m \le l.
$$

Then $rank M(l,m,n) = min\{m+1, n-l+1\}$ and $rank N(l,m,n) = min\{m+1, n-l+1\}$.

PROOF. Since $a_{i+1,j} + a_{i,j} = a_{i,j-1}$, we have, by elementary transforms of columns,

$$
M(l,m,n) \Rightarrow \begin{pmatrix}
a_{l,0} & 0 \cdots 0 \\
\vdots & \\
a_{l,l} & M(l, m-1, n-1) \\
0 &
\end{pmatrix}.
$$

By induction, we obtain $rank M(l,m,n) = mim\{m+1, n-l+1\}$. Similarly, we can prove that $rank N(l,m,n) = mim\{m+1, n-l+1\}$. $\square$

THEOREM 4.6.2. Let x_o is such an element in $\mathbf{t}$, as chosen above, that $[x_o, e] = e$ and $(m_1, m_2; n_1, n_2)_{\{\gamma_i\}}^{K_\theta^e}(p)$ the eigenspace of x_o in $(m_1, m_2; n_1, n_2)_{\{\gamma_i\}}^{K_\theta^e}$. Then we have
(a) If $(m_1 + m_2) \ge (n_1 + n_2) \ge (m_1 - m_2) \ge (n_1 - n_2)$, then the eigenvalues of x_o are

$$
p = 2m_1 - n_1 + n_2, 2m_1 - n_1 + n_2 + 2, \cdots, m_1 + m_2 + 2n_1
$$

and (a1) if $n_1 + n_2 - m_1 + m_2 < 2n_1 - 2n_2$, then

$$
\dim(m_1, m_2; n_1, n_2)_{\{\gamma_i\}}^{K_\theta^e}(p)
$$

$$
= \begin{cases}
1 + \frac{1}{2}(n_1 - n_2 - 2m_1 + p) & \text{if } 2m_1 - n_1 + n_2 \le p \le m_1 + m_2 + 2n_2, \\
1 + \frac{1}{2}(n_1 + n_2 - m_1 + m_2) & \text{if } m_1 + m_2 + 2n_2 \le p \le 2m_1 + n_1 - n_2, \\
1 + \frac{1}{2}(m_1 + m_2 + 2n_1 - p) & \text{if } 2m_1 + n_1 - n_2 \le p \le m_1 + m_2 + 2n_1.
\end{cases}
$$

(a2) if $n_1 + n_2 - m_1 + m_2 \geq 2n_1 - 2n_2$, then

$$\dim(m_1, m_2; n_1, n_2)_{\{\gamma_i\}}^{K_\theta^e}(p)$$

$$= \begin{cases} 1 + \frac{1}{2}(n_1 - n_2 - 2m_1 + p) & if \ 2m_1 - n_1 + n_2 \leq p \leq 2m_1 + n_1 - n_2, \\ 1 + (n_1 - n_2) & if \ 2m_1 + n_1 - n_2 \leq p \leq m_1 + m_2 + 2n_2, \\ 1 + \frac{1}{2}(m_1 + m_2 + 2n_1 - p) & if \ m_1 + m_2 + 2n_2 \leq p \leq m_1 + m_2 + 2n_1; \end{cases}$$

(b) If $(m_1 + m_2) \geq (n_1 + n_2) \geq (n_1 - n_2) > (m_1 - m_2)$, then the eigenvalues of $x_\circ$ are

$$p = 2m_2 + n_1 - n_2, 2m_2 + n_1 - n_2 + 2, \cdots, 2m_1 + n_1 + n_2$$

and (b1) if $n_2 < m_1 - m_2$, then

$$\dim(m_1, m_2; n_1, n_2)_{\{\gamma_i\}}^{K_\theta^e}(p)$$

$$= \begin{cases} 1 + \frac{1}{2}(n_2 - n_1 - 2m_2 + p) & if \ 2m_2 + n_1 - n_2 \leq p \leq 2m_2 + n_1 + n_2, \\ 1 + n_2 & if \ 2m_2 + n_1 + n_2 \leq p \leq 2m_1 + n_1 - n_2, \\ 1 + \frac{1}{2}(2m_1 + n_1 + n_2 - p) & if \ 2m_1 + n_1 - n_2 \leq p \leq 2m_1 + n_1 + n_2, \end{cases}$$

(b2) if $n_2 \geq m_1 - m_2$, then

$$\dim(m_1, m_2; n_1, n_2)_{\{\gamma_i\}}^{K_\theta^e}(p)$$

$$= \begin{cases} 1 + \frac{1}{2}(n_2 - n_1 - 2m_2 + p) & if \ 2m_2 + n_1 - n_2 \leq p \leq 2m_1 + n_1 - n_2, \\ 1 + (m_1 - m_2) & if \ 2m_1 + n_1 - n_2 \leq p \leq 2m_2 + n_1 + n_2, \\ 1 + \frac{1}{2}(2m_1 + n_1 + n_2 - p) & if \ 2m_2 + n_1 + n_2 \leq p \leq 2m_1 + n_1 + n_2; \end{cases}$$

(c) If $(n_1 + n_2) > (m_1 + m_2) \geq (m_1 - m_2) \geq (n_1 - n_2)$, then the eigenvalues of $x_\circ$ are

$$p = m_1 - m_2 + 2n_2, m_1 - m_1 + 2n_2 + 2, \cdots, m_1 + m_2 + 2n_1$$

and (c1) if $m_2 < n_1 - n_2$, then

$$\dim(m_1, m_2; n_1, n_2)_{\{\gamma_i\}}^{K_\theta^e}(p)$$

$$= \begin{cases} 1 + \frac{1}{2}(m_2 - m_1 - 2n_2 + p) & if \ m_1 - m_2 + 2n_2 \leq p \leq m_1 + m_2 + 2n_2, \\ 1 + m_2 & if \ m_1 + m_2 + 2n_2 \leq p \leq m_1 - m_2 + 2n_1, \\ 1 + \frac{1}{2}(m_1 + m_2 + 2n_1 - p) & if \ m_1 - m_2 + 2n_1 \leq p \leq m_1 + m_2 + 2n_1, \end{cases}$$

(c2) if $m_2 \geq n_1 - n_2$, then

$$\dim(m_1, m_2; n_1, n_2)_{\{\gamma_i\}}^{K_\theta^e}(p)$$

$$= \begin{cases} 1 + \frac{1}{2}(m_2 - m_1 - 2n_2 + p) & if \ m_1 - m_2 + 2n_2 \leq p \leq m_1 - m_2 + 2n_1, \\ 1 + (n_1 - n_2) & if \ m_1 - m_2 + n_1 \leq p \leq m_1 + m_2 + 2n_2, \\ 1 + \frac{1}{2}(m_1 + m_2 + 2n_1 - p) & if \ m_1 + m_1 + 2n_2 \leq p \leq m_1 + m_2 + 2n_1; \end{cases}$$

(d) If $(n_1 + n_2) > (m_1 + m_2) \geq (n_1 - n_2) > (m_1 - m_2)$, then the eigenvalues of $x_\circ$ are

$$p = 2n_2 + m_1 - m_2, 2n_2 + m_1 - m_2 + 2, \cdots, 2m_1 + n_1 + n_2$$

and (d1) if $m_1 + m_2 - n_1 + n_2 < 2m_1 - 2m_2$, then

$$\dim(m_1, m_2; n_1, n_2)_{\{\gamma_i\}}^{K_\theta^e}(p)$$

$$= \begin{cases} 1 + \frac{1}{2}(m_1 - m_2 - 2n_1 + p) & \text{if } 2n_1 - m_1 + m_2 \leq p \leq n_1 + n_2 + 2m_2, \\ 1 + \frac{1}{2}(m_1 + m_2 - n_1 + n_2) & \text{if } n_1 + n_2 + 2m_2 \leq p \leq 2n_1 + m_1 - m_2, \\ 1 + \frac{1}{2}(n_1 + n_2 + 2m_1 - p) & \text{if } 2n_1 + m_1 - m_2 \leq p \leq n_1 + n_2 + 2m_1, \end{cases}$$

(d2) if $m_1 + m_2 - n_1 + n_2 \geq 2m_1 - 2m_2$, then

$$\dim(m_1, m_2; n_1, n_2)_{\{\gamma_i\}}^{K_\theta^e}(p)$$

$$= \begin{cases} 1 + \frac{1}{2}(m_1 - m_2 - 2n_1 + p) & \text{if } 2n_1 - m_1 + m_2 \leq p \leq 2n_1 + m_1 - m_2, \\ 1 + (m_1 - m_2) & \text{if } 2n_1 + m_1 - m_2 \leq p \leq n_1 + n_2 + 2m_2, \\ 1 + \frac{1}{2}(n_1 + n_2 + 2m_1 - p) & \text{if } n_1 + n_2 + 2m_2 \leq p \leq n_1 + n_2 + 2m_1; \end{cases}$$

(e) If $(m_1 - m_2) > (n_1 + n_2)$ or $(n_1 - n_2) > (m_1 + m_2)$, then

$$\dim(m_1, m_2; n_1, n_2)_{\{\gamma_i\}}^{K_\theta^e}(p) = 0.$$

PROOF. Assume that $\varphi(x) = \sum_{i=1}^{q} a_i x^i$, $\tilde{\varphi}(x) = \sum_{j=1}^{q} b_j x^j \in \mathbb{C}[\overline{X}]$ such that $\tilde{\varphi}(x+1) = \varphi(x)$, i.e. $b_j = \sum_{i=j}^{q} (-1)^{i-j} \binom{i}{j} a_i$.

(a) $(m_1 + m_2) \geq (n_1 + n_2) \geq (m_1 - m_2) \geq (n_1 - n_2)$.

In this case, we have $2m_2 + n_1 + n_2 \leq m_1 + m_2 + 2n_2$, $m_1 + m_2 + 2n_1 \leq 2m_1 + n_1 + n_2$, and $m_1 - m_2 + 2n_1 \leq 2m_1 + n_1 - n_2$, and further we have $p \leq m_1 + m_2 + 2n_1$, $m_1 + m_2 + 2n_2 - p \leq 2i \leq m_1 + m_2 + 2n_1 - p$, and $b_j = 0$ if $2j < 2m_1 + n_1 - n_2 - p$ by Theorem 4.6.1.

(a1) If $n_1 + n_2 - m_1 + m_2 < 2n_1 - 2n_2$, then $2m_1 - n_1 + n_2 \leq m_1 + m_2 + 2n_2 < 2m_1 + n_1 - n_2$. we shall compute $\dim(m_1, m_2; n_1, n_2)_{\{\gamma_i\}}^{K_\theta^e}(p)$ case by case for p.

(a1i) Case $2m_1 + n_1 - n_2 \leq p \leq m_1 + m_2 + 2n_1$.

Since $2m_1 + n_1 - n_2 - p \leq 0$, there are no conditions for b_j and because $m_1 + m_2 + 2n_2 - p \leq 0$ we know that for such a p, the degree of the polynomial $\varphi(x)$ could be any integer between 0 and $\frac{1}{2}(m_1 + m_2 + 2n_1 - p)$. Thus

$$\dim(m_1, m_2; n_1, n_2)_{\{\gamma_i\}}^{K_\theta^e}(p) = 1 + \frac{1}{2}(m_1 + m_2 + 2n_1 - p).$$

(a1ii) Case $m_1 + m_2 + 2n_2 \leq p \leq 2m_1 + n_1 - n_2$.

In this case, the degree of $\varphi(x) = \sum_{i=1}^{q} a_i x^i$ will takes values of all integers between 0 and $\frac{1}{2}(m_1 + m_2 + 2n_1 - p)$ and there is the restriction for b_j, that is, $b_j = 0$ if $0 \leq j < \frac{1}{2}(2m_1 + n_1 - n_2 - p)$. This means that $\dim(m_1, m_2; n_1, n_2)_{\{\gamma_i\}}^{K_\theta^e}(p)$ is equal to the dimension of the solutions of the system of homogeneous linear equations $b_j = 0$ for $0 \leq j < \frac{1}{2}(2m_1 + n_1 - n_2 - p)$. Since the system of the equations

$$b_j = \sum_{i=j}^{q} (-1)^{i-j} \binom{i}{j} a_i = 0, \quad j = 0, 1, \cdots, \frac{1}{2}(2m_1 + n_1 - n_2 - p) - 1$$

has $1 + \frac{1}{2}(m_1 + m_2 + 2n_1 - p)$ variables and $\frac{1}{2}(2m_1 + n_1 - n_2 - p)$ equations and its rank is $v\frac{1}{2}(2m_1 + n_1 - n_2 - p)$, we know that the dimension of the solutions is $1 + \frac{1}{2}(m_1 + m_2 + 2n_1 - p) - \frac{1}{2}(2m_1 + n_1 - n_2 - p) = 1 + \frac{1}{2}(n_1 + n_2 - m_1 + m_2)$.

(a1iii) Case $p \leq m_1 + m_2 + 2n_2$

In this case, we have $a_i = 0$ if $2i < m_1 + m_2 + 2n_2 - p$. The system of homogeneous linear equations

$$b_j = \sum_{i=j}^{q} (-1)^{i-j} \binom{i}{j} a_i = 0, \quad j = 0, 1, \cdots, \frac{1}{2}(2m_1 + n_1 - n_2 - p) - 1$$

has $1 + (n_1 - n_2)$ variables and $\frac{1}{2}(2m_1 + n_1 - n_2 - p)$ equations and the corresponding matrix of the coefficients is $M(\frac{1}{2}(m_1 + m_2 + 2n_2 - p), \frac{1}{2}(2m_1 + n_1 - n_2 - p) - 1, \frac{1}{2}(m_1 + m_2 + 2n_1 - p))$. By Lemma 4.6.2, its rank is $mim\{\frac{1}{2}(2m_1 + n_1 - n_2 - p), 1 + (n_1 - n_2)\}$. Since $[\frac{1}{2}(2m_1 + n_1 - n_2 - p)] - [1 + (n_1 - n_2)] = \frac{1}{2}(2m_1 - n_1 + n_2 - p) - 1$, it is evident that if $p < 2m_1 - n_1 + n_2$, then the number of the variables equals the rank of the system of equations and is less than the number of the equations, and this means that the system of equations has no solutions, in other words, $\dim(m_1, m_2; n_1, n_2)_{\{\gamma_i\}}^{K_\theta^e}(p) = 0$; and if $2m_1 - n_1 + n_2 \leq p \leq m_1 + m_2 + 2n_2$, then the rank of the system of equations is $\frac{1}{2}(2m_1 + n_1 - n_2 - p)$ and the dimension of the solutions of the system of equations is $1 + \frac{1}{2}(n_1 - n_2 - 2m_1 + p)$. (a1) is proved.

(a2) If $n_1 + n_2 - m_1 + m_2 \geq 2n_1 - 2n_2$, then $m_1 + m_2 + 2n_2 \geq 2m_1 + n_1 - n_2 \geq 2m_1 - n_1 + n_2$. we shall compute $\dim(m_1, m_2; n_1, n_2)_{\{\gamma_i\}}^{K_\theta^e}(p)$ case by case for p.

(a2i) Case $m_1 + m_2 + 2n_2 \leq p \leq m_1 + m_2 + 2n_1$.

Since $2m_1 + n_1 - n_2 - p \leq 0$, there are no conditions for b_j and because $m_1 + m_2 + 2n_2 - p \leq 0$ we know that for such a p, the degree of the polynomial $\varphi(x)$ could be any integer between 0 and $\frac{1}{2}(m_1 + m_2 + 2n_1 - p)$. Thus

$$\dim(m_1, m_2; n_1, n_2)_{\{\gamma_i\}}^{K_\theta^e}(p) = 1 + \frac{1}{2}(m_1 + m_2 + 2n_1 - p).$$

(a2ii) Case $2m_1 + n_1 - n_2 \leq p \leq m_1 + m_2 + 2n_2$.

In this case, the degree of $\varphi(x) = \sum_{i=1}^{q} a_i x^i$ will takes values of all integers between $\frac{1}{2}(m_1 + m_2 + 2n_2 - p)$ and $\frac{1}{2}(m_1 + m_2 + 2n_1 - p)$, $a_i = 0$ if $2i < m_1 + m_2 + 2n_2 - p$, and there are no restrictions for b_j for $2m_1 + n_1 - n_2 - p \leq 0$. Therefore $\dim(m_1, m_2; n_1, n_2)_{\{\gamma_i\}}^{K_\theta^e}(p) = 1 + (n_1 - n_2)$.

(a2iii) Case $p \leq 2m_1 + n_1 - n_2$.

In this case, we have $a_i = 0$ if $2i < m_1 + m_2 + 2n_2 - p$. The system of homogeneous linear equations

$$b_j = \sum_{i=j}^{q} (-1)^{i-j} \binom{i}{j} a_i = 0, \quad j = 0, 1, \cdots, \frac{1}{2}(2m_1 + n_1 - n_2 - p) - 1$$

has $1 + (n_1 - n_2)$ variables and $\frac{1}{2}(2m_1 + n_1 - n_2 - p)$ equations and the corresponding matrix of the coefficients is $N(\frac{1}{2}(m_1 + m_2 + 2n_2 - p), \frac{1}{2}(2m_1 + n_1 - n_2 - p) - 1, \frac{1}{2}(m_1 + m_2 + 2n_1 - p))$. By Lemma 4.6.2, its rank is $mim\{\frac{1}{2}(2m_1 + n_1 - n_2 - p), 1 + (n_1 - n_2)\}$. Since $[\frac{1}{2}(2m_1 + n_1 - n_2 - p)] - [1 + (n_1 - n_2)] = \frac{1}{2}(2m_1 - n_1 + n_2 - p) - 1$, it is evident that if $p < 2m_1 - n_1 + n_2$, then the number of the variables equals the rank of the system of equations and is less than the number of the equations, and this means that the system of equations has no solutions, in other words, $\dim(m_1, m_2; n_1, n_2)_{\{\gamma_i\}}^{K_\theta^e}(p) = 0$; and if $2m_1 - n_1 + n_2 \leq p \leq 2m_1 + n_1 - n_2$, then the rank of the system of equations is $\frac{1}{2}(2m_1 + n_1 - n_2 - p)$ and the dimension of the solutions of the system of equations is $1 + \frac{1}{2}(n_1 - n_2 - 2m_1 + p)$. Case (a) is proved.

(b) $(m_1 + m_2) \geq (n_1 + n_2) \geq (n_1 - n_2) > (m_1 - m_2)$.

In this case, we have $2m_2 + n_1 + n_2 > m_1 + m_2 + 2n_2$, $m_1 + m_2 + 2n_1 > 2m_1 + n_1 + n_2$, and $m_1 - m_2 + 2n_1 \leq 2m_1 + n_1 - n_2$, and further we have $p \leq 2m_1 + n_1 + n_2$, $2m_2 + n_1 + n_2 - p \leq 2i \leq 2m_1 + n_1 + n_2 - p$, and $b_j = 0$ if $2j < 2m_1 + n_1 - n_2 - p$.

(b1) If $n_2 < 2m_1 - 2m_2$, then $2m_2 + n_1 - n_2 \leq 2m_2 + n_1 + n_2 < 2m_1 + n_1 - n_2$. we shall compute $\dim(m_1, m_2; n_1, n_2)_{\{\gamma_i\}}^{K_\theta^e}(p)$ case by case for p.

(b1i) Case $2m_1 + n_1 - n_2 \leq p \leq 2m_1 + n_1 + n_2$.

Since $2m_1 + n_1 - n_2 - p \leq 0$, there are no conditions for b_j and because $2m_2 + n_1 + n_2 - p \leq 0$ we know that for such a p, the degree of the polynomial $\varphi(x)$ could be any integer between 0 and $\frac{1}{2}(2m_1 + n_1 + n_2 - p)$. Thus

$$\dim(m_1, m_2; n_1, n_2)_{\{\gamma_i\}}^{K_\theta^e}(p) = 1 + \frac{1}{2}(2m_1 + n_1 + n_2 - p).$$

(b1ii) Case $2m_2 + n_1 + n_2 \leq p \leq 2m_1 + n_1 - n_2$.

In this case, the degree of $\varphi(x) = \sum_{i=1}^q a_i x^i$ will takes values of all integers between 0 and $\frac{1}{2}(2m_1 + n_1 + n_2 - p)$ and there is the restriction for b_j, that is, $b_j = 0$ if $0 \leq j < \frac{1}{2}(2m_1 + n_1 - n_2 - p)$. This means that $\dim(m_1, m_2; n_1, n_2)_{\{\gamma_i\}}^{K_\theta^e}(p)$ is equal to the dimension of the solutions of the system of homogeneous linear equations $b_j = 0$ for $0 \leq j < \frac{1}{2}(2m_1 + n_1 - n_2 - p)$. Since the system of the equations

$$b_j = \sum_{i=j}^q (-1)^{i-j} \binom{i}{j} a_i = 0, \quad j = 0, 1, \cdots, \frac{1}{2}(2m_1 + n_1 - n_2 - p) - 1$$

has $1 + \frac{1}{2}(2m_1 + n_1 + n_2 - p)$ variables and $\frac{1}{2}(2m_1 + n_1 - n_2 - p)$ equations and its rank is $\frac{1}{2}(2m_1 + n_1 - n_2 - p)$, we know that the dimension of the solutions is $1 + \frac{1}{2}(2m_1 + n_1 + n_2 - p) - \frac{1}{2}(2m_1 + n_1 - n_2 - p) = 1 + n_2$.

(b1iii) Case $p \leq 2m_2 + n_1 + n_2$

In this case, we have $a_i = 0$ if $2i < 2m_2 + n_1 + n_2 - p$. The system of homogeneous linear equations

$$b_j = \sum_{i=j}^{q} (-1)^{i-j} \binom{i}{j} a_i = 0, \quad j = 0, 1, \cdots, \frac{1}{2}(2m_1 + n_1 - n_2 - p) - 1$$

has $1 + (m_1 - m_2)$ variables and $\frac{1}{2}(2m_1 + n_1 - n_2 - p)$ equations and the corresponding matrix of the coefficients is $M(\frac{1}{2}(2m_2 + n_1 + n_2 - p), \frac{1}{2}(2m_1 + n_1 - n_2 - p) - 1, \frac{1}{2}(2m_1 + n_1 + n_2 - p))$. By Lemma 4.6.2, its rank is $mim\{\frac{1}{2}(2m_1 + n_1 - n_2 - p), 1 + (m_1 - m_2)\}$. Since $[1 + (m_1 - m_2)] - [\frac{1}{2}(2m_1 + n_1 - n_2 - p)] = 1 + \frac{1}{2}(-2m_2 - n_1 + n_2 + p)$, it is evident that if $p < 2m_2 + n_1 - n_2$, then the number of the variables equals the rank of the system of equations and is less than the number of the equations, and this means that the system of equations has no solutions, in other words, $\dim(m_1, m_2; n_1, n_2)_{\{\gamma_i\}}^{K_\theta^\varepsilon}(p) = 0$; and if $2m_2 + n_1 - n_2 \le p \le 2m_2 + n_1 + n_2$, then the rank of the system of equations is $\frac{1}{2}(2m_1 + n_1 - n_2 - p)$ and the dimension of the solutions of the system of equations is $1 + \frac{1}{2}(n_2 - n_1 - 2m_2 + p)$.

(b2) If $n_2 \ge 2m_1 - 2m_2$, then $2m_2 + n_1 + n_2 \ge 2m_1 + n_1 - n_2 \ge 2m_2 + n_1 - n_2$. we shall compute $\dim(m_1, m_2 | n_1, n_2)_{\{\gamma_i\}}^{K_\theta^\varepsilon}(p)$ case by case for p.

(b2i) Case $2m_2 + n_1 + n_2 \le p \le 2m_1 + n_1 + n_2$.

Since $2m_1 + n_1 - n_2 - p \le 0$, there are no conditions for b_j and because $2m_2 + n_1 + n_2 - p \le 0$ we know that for such a p, the degree of the polynomial $\varphi(x)$ could be any integer between 0 and $\frac{1}{2}(2m_1 + n_1 + n_2 - p)$. Thus

$$\dim(m_1, m_2; n_1, n_2)_{\{\gamma_i\}}^{K_\theta^\varepsilon}(p) = 1 + \frac{1}{2}(2m_1 + n_1 + n_2 - p).$$

(b2ii) Case $2m_1 + n_1 - n_2 \le p \le 2m_2 + n_1 + n_2$.

In this case, the degree of $\varphi(x) = \sum_{i=1}^{q} a_i x^i$ will takes values of all integers between $\frac{1}{2}(2m_2 + n_1 + n_2 - p)$ and $\frac{1}{2}(2m_1 + n_1 + n_2 - p)$, $a_i = 0$ if $2i < m_1 + m_2 + 2n_2 - p$, and there are no restrictions for b_j for $2m_1 + n_1 - n_2 - p \le 0$. Therefore $\dim(m_1, m_2; n_1, n_2)_{\{\gamma_i\}}^{K_\theta^\varepsilon}(p) = 1 + (m_1 - m_2)$.

(b2iii) Case $p \le 2m_1 + n_1 - n_2$.

In this case, we have $a_i = 0$ if $2i < 2m_2 + n_1 + n_2 - p$. The system of homogeneous linear equations

$$b_j = \sum_{i=j}^{q} (-1)^{i-j} \binom{i}{j} a_i = 0, \quad j = 0, 1, \cdots, \frac{1}{2}(2m_1 + n_1 - n_2 - p) - 1$$

has $1 + (m_1 - m_2)$ variables and $\frac{1}{2}(2m_1 + n_1 - n_2 - p)$ equations and the corresponding matrix of the coefficients is $N(\frac{1}{2}(2m_2 + n_1 + n_2 - p), \frac{1}{2}(2m_1 + n_1 - n_2 - p) - 1, \frac{1}{2}(2m_1 + n_1 + n_2 - p))$. By Lemma 4.6.2, its rank is $mim\{\frac{1}{2}(2m_1 + n_1 - n_2 - p), 1 + (m_1 - m_2)\}$. Since $[1 + (m_1 - m_2)] - [\frac{1}{2}(2m_1 + n_1 - n_2 - p)] = 1 + \frac{1}{2}(p - 2m_2 - n_1 + n_2)$, it is evident that if $p < 2m_2 + n_1 - n_2$, then the number of the variables equals the rank of the system of equations and is less than the number of the equations, and this means that the system of equations has no solutions, in other words, $\dim(m_1, m_2; n_1, n_2)^{K_\theta^e}_{\{\gamma_i\}}(p) = 0$; and if $2m_2 + n_1 - n_2 \le p \le 2m_1 + n_1 - n_2$, then the rank of the system of equations is $\frac{1}{2}(2m_1 + n_1 - n_2 - p)$ and the dimension of the solutions of the system of equations is $1 + \frac{1}{2}(n_2 - n_1 - 2m_2 + p)$.

(c) $(n_1 + n_2) > (m_1 + m_2) \ge (m_1 - m_2) \ge (n_1 - n_2)$.

In this case, we have $2m_2 + n_1 + n_2 \le m_1 + m_2 + 2n_2$, $m_1 + m_2 + 2n_1 \le 2m_1 + n_1 + n_2$, and $m_1 - m_2 + 2n_1 > 2m_1 + n_1 - n_2$, and further we have $0 \le p \le m_1 + m_2 + 2n_1$, $2n_2 + m_1 + m_2 - p \le 2i \le 2n_1 + m_1 + m_2 - p$, and $b_j = 0$ if $2j < 2n_1 + m_1 - m_2 - p$.

(c1) If $m_2 < 2n_1 - 2n_2$, then $2n_2 + m_1 - m_2 \le 2n_2 + m_1 + m_2 < 2n_1 + m_1 - m_2$. If we make a change of variables: $m_1 \leftrightarrow n_1, m_2 \leftrightarrow n_2$, then case (c1) will go to case (b1) when $(m_1 - m_2) > (n_1 - n_2)$, and to case (a1) when $(m_1 - m_2) = (n_1 - n_2)$. Hence if $(m_1 - m_2) > (n_1 - n_2)$,

$$\dim(m_1, m_2; n_1, n_2)^{K_\theta^e}_{\{\gamma_i\}}(p) =$$

$$= \begin{cases} 1 + \frac{1}{2}(m_2 - m_1 - 2n_2 + p) & if \ m_1 - m_2 + 2n_2 \le p \le m_1 + m_2 + 2n_2, \\ 1 + m_2 & if \ m_1 + m_2 + 2n_2 \le p \le m_1 - m_2 + 2n_1, \\ 1 + \frac{1}{2}(m_1 + m_2 + 2n_1 - p) & if \ m_1 - m_2 + 2n_1 \le p \le m_1 + m_2 + 2n_1, \end{cases}$$

and if $(m_1 - m_2) = (n_1 - n_2)$,

$$\dim(m_1, m_2; n_1, n_2)^{K_\theta^e}_{\{\gamma_i\}}(p) =$$

$$= \begin{cases} 1 + \frac{1}{2}(m_1 - m_2 - 2n_1 + p) & if \ 2n_1 - m_1 + m_2 \le p \le n_1 + n_2 + 2m_2, \\ 1 + \frac{1}{2}(m_1 + m_2 - n_1 + n_2) & if \ n_1 + n_2 + 2m_2 \le p \le 2n_1 + m_1 - m_2, \\ 1 + \frac{1}{2}(n_1 + n_2 + 2m_1 - p) & if \ 2n_1 + m_1 - m_2 \le p \le n_1 + n_2 + 2m_1. \end{cases}$$

However, both expressions are actually same since $m_2 - m_1 - 2n_2 = m_1 - m_2 - 2n_1$, $m_1 - m_2 + 2n_2 = 2n_1 - m_1 + m_2$, and $m_1 + m_2 + 2n_2 = n_1 + n_2 + 2m_2$; $2m_2 = m_1 + m_2 - n_1 + n_2$; and $m_1 + m_2 + 2n_1 = n_1 + n_2 + 2m_1$. We are done in this case.

(c2) If $m_2 \ge n_1 - n_2$, then $m_1 + m_2 + 2n_2 \ge 2n_1 + m_1 - m_2 \ge m_1 - m_2 + 2n_2$. Similarly, if the variables are changed by $m_1 \leftrightarrow n_1$ and $m_2 \leftrightarrow n_2$, then case (c2) goes to case (b2) if $(m_1 - m_2) > (n_1 - n_2)$, and to case (a2) if $(m_1 - m_2) = (n_1 - n_2)$. Just as case (c1), we will get our results for case (c2) as stated in the theorem.

(d) $(n_1 + n_2) > (m_1 + m_2) \ge (n_1 - n_2) > (m_1 - m_2)$.

In this case, we will make the same change of variables: $m_1 \leftrightarrow n_1$ and $m_2 \leftrightarrow n_2$, and case (d) is switched to case (a). The results for case (d) follow from the same argument as in case (c).

(e) $(m_1 - m_2) > (n_1 + n_2)$ or $(n_1 - n_2) > (m_1 + m_2)$.

If $(m_1 - m_2) > (n_1 + n_2)$, then $(m_1 + m_2 + 2n_1) < (2m_1 + n_1 - n_2)$ and $m_1 + m_2 + 2n_2 - p \leq 2i \leq m_1 + m_2 + 2n_1 - p$. We have $b_j = 0$ for $0 \leq j \leq \frac{1}{2}(m_1 + m_2 + 2n_1 - p) = q$. Therefore this system of equations has no solutions since the rank of the system is equal to the number of the variables and less than the number of the equations, the result follows.

If $(n_1 - n_2) > (m_1 + m_2)$, just changing the variables by $m_1 \leftrightarrow n_1$ and $m_2 \leftrightarrow n_2$, we will obtain the result.

We are done. $\qquad\square$

From the proof above, the eigenvalue p of $x_\circ$ must be nonnegative.

COROLLARY 4.6.1. *The dimension of the space* $(m_1, m_2; n_1, n_2))_{\{\gamma_i\}}^{K_\theta^e}$ *is determined as follows:*

(a) *If* $(m_1 + m_2) \geq (n_1 + n_2) \geq (m_1 - m_2) \geq (n_1 - n_2)$, *then*
$$\dim(m_1, m_2; n_1, n_2))^{K_\theta^e} = \frac{1}{2}(n_1 + n_2 - m_1 + m_2 + 2)(n_1 - n_2 + 1);$$

(b) *If* $(m_1 + m_2) \geq (n_1 + n_2) \geq (n_1 - n_2) > (m_1 - m_2)$, *then*
$$\dim(m_1, m_2; n_1, n_2))^{K_\theta^e} = (n_2 + 1)(m_1 - m_2 + 1);$$

(c) *If* $(n_1 + n_2) > (m_1 + m_2) \geq (m_1 - m_2) \geq (n_1 - n_2)$, *then*
$$\dim(m_1, m_2; n_1, n_2))^{K_\theta^e} = (m_2 + 1)(n_1 - n_2 + 1);$$

(d) *If* $(n_1 + n_2) > (m_1 + m_2) \geq (n_1 - n_2) > (m_1 - m_2)$, *then*
$$\dim(m_1, m_2; n_1, n_2))^{K_\theta^e} = \frac{1}{2}(m_1 + m_2 - n_1 + n_2 + 2)(m_1 - m_2 + 1);$$

(e) *If* $(m_1 - m_2) > (n_1 + n_2)$ *or* $(n_1 - n_2) > (m_1 + m_2)$, *then*
$$\dim(m_1, m_2; n_1, n_2))^{K_\theta^e} = 0.$$

PROOF. According to the Theorem, we have
$$\dim(m_1, m_2; n_1, n_2)_{\{\gamma_i\}}^{K_\theta^e} = \sum_p \dim(m_1, m_2; n_1, n_2)_{\{\gamma_i\}}^{K_\theta^e}(p), \tag{66}$$

where the summation takes over the sets of integers of different types subject to the conditions on m_1, m_2, n_1, and n_2, which are

(a) If $(m_1 + m_2) \geq (n_1 + n_2) \geq (m_1 - m_2) \geq (n_1 - n_2)$, then
$$p \in \{2m_1 - n_1 + n_2, 2m_1 - n_1 + n_2 + 2, \cdots, m_1 + m_2 + 2n_1\}.$$

(b) If $(m_1 + m_2) \geq (n_1 + n_2) \geq (n_1 - n_2) > (m_1 - m_2)$, then
$$p \in \{2m_2 + n_1 - n_2, 2m_2 + n_1 - n_2 + 2, \cdots, 2m_1 + n_1 + n_2\}.$$

(c) If $(n_1 + n_2) > (m_1 + m_2) \geq (m_1 - m_2) \geq (n_1 - n_2)$, then
$$p \in \{m_1 - m_2 + 2n_2, m_1 - m_2 + 2n_2 + 2, \cdots, m_1 + m_2 + 2n_1\}.$$

(d) If $(n_1 + n_2) > (m_1 + m_2) \geq (n_1 - n_2) > (m_1 - m_2)$, then

$$p \in \{2n_2 + m_1 - m_2, 2n_2 + m_1 - m_2 + 2, \cdots, 2m_1 + n_1 + n_2\}.$$

(e) If $(m_1 - m_2) > (n_1 + n_2)$ or $(n_1 - n_2) > (m_1 + m_2)$, then the summation is over an empty set. Thus the dimension of the space is zero.

By means of the conditions of congruence on those integers m_1, m_2, n_1, n_2, and p:

$$m_1 + m_2 + n_1 + n_2 \equiv 0 (mod2) \text{ and } m_1 + m_2 \equiv p(mod2),$$

it is not difficult to calculate the dimension of the space, which is exactly as same as stated in the Corollary.   □

In the subsection 4.4, we used the Branching Rule to compute the multiplicity of the representation $(m_1, m_2; n_1, n_2)$ occurring in the space $\mathcal{H}$ of all harmonic polynomial functions on $\mathbf{p}$, which is $l((m_1, m_2; n_1, n_2))$ and is equal to $\dim(m_1, m_2; n_1, n_2)(z)$ for some regular semisimple element z in $\mathbf{p}$. Following Kostant-Rallis [**KoRa**], the multiplicity $l((m_1, m_2; n_1, n_2))$ is also equal to the dimension of $(m_1, m_2; n_1, n_2)(e)$ for any regular nilpotent element $e \in \mathbf{p}$. Following Corollary 4.6.1, we obtain

COROLLARY 4.6.2. *Notations are as above.*

(a) *We have following equalities*

$$l((m_1, m_2; n_1, n_2)) = \dim(m_1, m_2; n_1, n_2)(e) = \dim(m_1, m_2; n_1, n_2)_{\{\gamma_i\}}^{K_\theta^e}.$$

(b) *We have* $(m_1, m_2; n_1, n_2)(e) = (m_1, m_2; n_1, n_2)^{K_\theta^e}$.

PROOF. Part (a) follows from Corollary 4.4.3 in subsection 4.4 and Corollary 4.6.1 in this subsection. Part (b) holds because we already know the following filtration of subspaces:

$$(m_1, m_2; n_1, n_2)(e) \subset (m_1, m_2; n_1, n_2)^{K_\theta^e} \subset (m_1, m_2; n_1, n_2)_{\{\gamma_i\}}^{K_\theta^e}.$$

□

Finally we can write down explicitly the eigenvalues and the dimensions of the corresponding eigensubspaces of x_o in $(m_1, m_2; n_1, n_2)(e)$.

THEOREM 4.6.3. *Let* $(m_1, m_2; n_1, n_2)$ *be the finitely dimensional complex representation of* K_θ *with highest weight* $(m_1, m_2) \oplus (n_1, n_2)$. *Let* e *be the regular nilpotent element in* $\mathbf{p}$ *as in proposition 1.1 and* x_o *in* $\mathbf{t}$ *such that* $[x_o, e] = e$. *Then the eigenvalues* p *and the dimensions of the corresponding eigensubspaces* $(m_1, m_2; n_1, n_2)(e)(p)$ *of* x_o *in the subspace* $(m_1, m_2; n_1, n_2)(e)$ *(which is defined in* [**KoRa**]*) can be determined as follows:*

(a) *If* $(m_1 + m_2) \geq (n_1 + n_2) \geq (m_1 - m_2) \geq (n_1 - n_2)$, *then the eigenvalues of* x_o are

$$p = 2m_1 - n_1 + n_2, 2m_1 - n_1 + n_2 + 2, \cdots, m_1 + m_2 + 2n_1$$

and (a1) if $n_1 + n_2 - m_1 + m_2 < 2n_1 - 2n_2$, then

$$\dim(m_1, m_2; n_1, n_2)(e)(p)$$
$$= \begin{cases} 1 + \frac{1}{2}(n_1 - n_2 - 2m_1 + p) & if\ 2m_1 - n_1 + n_2 \le p \le m_1 + m_2 + 2n_2, \\ 1 + \frac{1}{2}(n_1 + n_2 - m_1 + m_2) & if\ m_1 + m_2 + 2n_2 \le p \le 2m_1 + n_1 - n_2, \\ 1 + \frac{1}{2}(m_1 + m_2 + 2n_1 - p) & if\ 2m_1 + n_1 - n_2 \le p \le m_1 + m_2 + 2n_1, \end{cases}$$

(a2) if $n_1 + n_2 - m_1 + m_2 \ge 2n_1 - 2n_2$, then

$$\dim(m_1, m_2; n_1, n_2)(e)(p)$$
$$= \begin{cases} 1 + \frac{1}{2}(n_1 - n_2 - 2m_1 + p) & if\ 2m_1 - n_1 + n_2 \le p \le 2m_1 + n_1 - n_2, \\ 1 + (n_1 - n_2) & if\ 2m_1 + n_1 - n_2 \le p \le m_1 + m_2 + 2n_2, \\ 1 + \frac{1}{2}(m_1 + m_2 + 2n_1 - p) & if\ m_1 + m_2 + 2n_2 \le p \le m_1 + m_2 + 2n_1; \end{cases}$$

(b) If $(m_1 + m_2) \ge (n_1 + n_2) \ge (n_1 - n_2) > (m_1 - m_2)$, then the eigenvalues of $x_\circ$ are

$$p = 2m_2 + n_1 - n_2, 2m_2 + n_1 - n_2 + 2, \cdots, 2m_1 + n_1 + n_2$$

and (b1) if $n_2 < m_1 - m_2$, then

$$\dim(m_1, m_2; n_1, n_2)(e)(p)$$
$$= \begin{cases} 1 + \frac{1}{2}(n_2 - n_1 - 2m_2 + p) & if\ 2m_2 + n_1 - n_2 \le p \le 2m_2 + n_1 + n_2, \\ 1 + n_2 & if\ 2m_2 + n_1 + n_2 \le p \le 2m_1 + n_1 - n_2, \\ 1 + \frac{1}{2}(2m_1 + n_1 + n_2 - p) & if\ 2m_1 + n_1 - n_2 \le p \le 2m_1 + n_1 + n_2, \end{cases}$$

(b2) if $n_2 \ge m_1 - m_2$, then

$$\dim(m_1, m_2; n_1, n_2)(e)(p)$$
$$= \begin{cases} 1 + \frac{1}{2}(n_2 - n_1 - 2m_2 + p) & if\ 2m_2 + n_1 - n_2 \le p \le 2m_1 + n_1 - n_2, \\ 1 + (m_1 - m_2) & if\ 2m_1 + n_1 - n_2 \le p \le 2m_2 + n_1 + n_2, \\ 1 + \frac{1}{2}(2m_1 + n_1 + n_2 - p) & if\ 2m_2 + n_1 + n_2 \le p \le 2m_1 + n_1 + n_2; \end{cases}$$

(c) If $(n_1 + n_2) > (m_1 + m_2) \ge (m_1 - m_2) \ge (n_1 - n_2)$, then the eigenvalues of $x_\circ$ are

$$p = m_1 - m_2 + 2n_2, m_1 - m_1 + 2n_2 + 2, \cdots, m_1 + m_2 + 2n_1$$

and (c1) if $m_2 < n_1 - n_2$, then

$$\dim(m_1, m_2; n_1, n_2)(e)(p)$$
$$= \begin{cases} 1 + \frac{1}{2}(m_2 - m_1 - 2n_2 + p) & if\ m_1 - m_2 + 2n_2 \le p \le m_1 + m_2 + 2n_2, \\ 1 + m_2 & if\ m_1 + m_2 + 2n_2 \le p \le m_1 - m_2 + 2n_1, \\ 1 + \frac{1}{2}(m_1 + m_2 + 2n_1 - p) & if\ m_1 - m_2 + 2n_1 \le p \le m_1 + m_2 + 2n_1, \end{cases}$$

(c2) if $m_2 \geq n_1 - n_2$, then

$$\dim(m_1, m_2; n_1, n_2)(e)(p)$$
$$= \begin{cases} 1 + \frac{1}{2}(m_2 - m_1 - 2n_2 + p) & if\ m_1 - m_2 + 2n_2 \leq p \leq m_1 - m_2 + 2n_1, \\ 1 + (n_1 - n_2) & if\ m_1 - m_2 + n_1 \leq p \leq m_1 + m_2 + 2n_2, \\ 1 + \frac{1}{2}(m_1 + m_2 + 2n_1 - p) & if\ m_1 + m_1 + 2n_2 \leq p \leq m_1 + m_2 + 2n_1; \end{cases}$$

(d) If $(n_1 + n_2) > (m_1 + m_2) \geq (n_1 - n_2) > (m_1 - m_2)$, then the eigenvalues of $x_\circ$ are

$$p = 2n_2 + m_1 - m_2, 2n_2 + m_1 - m_2 + 2, \cdots, 2m_1 + n_1 + n_2$$

and (d1) if $m_1 + m_2 - n_1 + n_2 < 2m_1 - 2m_2$, then

$$\dim(m_1, m_2; n_1, n_2)(e)(p)$$
$$= \begin{cases} 1 + \frac{1}{2}(m_1 - m_2 - 2n_1 + p) & if\ 2n_1 - m_1 + m_2 \leq p \leq n_1 + n_2 + 2m_2, \\ 1 + \frac{1}{2}(m_1 + m_2 - n_1 + n_2) & if\ n_1 + n_2 + 2m_2 \leq p \leq 2n_1 + m_1 - m_2, \\ 1 + \frac{1}{2}(n_1 + n_2 + 2m_1 - p) & if\ 2n_1 + m_1 - m_2 \leq p \leq n_1 + n_2 + 2m_1, \end{cases}$$

(d2) if $m_1 + m_2 - n_1 + n_2 \geq 2m_1 - 2m_2$, then

$$\dim(m_1, m_2; n_1, n_2)(e)(p)$$
$$= \begin{cases} 1 + \frac{1}{2}(m_1 - m_2 - 2n_1 + p) & if\ 2n_1 - m_1 + m_2 \leq p \leq 2n_1 + m_1 - m_2, \\ 1 + (m_1 - m_2) & if\ 2n_1 + m_1 - m_2 \leq p \leq n_1 + n_2 + 2m_2, \\ 1 + \frac{1}{2}(n_1 + n_2 + 2m_1 - p) & if\ n_1 + n_2 + 2m_2 \leq p \leq n_1 + n_2 + 2m_1; \end{cases}$$

(e) If $(m_1 - m_2) > (n_1 + n_2)$ or $(n_1 - n_2) > (m_1 + m_2)$, then

$$\dim(m_1, m_2; n_1, n_2)(e)(p) = 0.$$

PROOF. By the filtration of those three subspaces:

$$(m_1, m_2; n_1, n_2)(e) \subset (m_1, m_2; n_1, n_2)^{K_\theta^\varepsilon} \subset (m_1, m_2; n_1, n_2)^{K_\theta^\varepsilon}_{\{\gamma_i\}},$$

we have

$$\dim(m_1, m_2; n_1, n_2)(e) \leq \dim(m_1, m_2; n_1, n_2)^{K_\theta^\varepsilon} \leq \dim(m_1, m_2; n_1, n_2)^{K_\theta^\varepsilon}_{\{\gamma_i\}}.$$

By corollary 11, we have

$$\sum_p \dim(m_1, m_2; n_1, n_2)(e)(p) = \dim(m_1, m_2; n_1, n_2)(e)$$
$$= \dim(m_1, m_2; n_1, n_2)^{K_\theta^\varepsilon}_{\{\gamma_i\}}$$
$$= \sum_p \dim(m_1, m_2; n_1, n_2)^{K_\theta^\varepsilon}_{\{\gamma_i\}}(p).$$

Therefore, for each eigenvalue p, we have

$$\dim(m_1, m_2; n_1, n_2)(e)(p) = \dim(m_1, m_2; n_1, n_2)^{K_\theta^\varepsilon}_{\{\gamma_i\}}(p).$$

The theorem follows from Theorem 4.6.2. $\square$

4.7. Application of Kostant-Rallis' Algorithm. We are able to apply the algorithm of Kostant-Rallis to compute the polynomials $P(m_1, m_2, n_1, n_2; X)$, which is the main part of our computation of the local Langlands factor $L_v(\frac{s+1}{2}, \pi_{1v} \otimes \pi_{2v}, (\rho_1 \otimes \rho_1)_v)$.

THEOREM 4.7.1. *Let m_1, m_2, n_1, and n_2 be any four nonnegative integers. Then the polynomial $P(m_1, m_2, n_1, n_2; X)$ can be expressed as follows:*

(a) If $(m_1 + m_2) \geq (n_1 + n_2) \geq (m_1 - m_2) \geq (n_1 - n_2)$, then we have

$$P(m_1, m_2, n_1, n_2; X) = X^{2m_1 - n_1 + n_2} \frac{(1 - X^{n_1 + n_2 - m_1 + m_2 + 2})}{(1 - X^2)} \frac{(1 - X^{2(n_1 - n_2 + 1)})}{(1 - X^2)}.$$

(b) If $(m_1 + m_2) \geq (n_1 + n_2) \geq (n_1 - n_2) > (m_1 - m_2)$, then

$$P(m_1, m_2, n_1, n_2; X) = X^{2m_2 + n_1 - n_2} \frac{(1 - X^{2(n_2 + 1)})}{(1 - X^2)} \frac{(1 - X^{2(m_1 - m_2 + 1)})}{(1 - X^2)}.$$

(c) If $(n_1 + n_2) > (m_1 + m_2) \geq (m_1 - m_2) \geq (n_1 - n_2)$, then

$$P(m_1, m_2, n_1, n_2; X) = X^{m_1 - m_2 + 2n_2} \frac{(1 - X^{2(m_2 + 1)})}{(1 - X^2)} \frac{(1 - X^{2(n_1 - n_2 + 1)})}{(1 - X^2)}.$$

(d) If $(n_1 + n_2) > (m_1 + m_2) \geq (n_1 - n_2) > (m_1 - m_2)$, then

$$P(m_1, m_2, n_1, n_2; X) = X^{2n_1 - m_1 + m_2} \frac{(1 - X^{m_1 + m_2 - n_1 + n_2 + 2})}{(1 - X^2)} \frac{(1 - X^{2(m_1 - m_2 + 1)})}{(1 - X^2)}.$$

(e) If $(m_1 - m_2) > (n_1 + n_2)$ or $(n_1 - n_2) > (m_1 + m_2)$, then $P(m_1, m_2, n_1, n_2; X) = 0$.

PROOF. By Proposition 4.2.4, one has $P(m_1, m_2, n_1, n_2; X) = \sum_{l=0}^{q} p_l X^l$, each term of which can be interpreted as follows: the term $p_l X^l$ indicates that the irreducible representation $(m_1, m_2) \oplus (n_1, n_2)$ occurs, with multiplicity p_l, in the K_θ-submodule of $\mathcal{H}$ consisting of all homogeneous functions of degree l. In other words, l is an eigenvalue of $x_\circ$ in $(m_1, m_2; n_1, n_2)(e)$ and $p_l = \dim(m_1, m_2; n_1, n_2)(e)(l)$. By Theorem 4.6.3, we will get the expression of the polynomial $P(m_1, m_2, n_1, n_2; X)$ by means of m_1, m_2, n_1, and n_2. Note that $m_1 + m_2 + n_1 + n_2 \equiv 0 \pmod 2$ and $l \equiv m_1 + m_2 \pmod 2$.

(a) If $(m_1 + m_2) \geq (n_1 + n_2) \geq (m_1 - m_2) \geq (n_1 - n_2)$ and if $n_1 + n_2 - m_1 + m_2 < 2n_1 - 2n_2$, then the multiplicity p_l can be expressed as

$$p_l = \begin{cases} 1 + \frac{1}{2}(n_1 - n_2 - 2m_1 + l) & \text{if } 2m_1 - n_1 + n_2 \leq l \leq m_1 + m_2 + 2n_2, \\ 1 + \frac{1}{2}(n_1 + n_2 - m_1 + m_2) & \text{if } m_1 + m_2 + 2n_2 \leq l \leq 2m_1 + n_1 - n_2, \\ 1 + \frac{1}{2}(m_1 + m_2 + 2n_1 - l) & \text{if } 2m_1 + n_1 - n_2 \leq l \leq m_1 + m_2 + 2n_1. \end{cases}$$

Thus the polynomial $P(m_1, m_2, n_1, n_2; X)$ can be deduced as

$$
\begin{aligned}
= & \ X^{2m_1-n_1+n_2} + 2X^{2m_1-n_1+n_2+2} + \cdots + \frac{1}{2}(n_1 + n_2 - m_1 + m_2)X^{m_1+m_2+2n_2-2} \\
& + [1 + \frac{1}{2}(n_1 + n_2 - m_1 + m_2)](X^{m_1+m_2+2n_2} + \cdots + X^{2m_1+n_1-n_2}) \\
& + \frac{1}{2}(n_1 + n_2 - m_1 + m_2)X^{2m_1+n_1-n_2+2} + 2X^{m_1+m_2+2n_1-2} + X^{m_1+m_2+2n_1}.
\end{aligned}
$$

It is not difficult to see that $P(m_1, m_2, n_1, n_2; X) = X^{2m_1-n_1+n_2} \cdot P'(X)$ where the polynomial factor $P'(X)$ has form

$$
\begin{aligned}
P'(X) = & \ 1 + 2X^2 + \cdots + \frac{1}{2}(n_1 + n_2 - m_1 + m_2)X^{n_1+n_2-m_1+m_2-2} \\
& + [1 + \frac{1}{2}(n_1 + n_2 - m_1 + m_2)](X^{n_1+n_2-m_1+m_2} + \cdots + X^{2n_1-2n_2}) \\
& + \frac{1}{2}(n_1 + n_2 - m_1 + m_2)X^{2n_1-2n_2+2} \\
& + 2X^{3n_1-n_2-m_1+m_2-2} + X^{3n_1-n_2-m_1+m_2}.
\end{aligned}
$$

Factoring the polynomial $P'(X)$, we obtain that

$$
\begin{aligned}
& P(m_1, m_2, n_1, n_2; X) \\
= & \ X^{2m_1-n_1+n_2} \cdot P'(X) \\
= & \ X^{2m_1-n_1+n_2} \frac{(1 - X^{n_1+n_2-m_1+m_2+2})(1 - X^{2(n_1-n_2+1)})}{(1 - X^2)^2}.
\end{aligned}
$$

Similarly, we can show that if $n_1 + n_2 - m_1 + m_2 \geq 2n_1 - 2n_2$, the polynomial $P(m_1, m_2, n_1, n_2; X)$ is also equal to

$$
= X^{2m_1-n_1+n_2} \frac{(1 - X^{n_1+n_2-m_1+m_2+2})}{(1 - X^2)} \frac{(1 - X^{2(n_1-n_2+1)})}{(1 - X^2)}.
$$

This proves (a).

By using the same argument, we will obtain the following identities:

(b) If $(m_1 + m_2) \geq (n_1 + n_2) \geq (n_1 - n_2) > (m_1 - m_2)$, then

$$
P(m_1, m_2, n_1, n_2; X) = X^{2m_2+n_1-n_2} \frac{(1 - X^{2(n_2+1)})}{(1 - X^2)} \frac{(1 - X^{2(m_1-m_2+1)})}{(1 - X^2)}.
$$

(c) If $(n_1 + n_2) > (m_1 + m_2) \geq (m_1 - m_2) \geq (n_1 - n_2)$, then

$$
P(m_1, m_2, n_1, n_2; X) = X^{m_1-m_2+2n_2} \frac{(1 - X^{2(m_2+1)})}{(1 - X^2)} \frac{(1 - X^{2(n_1-n_2+1)})}{(1 - X^2)}.
$$

(d) If $(n_1 + n_2) > (m_1 + m_2) \geq (n_1 - n_2) > (m_1 - m_2)$, then

$$P(m_1, m_2, n_1, n_2; X) = X^{2n_1 - m_1 + m_2} \frac{(1 - X^{m_1 + m_2 - n_1 + n_2 + 2})}{(1 - X^2)} \frac{(1 - X^{2(m_1 - m_2 + 1)})}{(1 - X^2)}.$$

(e) If $(m_1 - m_2) > (n_1 + n_2)$ or $(n_1 - n_2) > (m_1 + m_2)$, then $P(m_1, m_2, n_1, n_2; X) = 0$. The theorem is proved. $\qquad \square$

Combining Theorem 4.7.1 and Proposition 4.2.4, we obtain

COROLLARY 4.7.1. *Let (m_1, m_2), (n_1, n_2) be the finite-dimensional complex representations of $Sp(2)$ with highest weight $m_1 \varepsilon_1 + m_2 \varepsilon_2$, $m_1 \geq m_2 \geq 0$, $n_1 \varepsilon_1 + n_2 \varepsilon_2$, $n_1 \geq n_2 \geq 0$, respectively. Let $(m_1, m_2 | n_1, n_2)$ be the trace of the tensor product of representations (m_1, m_2) and (n_1, n_2) evaluated at the matrix $(\rho_1 \otimes \rho_1)_v (\tau_{1v} \times \tau_{2v})$. Then*

$$\sum_{l=0}^{\infty} trace(Sym_{\mathcal{H}}^l((\rho_1 \otimes \rho_1)_v(t_{1v} \times t_{2v}))) X^l$$

$$= \sum_{\substack{m_1 + m_2 \geq n_1 + n_2 \geq \\ m_1 - m_2 \geq n_1 - n_2}} (m_1, m_2 | n_1, n_2) Q_1(m_1, m_2, n_1, n_2; X)$$

$$+ \sum_{\substack{m_1 + m_2 \geq n_1 + n_2 \\ n_1 - n_2 > m_1 - m_2}} (m_1, m_2 | n_1, n_2) Q_2(m_1, m_2, n_1, n_2; X)$$

$$+ \sum_{\substack{n_1 + n_2 > m_1 + m_2 \\ m_1 - m_2 \geq n_1 - n_2}} (m_1, m_2 | n_1, n_2) Q_3(m_1, m_2, n_1, n_2; X)$$

$$+ \sum_{\substack{n_1 + n_2 > m_1 + m_2 \geq \\ n_1 - n_2 > m_1 - m_2}} (m_1, m_2 | n_1, n_2) Q_4(m_1, m_2, n_1, n_2; X),$$

where

$$Q_1(m_1, m_2, n_1, n_2; X) = X^{2m_1 - n_1 + n_2} \frac{(1 - X^{n_1 + n_2 - m_1 + m_2 + 2})(1 - X^{2(n_1 - n_2 + 1)})}{(1 - X^2)^2},$$

$$Q_2(m_1, m_2, n_1, n_2; X) = X^{2m_2 + n_1 - n_2} \frac{(1 - X^{2(n_2 + 1)})(1 - X^{2(m_1 - m_2 + 1)})}{(1 - X^2)^2},$$

$$Q_3(m_1, m_2, n_1, n_2; X) = X^{m_1 - m_2 + 2n_2} \frac{(1 - X^{2(m_2 + 1)})(1 - X^{2(n_1 - n_2 + 1)})}{(1 - X^2)^2},$$

and

$$Q_4(m_1, m_2, n_1, n_2; X) = X^{2n_1 - m_1 + m_2} \frac{(1 - X^{m_1 + m_2 - n_1 + n_2 + 2})(1 - X^{2(m_1 - m_2 + 1)})}{(1 - X^2)^2}.$$

4.8. The Proof of Theorem 3.3.3. Recall from Theorem 3.3.2 and Proposition 4.2.3 that

$$\mathcal{Z}_v^*(s, W_{1v}^\circ, W_{2v}^\circ, f_v^\circ) = \frac{P(X)}{(1-X^2)(1-X^4)}.$$

and

$$L_v(\frac{s+1}{2}, \pi_{1v} \otimes \pi_{2v}, (\rho_1 \otimes \rho_1)_v) = \frac{\sum_{l=0}^{\infty} trace(Sym_{\mathcal{H}}^l((\rho_1 \otimes \rho_1)_v(t_{1v} \times t_{2v})))X^l}{(1-X^2)(1-X^4)}.$$

Theorem 3.3.3 will be proved if we prove the following identity:

$$P(X) = \sum_{l=0}^{\infty} trace(Sym_{\mathcal{H}}^l((\rho_1 \otimes \rho_1)_v(t_{1v} \times t_{2v})))X^l. \tag{67}$$

In Theorem 3.3.2, we use (m_1, m_2) to indicate the irreducible complex representation of $Sp(2)$ with highest weight $m_1\varepsilon_1 + m_2(\varepsilon_1 + \varepsilon_2)$, $m_1, m_2 \geq 0$. In order to make the notations compatible with these used in this section, we have to set some substitutions and rewrite the rational function $P(X)$.

Let $k_1 = m_1 + m_2$, $k_2 = m_2$; $l_1 = 2n_1 - m_1 + n_2$, $l_2 = n_2$. Then

(0) $k_1 \geq k_2 \geq 0$, $l_1 \geq l_2 \geq 0$ and $k_1 + k_2 + l_1 + l_2 = 2(m_2 + n_1 + n_2) \equiv 0 \pmod{2}$;

(a) $m_1 + m_2 \geq n_1 + n_2 \geq m_1$ and $2n_1 \geq m_1 \geq n_1$ are equivalent to $k_1 \geq \frac{1}{2}(l_1 + l_2 + k_1 - k_2) \geq k_1 - k_2$ and $l_1 - l_2 + k_1 - k_2 \geq k_1 - k_2 \geq \frac{1}{2}(l_1 - l_2 + k_1 - k_2)$, that is, $k_1 + k_2 \geq l_1 + l_2 \geq k_1 - k_2 \geq l_1 - l_2$, and also $3m_1 + 2m_2 - 2n_1 = 2k_1 - l_1 + l_2$, $2(n_1 + n_2 - m_1 + 1) = l_1 + l_2 - k_1 + k_2 + 2$, and $2(2n_1 - m_1 + 1) = 2(l_1 - l_2 + 1)$;

(b) $m_1 + m_2 \geq n_1 + n_2$ and $n_1 > m_1$ are equivalent to $k_1 + k_2 \geq l_1 + l_2$ and $l_1 - l_2 > k_1 - k_2$, and also $2m_2 - m_1 + 2n_1 = 2k_2 + l_1 - l_2$, $2(n_2 + 1) = 2(l_2 + 1)$, and $2(m_1 + 1) = 2(k_1 - k_2 + 1)$;

(c) $n_1 + n_2 > m_1 + m_2$ and $2n_1 \geq m_1 \geq n_1$ are equivalent to $n_1 + n_2 > m_1 + m_2$ and $m_1 - m_2 \geq n_1 - n_2$, and also $m_1 + 2n_2 = k_1 - k_2 + 2l_2$, $2(m_2 + 1) = 2(k_2 + 1)$, and $2(2n_1 - m_1 + 1) = 2(l_1 - l_2 + 1)$; and

(d) $n_1 + n_2 > m_1 + m_2 \geq n_1$ and $n_1 > m_1$ are equivalent to $n_1 + n_2 > m_1 + m_2 \geq n_1 - n_2 > m_1 - m_2$, and also $4n_1 + 2n_2 - 3m_1 = 2l_1 - k_1 + k_2$, $2(m_1 + m_2 - n_1 + 1) = k_1 + k_2 - l_1 + l_2 + 2$, and $2(m_1 + 1) = 2(k_1 - k_2 + 1)$.

After substituting those data into $P(X)$, identity (63) follows easily. Therefore Theorem 3.3.3 is finally proved.

5. The Fundamental Identity

According to the results in previous sections, our degree 16 L-function $L^S(s', \pi_1 \otimes \pi_2, \rho_1 \otimes \rho_1)$ can be expressed by the normalized global zeta integral $\mathcal{Z}^*(s, \phi_1, \phi_2, f)$ with a product of finite ramified local zeta integrals. More precisely, let $\pi_1 = \otimes_v \pi_{1,v}$ and $\pi_2 = \otimes_v \pi_{2,v}$ be irreducible automorphic cuspidal representations of $GSp(2, \mathbb{A})$. The (global) degenerate principal series representation $I_3^4(s)$ is defined as in section

2.2. ψ is a generic additive unitary character of the adelic group $N^2(\mathbb{A})$ of the standard maximal unipotent subgroup N^2 of $GSp(2)$. We have the following global result.

THEOREM 5.0.1 (Fundamental Identity). *Assume that the data* $(\phi_1, \phi_2, f, \psi)$ *are factorizable, i.e.,* $\phi_1 = \otimes_v \phi_{1,v} \in \pi_1$, $\phi_2 = \otimes_v \phi_{2,v} \in \pi_2$, $f(\cdot, s) = \otimes_v f_v(\cdot, s) \in I_3^4(s)$, *and* $\psi = \otimes_v \psi_v$. *Let* S *be such a finite subset of places of the totally real number field* F *that* S *contains all archimedean places of* F *and at any finite place* v *which is not in* S, *the local representations* $\pi_{1,v}$ *and* $\pi_{2,v}$ *are unramified (or spherical) and the local character* ψ_v *is unramified and generic. Then the global zeta integral can be factorized as*

$$\mathcal{Z}^*(s, \phi_1, \phi_2, f) = L^S(s', \pi_1 \otimes \pi_2, \rho_1 \otimes \rho_1) \cdot \prod_{v \in S} \mathcal{Z}_v(s, W_{\phi_{1v}}^{\psi_v}, W_{\phi_{2v}}^{\overline{\psi_v}}, f_v) \qquad (68)$$

where ρ_1 *is the standard representation of the algebraic reductive group* $GSp(2)$ *and* $s' = \frac{s+1}{2}$.

This fundamental identity is our starting point to study the degree 16 standard L-function $L^S(s', \pi_1 \otimes \pi_2, \rho_1 \otimes \rho_1)$. Next we have to study the Eisenstein series, the analytic properties of which will determine those of the global integral, and the ramified local integrals.

CHAPTER 3

Poles of Eisenstein Series of $Sp(n)$

In this Chapter, we will apply the methods, developed by Piatetski-Shapiro and Rallis in [**PSRa**] and [**PSRa1**], and Kudla and Rallis [**KuRa**] and [**KuRa1**], to determine the possible poles of a family of non-Siegel type Eisenstein series of $Sp(n)$. We shall first compute the constant term of the Eisenstein series under consideration, which leads an inductive formula. Section 2 is devoted to the determination of the poles of certain intertwining operators. The main result on the poles of the Eisenstein series will be proved in section 4. The proof is in principal reduced to the case $n \leq 3$, which is done in section 3. Two technical lemmas are proved in the last section.

1. The P_1^n-constant Term of Eisenstein Series

Let $G_n := Sp(n)$ be the symplectic group of rank n over the totally real number field F. As usual, $G_n(\mathbb{A})$ is the adelic group of G_n and $G_n(F)$ is the group of F-rational points of G_n. Let $P_r^n = M_r^n N_r^n$ be the standard maximal parabolic subgroup of G_n with its Levi factor M_r^n and its unipotent radical N_r^n, that is,

$$M_r^n = \{m = \begin{pmatrix} a & 0 & 0 & 0 \\ 0 & x_1 & 0 & x_2 \\ 0 & 0 & {}^t a^{-1} & 0 \\ 0 & x_3 & 0 & x_4 \end{pmatrix} \in GL(r) \times G_{n-r}\}$$

and

$$N_r^n = \{n(x, w, y) = \begin{pmatrix} I_r & x & w & y \\ 0 & I_{n-r} & {}^t y & 0 \\ 0 & 0 & I_r & 0 \\ 0 & 0 & -{}^t x & I_{n-r} \end{pmatrix}\}.$$

We will use the notation that $a(p) = a(m) = \det(a)$. Then the modulus character $\delta_{P_r^n} = |a(m)|^{2n-r+1}$.

For each place v of the number field F, we choose a maximal compact subgroup $K_{n,v}$ of $G_n(F_v)$ as follows: If v is a non-archimedean place of F, we let $K_{n,v} = Sp(n, \mathcal{O}_v)$, and if real archimedean place v, we let $K_{n,v} \cong U(n)$ via the canonical embedding from $U(n)$ into $G_n(F_v)$. We let $K_n = \prod_v K_{n,v}$, which is a maximal compact subgroup of $G_n(\mathbb{A})$. By Iwasawa decomposition, we have $G_n(F_v) = P_r^n(F_v)K_{n,v}$ for each place v and $G_n(\mathbb{A}) = P_r^n(\mathbb{A})K_n$ globally.

We denote by $I_r^n(s) = Ind_{P_r^n(\mathbb{A})}^{G_n(\mathbb{A})}(|a(m)|^s)$ the degenerate principal series representation of $G_n(\mathbb{A})$, which consists of smooth functions $f(\cdot, s) : G_n(\mathbb{A}) \to \mathbb{C}$ satisfying the following condition:

$$f(pg, s) = |a(p)|^{s + \frac{2n-r+1}{2}} f(g, s) \tag{69}$$

for $p \in P_r^n(\mathbb{A})$, $g \in G_n(\mathbb{A})$. The group action is the right translation. By smoothness of the function f we mean that the function f is locally constant as a function over the non-archimedean local variables and smooth in the usual sense as a function over the archimedean local variables. A section $f(g, s)$ is called holomorphic or entire if the section $f(g, s)$, as a function of one complex variable s, is holomorphic or entire. We also assume that sections $f(g, s)$ is right K-finite. This implies that $I_r^n(s)$ is, in fact, a representation of $(\mathfrak{g}_{n,\infty}, K_{n,\infty}) \times G(\mathbb{A}_f)$, where $\mathfrak{g}_{n,\infty}$ is the Lie algebra of $G_{n,\infty}$ and $\mathbb{A}_f$ indicates the group of finite adeles.

We define, for any section $f_s = f(\cdot, s) \in I_r^n(s)$, an Eisenstein series as follows:

$$E_r^n(g, s; f_s) = \sum_{\gamma \in P_r^n(F) \backslash G_n(F)} f(\gamma g; s). \tag{70}$$

According to the general theory of Eisenstein series [**Lan**] and [**Art**], Such Eisenstein series is absolutely convergent for $Re(s)$ large and has meromorphic continuation to the whole complex plane as function of s and satisfies certain functional equation. In practise, the knowledge about the precise location and the exact order of the poles of Eisenstein series is important and interesting.

In the rest of this Chapter, we concentrate our study on the family of Eisenstein series $E_{n-1}^n(g, s; f_s)$. We will determine specifically the location and the order of the Eisenstein series.

From the general theory of Eisenstein series [**Lan**], [**Art**], and [**MoWa**]. Eisenstein series and its certain constant terms share the same analytic properties. We start with computing the constant term of $E_{n-1}^n(g, s; f_s)$ along the maximal parabolic subgroup P_1^n. We can do this because the Eisenstein series $E_{n-1}^n(g, s; f_s)$ is concentrated on the Borel subgroup of G_n in the terminology of Langlands [**Lan**] and Jacquet [**Jac**].

The unipotent radical of P_1^n is

$$N_1^n = \{ \begin{pmatrix} 1 & x_{n-1} & z & y \\ 0 & I_{n-1} & y & 0 \\ 0 & 0 & 1 & 0 \\ 0 & 0 & -{}^t x_{n-1} & 1 \end{pmatrix} \}.$$

As usual, the P_1^n-constant term of the Eisenstein series $E_{n-1}^n(g, s; f_s)$ is defined as

$$E_{n-1,P_1^n}^n(g, s; f_s) = \int_{N_1^n(F) \backslash N_1^n(\mathbb{A})} E_{n-1}^n(ng, s; f_s) dn. \tag{71}$$

After the standard unfolding, we obtain

$$E_{n-1,P_1^n}^n(g, s; f_s) = \sum_{\gamma \in P_{n-1}^n(F) \backslash Sp(n,F)/N_1^n(F)} \int_{N_1^{n,\gamma}(F) \backslash N_1^n(\mathbb{A})} f_s(\gamma n g) dn. \tag{72}$$

Considering the P_1^n-orbit decomposition on the flag variety $P_{n-1}^n \backslash Sp(n)$, we have

LEMMA 1.0.1. *Let w_1, w_2 be Weyl group elements of following type:*

$$w_1 = \begin{pmatrix} 0 & 0 & 1 & & & & \\ 0 & I_{n-2} & 0 & & 0 & & \\ 1 & 0 & 0 & & & & \\ & & & 0 & 0 & 1 \\ & 0 & & 0 & I_{n-2} & 0 \\ & & & 1 & 0 & 0 \end{pmatrix} ; \quad w_2 = \begin{pmatrix} 0 & 0 & 1 & 0 \\ 0 & I_{n-1} & 0 & 0 \\ -1 & 0 & 0 & 0 \\ 0 & 0 & 0 & I_{n-1} \end{pmatrix}. \tag{73}$$

Then we have

(a) *The double coset space $P_{n-1}^n(F) \setminus Sp(n, F)/N_1^n(F)$ is decomposed as a disjoint union of $[P_{n-1}^n \setminus P_{n-1}^n P_1^n/N_1^n]$, $[P_{n-1}^n \setminus P_{n-1}^n w_1 P_1^n/N_1^n]$, and $[P_{n-1}^n \setminus P_{n-1}^n w_2 P_1^n/N_1^n]$.*

(b) *As algebraic varieties, we have*

$$[P_{n-1}^n \setminus P_{n-1}^n P_1^n/N_1^n] = P_{n-2}^{n-1} \setminus Sp(n-1),$$

$$[P_{n-1}^n \setminus P_{n-1}^n w_1 P_1^n/N_1^n] = P_{n-1}^{n-1} \setminus Sp(n-1),$$

and

$$[P_{n-1}^n \setminus P_{n-1}^n w_2 P_1^n/N_1^n] = P_{n-2}^{n-1} \setminus Sp(n-1).$$

In order to study the P_1^n-constant term $E_{n-1,P_1^n}^n(g, s; f_s)$, we have to introduce some intertwining operators i_{n-1}^*, $i_{n-1}^* \circ \mathcal{U}_{w_1}^n(s)$, $i_{n-1}^* \circ \mathcal{U}_{w_2}^n(s)$, and $\mathcal{U}_{w_o}^n(s)$, which are defined in the following ways: By means of the canonical embedding of $Sp(n-1)$ into the Levi factor M_1^n of P_1^n, that is, $Sp(n-1) = Sp(V_{1,n-1})$, where $V_{1,n-1} = (Fe_1 \oplus Fe_1^*)^\perp$ in V, we define, for any section $f_s \in I_{n-1}^n(s)$, $i_{n-1}^*(f_s)(g)$ to be simply the restriction of f_s to the subgroup $Sp(n-1)$, and

$$\mathcal{U}_{w_1}^n(s)(f_s)(g) = \int_{N_{w_1}^n(\mathbb{A})} f_s(w_1 n g)dn,$$

$$\mathcal{U}_{w_2}^n(s)(f_s)(g) = \int_{N_{w_2}^n(\mathbb{A})} f_s(w_2 n g)dn, \tag{74}$$

$$\mathcal{U}_{w_o}^n(s)(f_s)(g) = \int_{N_{n-1}^n(\mathbb{A})} f_s(w_o n g)dn.$$

where w_1 and w_2 are the Weyl group elements defined as in (69), which represents two different P_{n-1}^n-orbits on $P_{n-1}^n \setminus Sp(n)$, and w_o is the longest Weyl group element; $N_{w_i}^n$, $i = 1, 2$ are such subgroups of N^n, the standard maximal unipotent subgroup of G_n, that $N^n = N^{w_i} N_{w_i}^n$, $N^{w_i} = (w_i^{-1} P_{n-1}^n w_i) \cap N^n$. By [**Sch**] and [**Jac1**], those intertwining operators converge absolutely for $Re(s)$ large and have meromorphic continuation to the whole s-plane. Those intertwining operators enjoy following properties. In the next section, we shall locate the poles of those operators.

LEMMA 1.0.2. *For $m_1(t_1, g') \in M_1^n$ ($M_1^n = GL(1) \times Sp(n-1)$) and $f_s \in I_{n-1}^n(s)$, we have*

(a) $f_s(m_1(t_1, g')) = |t_1|^{s + \frac{n+2}{2}} f_s(m_1(1, g'))$,

(b) $\mathcal{U}_{w_1}^n(s)(f_s)(m_1(t_1, g')) = |t_1|^{n-1} \mathcal{U}_{w_1}^n(s)(f_s)(m_1(1, g'))$, *and*

(c) $\mathcal{U}_{w_2}^n(s)(f_s)(m_1(t_1,g')) = |t_1|^{-s+\frac{n+2}{2}}\mathcal{U}_{w_2}^n(s)(f_s)(m_1(1,g')).$

PROPOSITION 1.0.1. *For generic complex values of s, we have the following intertwining operators: If $n \geq 3$, then*

(a) $i_{n-1}^* : I_{n-1}^n(s) \longrightarrow I_{n-2}^{n-1}(s+\frac{1}{2}),$

(b) $i_{n-1}^* \circ \mathcal{U}_{w_1}^n(s) : I_{n-1}^n(s) \longrightarrow I_{n-1}^{n-1}(s),$ *and*

(c) $i_{n-1}^* \circ \mathcal{U}_{w_2}^n(s) : I_{n-1}^n(s) \longrightarrow I_{n-2}^{n-1}(s-\frac{1}{2}).$

(d) $\mathcal{U}_{w_0}^n(s) : I_{n-1}^n(s) \longrightarrow I_{n-1}^n(-s).$

If $n = 2$, then $i_1^* \circ \mathcal{U}_{w_1}^2(s) : I_1^2(s) \longrightarrow I_1^1(s),$ and i_1^*, $i_1^* \circ \mathcal{U}_{w_2}^2(s)$ map $I_1^2(s)$ to the trivial representation of $Sp(1)$.

By the generic complex values of s we mean that for such values of s those intertwining operators are holomorphic. We will prove, in the next section, that there exists a normalizing factor to each intertwining operator so that the normalized intertwining operator will be holomorphic for Re(s) positive. Our main result in this section is the following theorem, which suggests an inductive way to study the analytic properties of Eisenstein series.

THEOREM 1.0.2. *For generic complex values of s, the constant term of the Eisenstein series $E_{n-1}^n(g,s;f_s)$ along the maximal parabolic subgroup P_1^n can be written, if $n \geq 3$, as a sum of Eisenstein series of G_{n-1}, that is,*

$$
\begin{aligned}
E_{n-1,P_1^n}^n(m_1(t_1,g'),s;f_s) &= |t_1|^{s+\frac{n+2}{2}}E_{n-2}^{n-1}(g',s+\frac{1}{2};i_{n-1}^*(f_s)) \\
&\quad + |t_1|^{n-1}E_{n-1}^{n-1}(g',s;i_{n-1}^*\circ\mathcal{U}_{w_1}^n(s)(f_s)) \qquad (75) \\
&\quad + |t_1|^{-s+\frac{n+2}{2}}E_{n-2}^{n-1}(g',s-\frac{1}{2};i_{n-1}^*\circ\mathcal{U}_{w_2}^n(s)(f_s)),
\end{aligned}
$$

where $m_1(t_1,g') \in M_1^n$. If $n = 2$, then the constant term $E_{1,P_1^2}^2(m_1(t_1,g'),s;f_s)$ can be expressed as follows:

$$
\begin{aligned}
&E_{1,P_1^2}^2(m_1(t_1,g'),s;f_s) \qquad\qquad\qquad\qquad\qquad\qquad\qquad (76) \\
&= |t_1|^{s+2}f_s(g') + |t_1|E_1^1(g',s;i_1^*\circ\mathcal{U}_{w_1}^2(s)(f_s)) + |t_1|^{-s+2}\mathcal{U}_{w_2}^2(s)(f_s)(g').
\end{aligned}
$$

PROOF. The Haar measure on $N_1^n(F) \setminus N_1^n(\mathbb{A})$ is so normalized that the volume of $N_1^{n,w}(F) \setminus N_1^{n,w}(\mathbb{A})$ is 1. We first consider the case of $n \geq 3$. According to (68), we have

$$E_{n-1,P_1^n}^n(g,s;f_s) = \sum_{\gamma \in (P_{n-1}^n \backslash P_{n-1}^n P_1^n / N_1^n)(F)} \int_{N_1^{n,\gamma}(F) \backslash N_1^n(\mathbb{A})} f_s(\gamma n g) dn$$

$$+ \sum_{\gamma \in (P_{n-1}^n \backslash P_{n-1}^n w_1 P_1^n / N_1^n)(F)} \int_{N_1^{n,\gamma}(F) \backslash N_1^n(\mathbb{A})} f_s(\gamma n g) dn$$

$$+ \sum_{\gamma \in (P_{n-1}^n \backslash P_{n-1}^n w_2 P_1^n / N_1^n)(F)} \int_{N_1^{n,\gamma}(F) \backslash N_1^n(\mathbb{A})} f_s(\gamma n g) dn$$

$$= E_r + E_{w_1} + E_{w_2}.$$

Applying Lemma 1.0.1, we can compute each term as follows: The first term E_r is easy to be deduced as follows:

$$E_r = \sum_{\gamma \in [P_{n-1}^n \backslash P_{n-1}^n P_1^n / N_1^n](F)} \int_{N_1^{n,\gamma}(F) \backslash N_1^n(\mathbb{A})} f_s(\gamma n g) dn$$

$$= \sum_{\gamma \in [P_{n-2}^{n-1} \backslash Sp(n-1)](F)} \int_{N_1^n(F) \backslash N_1^n(\mathbb{A})} f_s(m_1(1,\gamma) n g) dn$$

$$= \sum_{\gamma \in [P_{n-2}^{n-1} \backslash Sp(n-1)](F)} f_s(m_1(1,\gamma) g) \int_{N_1^n(F) \backslash N_1^n(\mathbb{A})} dn$$

$$= \sum_{\gamma \in [P_{n-2}^{n-1} \backslash Sp(n-1)](F)} f_s(m_1(1,\gamma) g),$$

since f_s is left $N_1^n(\mathbb{A})$-invariant. The computation of the term E_{w_1} will be a little longer.

$$E_{w_1} = \sum_{\gamma \in [P_{n-1}^n \backslash P_{n-1}^n w_1 P_1^n / N_1^n](F)} \int_{N_1^{n,\gamma}(F) \backslash N_1^n(\mathbb{A})} f_s(\gamma n g) dn$$

$$= \sum_{\gamma \in [P_{n-1}^{n-1} \backslash Sp(n-1)](F)} \int_{N_1^{n,w_1}(F) \backslash N_1^n(\mathbb{A})} f_s(w_1 m_1(1,\gamma) n g) dn$$

$$= \sum_{\gamma \in [P_{n-1}^{n-1} \backslash Sp(n-1)](F)} \int_{N_1^{n,w_1}(F) \backslash N_1^{n,w_1}(\mathbb{A})} \int_{N_{1,w_1}^n(\mathbb{A})} f_s(w_1 n n^- m_1(1,\gamma) g) dn^- dn$$

$$= \sum_{\gamma \in [P_{n-1}^{n-1} \backslash Sp(n-1)](F)} \int_{N_{1,w_1}^n(\mathbb{A})} f_s(w_1 n m_1(1,\gamma) g) dn,$$

since $N_1^n = N_1^{n,w_1} N_{1,w_1}^n$ and f_s is left $N_1^{n,w_1}(\mathbb{A})$- invariant and the volume of $N_1^{n,w_1}(F) \backslash N_1^{n,w_1}(\mathbb{A})$ is one. It is easy to check that $N_{1,w_1}^n = N_{w_1}$. Hence we can write the term E_{w_1} in terms of the intertwining operator $\mathcal{U}_{w_1}^n(s)$, that is,

$$E_{w_1} = \sum_{\gamma \in [P_{n-1}^{n-1} \backslash Sp(n-1)](F)} \mathcal{U}_{w_1}^n(s)(f_s)(m_1(1,\gamma) g).$$

By the similar reason, we can deduce the last term E_{w_2} as follows:

$$
\begin{aligned}
E_{w_2} &= \sum_{\gamma \in [P_{n-1}^n \backslash P_{n-1}^n w_2 P_1^n / N_1^n](F)} \int_{N_1^{n,\gamma}(F) \backslash N_1^n(\mathbb{A})} f_s(\gamma n g) dn \\
&= \sum_{\gamma \in [P_{n-2}^{n-1} \backslash Sp(n-1)](F)} \int_{N_1^{n,w_2}(F) \backslash N_1^n(\mathbb{A})} f_s(w_2 m_1(1,\gamma) n g) dn \\
&= \sum_{\gamma \in [P_{n-2}^{n-1} \backslash Sp(n-1)](F)} \int_{N_1^{n,w_2}(F) \backslash N_1^{n,w_2}(\mathbb{A})} \int_{N_{1,w_2}^n(\mathbb{A})} f_s(w_2 n n^- m_1(1,\gamma) g) dn^- dn \\
&= \sum_{\gamma \in [P_{n-2}^{n-1} \backslash Sp(n-1)](F)} \int_{N_{1,w_2}^n(\mathbb{A})} f_s(w_2 n m_1(1,\gamma) g) dn \\
&= \sum_{\gamma \in [P_{n-2}^{n-1} \backslash Sp(n-1)](F)} \mathcal{U}_{w_2}^n(s)(f_s)(m_1(1,\gamma)g).
\end{aligned}
$$

Therefore the constant term $E_{n-1,P_1^n}^n(g,s;f_s)$ can be expressed in the following way:

$$
\begin{aligned}
E_{n-1,P_1^n}^n(g,s;f_s) &= \sum_{\gamma \in [P_{n-2}^{n-1} \backslash Sp(n-1)](F)} f_s(m_1(1,\gamma)g) \\
&+ \sum_{\gamma \in [P_{n-1}^{n-1} \backslash Sp(n-1)](F)} \mathcal{U}_{w_1}^n(s)(f_s)(m_1(1,\gamma)g) \\
&+ \sum_{\gamma \in [P_{n-2}^{n-1} \backslash Sp(n-1)](F)} \mathcal{U}_{w_2}^n(s)(f_s)(m_1(1,\gamma)g).
\end{aligned}
$$

After restricting $E_{n_1,P_1^n}^n(g,s;f_s)$ to the Levi factor M_1^n of the parabolic subgroup P_1^n and using Lemma 1.0.2, we obtain that

$$
\begin{aligned}
E_{n-1,P_1^n}^n(m_1(t_1,g'),s;f_s) &= |t_1|^{s+\frac{n+2}{2}} E_{n-2}^{n-1}\left(g',s+\frac{1}{2};i_{n-1}^*(f_s)\right) \\
&+ |t_1|^{n-1} E_{n-1}^{n-1}\left(g',s;i_{n-1}^* \circ \mathcal{U}_{w_1}^n(s)(f_s)\right) \\
&+ |t_1|^{-s+\frac{n+2}{2}} E_{n-2}^{n-1}\left(g',s-\frac{1}{2};i_{n-1}^* \circ \mathcal{U}_{w_2}^n(s)(f_s)\right),
\end{aligned}
$$

for generic complex values of s. In the case of $n = 2$, one has that

$$
P_1^2 \backslash P_1^2 P_1^2 / N_1^2 = P_1^2 \backslash P_1^2 w_2 P_1^2 / N_1^2 = 1.
$$

Following the same calculation, we will obtain the expression for the constant term $E_{1,P_1^2}^2(m_1(t_1,g'),s;f_s)$. $\qquad\square$

2. Local Analyses of Intertwining Operators

In this section, we assume that F_v is the local completion of the number field F at the prime v, finite or real. As usual, $G_{n,v} = G_n(F_v)$ is the F_v-rational point of the group G_n.

We are going to determine the holomorphy of those intertwining operators $\mathcal{U}^n_{w_1}(s)$ and $\mathcal{U}^n_{w_2}(s)$ involved in our inductive formula for the constant term of the Eisenstein series, and that of the intertwining operator $\mathcal{U}^n_{w_o}(s)$ attached to the longest Weyl group element w_o, which will be used to determine the poles of Eisenstein series on the negative half-plan. Since those intertwining operators are eulerian, we can deal with the problem for each corresponding local intertwining operator $\mathcal{U}^n_{w_1,v}(s)(f_s)$, $\mathcal{U}^n_{w_2,v}(s)(f_s)$, and $\mathcal{U}^n_{w_o,v}(s)$. These intertwining operators are well known to be absolutely convergent for $Re(s)$ large and have a meromorphic continuation to the whole complex plane. The point here is to determine the exact location of their possible poles. The methods used in this section were developed by Piatetski-Shapiro and Rallis in [**PSRa**] and [**PSRa1**], and by Kudla and Rallis in [**KuRa**] and [**KuRa1**]. It should be mentioned that various version of the useful Lemma of Rallis play a very important role in our concrete computation. Note that the intertwining operators under consideration are defined over nonabelian unipotent subgroups. Therefore the situation will be more complicated. Our process consists of two steps: (1) work with general holomorphic sections in $I^n_{n-1}(s)$ and (2) work with the spherical sections in $I^n_{n-1}(s)$. Since in step (1) we need to decompose our intertwining operators, we may gain extra poles for our original intertwining operators. This is the reason we need step (2) to confirm the poles of our original intertwining operators.

2.1. Intertwining Operator $\mathcal{U}^n_{w_1,v}(s)$: general sections. We recall some notations from previous sections. Let

$$w_1 = \begin{pmatrix} 0 & 0 & 1 & & & \\ 0 & I_{n-2} & 0 & & 0 & \\ 1 & 0 & 0 & & & \\ & & & 0 & 0 & 1 \\ & 0 & & 0 & I_{n-2} & 0 \\ & & & 1 & 0 & 0 \end{pmatrix}, \quad N_{w_1} = \left\{ \begin{pmatrix} 1 & x_{n-2} & z & & & \\ 0 & I_{n-2} & 0 & & 0 & \\ 0 & 0 & 1 & & & \\ & & & 1 & 0 & 0 \\ & 0 & & -{}^t x_{n-2} & I_{n-2} & 0 \\ & & & -z & 0 & 1 \end{pmatrix} \right\},$$

and for any section $f_s \in I^n_{n-1,v}(s)$, the local component $\mathcal{U}^n_{w_1,v}(s)$ of the intertwining operator $\mathcal{U}^n_{w_1}(s)$ attached to w_1 is defined as

$$\mathcal{U}^n_{w_1,v}(s)(f_s)(g) = \int_{N_{w_1}(F_v)} f_s(w_1 n g) dn, \tag{77}$$

which is a $G_{n,v}$-intertwining operator from $I^n_{n-1,v}(s)$ to the induced representation $ind^{Sp(n,F_v)}_{P^n_{1,n-1}(F_v)}(|t|^{n-1}|\det m|^{s+\frac{n}{2}})$, the unnormalized induced representation of $G_{n,v} = Sp(n,F_v)$ from a parabolic (not maximal) subgroup $P^n_{1,n-1}$, which is in the following form:

$$P^n_{1,n-1} = \left\{ \begin{pmatrix} t & * & & * & \\ 0 & m & & & \\ & & t^{-1} & 0 & \\ 0 & & * & {}^t m^{-1} \end{pmatrix} : m \in GL(n-1, F_v) \right\}. \tag{78}$$

Since the analytic properties in s of the family

$$\{\mathcal{U}^n_{w_1,v}(s)(f_s) \ : \ f_s \text{ varies as holomorphic sections in } I^n_{n-1,v}(s)\}$$

are independent of the evaluation of $\mathcal{U}^n_{w_1,v}(s)(f_s)$ at g, we know that the analytic properties in s of the family $\{\mathcal{U}^n_{w_1,v}(f_s)\}$ is determined by that of the restriction of the family $\{\mathcal{U}^n_{w_1,v}(f_s)\}$ to $GL(n, F_v)$, which is canonically embedded into $Sp(n, F_v)$ via $g \mapsto \begin{pmatrix} g & 0 \\ 0 & {}^tg^{-1} \end{pmatrix}$. Then $\mathcal{U}^n_{w_1,v}(s)$ defines a $GL(n, F_v)$-intertwining operator from $ind^{GL(n,F_v)}_{P_{n-1,1}(F_v)}(|\det m|^{s+\frac{n+2}{2}})$ to $ind^{GL(n,F_v)}_{P_{1,n-1}(F_v)}(|t|^{n-1}|\det m|^{s+\frac{n}{2}})$, where $P_{n-1,1}$ and $P_{1,n-1}$ are maximal parabolic subgroups of $GL(n)$ in the following form:

$$P_{n-1,1} = \{\begin{pmatrix} m & * \\ 0 & t \end{pmatrix}\} \ ; \ P_{1,n-1} = \{\begin{pmatrix} t & * \\ 0 & m \end{pmatrix}\}.$$

More precisely, for any section $f_s \in ind^{GL(n,F_v)}_{P_{n-1,1}(F_v)}(|\det m|^{s+\frac{n+2}{2}})$, the restriction of the intertwining operator $\mathcal{U}^n_{w_1,v}(s)$ is

$$\mathcal{U}^n_{w_1,v}(s)(f_s)(g) = \int_{F^{n-1}_v} f_s(\overline{w_1}n(x_{n-2}, z)g)dx_{n-2}dz$$

with $\overline{w_1} = \begin{pmatrix} & & 1 \\ & I_{n-2} & \\ 1 & & \end{pmatrix}$ and $n(x_{n-2}, z) = \begin{pmatrix} 1 & x_{n-2} & z \\ 0 & I_{n-2} & 0 \\ 0 & 0 & 1 \end{pmatrix}$.

The following lemma is a $GL(n)$-version of the Rallis Lemma [**PSRa1**]. Since the proof is similar, we omit it here.

LEMMA 2.1.1. *Let $\overline{w}_\circ$ be the longest element of the Weyl group of $GL(n)$ i.e.*

$$\overline{w}_\circ = J_n = \begin{pmatrix} & & 1 \\ & J_{n-2} & \\ 1 & & \end{pmatrix} \text{ and } S := \{\phi \in I^n_{n-1,v}(s) \ : \ supp(\phi) \subset P_{n-1,1}\overline{w}_\circ P_{n-1,1}\}.$$

Then the analytic properties in s of the family

$$\{\mathcal{U}^n_{w_1,v}(s)(f_s) \ : \ f_s \text{ varies as holomorphic sections in } I^n_{n-1,v}(s)\}$$

coincide with that of the family $\{\mathcal{U}^n_{w_1,v}(s)(\phi)(\overline{w}_\circ)\}$ with $\phi \in S$.

According to this Lemma, it is sufficient to determine the holomorphy of the family $\{\mathcal{U}^n_{w_1,v}(s)(\phi)(\overline{w}_\circ) \ : \ \phi \in S\}$. By Bruhat decomposition,

$$\overline{w}_1 n(x_{n-2}, z)\overline{w}_\circ = \begin{pmatrix} -z^{-1} & -x_{n-2}z^{-1} & 1 \\ 0 & I_{n-2} & 0 \\ 0 & 0 & z \end{pmatrix}\overline{w}_\circ\begin{pmatrix} 1 & x_{n-2}J_{n-2}z^{-1} & z^{-1} \\ 0 & I_{n-2} & 0 \\ 0 & 0 & 1 \end{pmatrix}.$$

The analytic property of $\mathcal{U}^n_{w_1,v}(s)(\phi)(\overline{w}_\circ)$ can be determined in the following standard way:

$$\begin{aligned} \mathcal{U}^n_{w_1,v}(s)(\phi)(\overline{w}_\circ) &= \int_{F^{n-1}_v} \phi(\overline{w}_1 n(x_{n-2}, z)\overline{w}_\circ)dx_{n-2}dz \\ &= \int_{F^{n-1}_v} |z^{-1}|^{s+\frac{n+2}{2}}\phi(n(x_{n-2}J_{n-2}z^{-1}, z^{-1}))dx_{n-2}dz. \end{aligned}$$

By changing the variable $x_{n-2}J_{n-2}z^{-1} \mapsto x_{n-2}$, and then $z^{-1} \mapsto z$, and using the rule of separation variables for smooth functions ϕ, i.e., $\phi(x_{n-2}, z) = \sum_i \phi'_i(x_{n-2})\phi''_i(z)$, we deduce that

$$\begin{aligned}
\mathcal{U}^n_{w_1,v}(s)(\phi)(\overline{w}_\circ) &= \int_{F_v^\times} \int_{F_v^{n-2}} |z|^{s-\frac{n}{2}+2} \phi(n(x_{n-2}, z))dx_{n-2}d^\times z \\
&= \sum_i \int_{F_v^\times} |z|^{s-\frac{n}{2}+2} \phi''_i(z)dz^\times \cdot \int_{F_v^{n-2}} \phi'_i(x_{n-2})dx_{n-2}.
\end{aligned}$$

Since the integration over the variable x_{n-2} is independent of s and the integration over the variable z equals, by standard computation, to a product of $\zeta_v(s - \frac{n}{2} + 2)$ times a polynomial of s. Therefore we can express $\mathcal{U}^n_{w_1,v}(s)(\phi)(\overline{w}_\circ)$ in the following way:

$$\mathcal{U}^n_{w_1,v}(s)(\phi)(\overline{w}_\circ) = \zeta_v\left(s - \frac{n}{2} + 2\right) \times \text{ an entire function in } s.$$

This proves the following Proposition.

PROPOSITION 2.1.1. *The modified intertwining operator* $\frac{1}{\zeta_v(s-\frac{n}{2}+2)}\mathcal{U}^n_{w_1,v}(s)$ *is holomorphic. In other words, for any holomorphic section* f_s *in* $I^n_{n-1,v}(s)$,

$$\frac{1}{\zeta_v(s - \frac{n}{2} + 2)}\mathcal{U}^n_{w_1,v}(s)(f_s)$$

is a holomorphic section in $\text{ind}^{Sp(n,F_v)}_{P^n_{1,n-1}(F_v)}(|t|^{n-1}| \det m|^{s+\frac{n}{2}})$.

2.2. Intertwining Operator $\mathcal{U}^n_{w_2,v}(s)$: general sections. In this case, we will deal with the following Weyl group element w_2 and the corresponding unipotent subgroup N_{w_2}:

$$w_2 = \begin{pmatrix} 0 & 0 & 1 & 0 \\ 0 & I_{n-1} & 0 & 0 \\ -1 & 0 & 0 & 0 \\ 0 & 0 & 0 & I_{n-1} \end{pmatrix} \text{ and } N^n_{w_2} = \left\{ \begin{pmatrix} 1 & x_{n-2} & x & z & 0 & y \\ 0 & I_{n-2} & 0 & 0 & 0 & 0 \\ 0 & 0 & 1 & y & 0 & 0 \\ 0 & 0 & 0 & 1 & 0 & 0 \\ 0 & 0 & 0 & -{}^tx_{n-2} & I_{n-2} & 0 \\ 0 & 0 & 0 & -x & 0 & 1 \end{pmatrix} \right\}.$$

The local component $\mathcal{U}^n_{w_2,v}(s)$ of the intertwining operator $\mathcal{U}^n_{w_2}(s)$ is defined as follows: for any section $f_s \in I^n_{n-1,v}(s)$ and generic value of s,

$$\mathcal{U}^n_{w_2,v}(s)(f_s)(g) = \int_{N^n_{w_2}(F_v)} f_s(w_2ng)dn, \tag{79}$$

which takes, if $n \geq 3$, sections in $I^n_{n-1,v}(s)$ to sections in the induced representation $\text{ind}^{Sp(n,F_v)}_{P^n_{1,n-2}(F_v)}(|t_1|^{-s+\frac{n+2}{2}}| \det m|^{s+\frac{n}{2}})$, which is the unnormalized induced representation of $G_{n,v}$ from the standard parabolic subgroup $P^n_{1,n-2}$ of form:

$$P^n_{1,n-2} = \{ \begin{pmatrix} t_1 & * & * & * & & * & * \\ & m & * & * & & * & * \\ & & * & * & * & * & * \\ & & & t_1^{-1} & & & \\ & & & * & {}^t m^{-1} & & \\ & & & * & * & * & * \end{pmatrix} \} = [GL(1) \times GL(n-2) \times Sp(1)]N^n_{1,n-2}.$$

(80)

If $n = 2$, $\mathcal{U}^2_{w_2,v}(s)$ intertwines $I^2_1(s)$ and $I^2_1(-s)$ for generic value of s.

For general section f_s, the analytic property of the local intertwining operator $\mathcal{U}^n_{w_2,v}(s)$ is in some sense hard to deal with. Our idea is to decompose this intertwining operator into two intertwining operators, the analytic properties of which are easy to determine, and then to verify that our method does not create extra poles for the original intertwining operator $\mathcal{U}^n_{w_2,v}(s)$, which will be done in the last subsection of the present section.

Any element $n(x_{n-2}, x, y, z) \in N_{w_2}$ can be written in the following form:

$$n(x_{n-2}, x, y, z) = \begin{pmatrix} 1 & 0 & x & z & 0 & y \\ 0 & I_{n-2} & 0 & 0 & 0 & 0 \\ 0 & 0 & 1 & y & 0 & 0 \\ 0 & 0 & 0 & 1 & 0 & 0 \\ 0 & 0 & 0 & 0 & I_{n-2} & 0 \\ 0 & 0 & 0 & -x & 0 & 1 \end{pmatrix} \begin{pmatrix} 1 & x_{n-2} & 0 & & & \\ 0 & I_{n-2} & 0 & & 0 & \\ 0 & 0 & 1 & & & \\ & & & 1 & 0 & 0 \\ & 0 & & -{}^t x_{n-2} & I_{n-2} & 0 \\ & & & 0 & 0 & 1 \end{pmatrix}$$

$$= n(x, y, z)n(x_{n-2}).$$

The later factor is a subgroup in the center of N_{w_2}. Then for any $f_s \in I^n_{n-1,v}(s)$, our intertwining operator $\mathcal{U}^n_{w_2,v}(s)$ can be decomposed in the following way:

$$\begin{aligned} \mathcal{U}^n_{w_2,v}(s)(f_s)(g) &= \int_{N_{w_2}(F_v)} f_s(w_2 ng)dn \\ &= \int_{F_v^{n+1}} f_s(w_2 n(x,y,z)n(x_{n-2})g)dxdydzdx_{n-2} \\ &= \int_{F_v^{n-2}} \mathcal{M}_s(f_s)(n(x_{n-2})g)dx_{n-2} \\ &= \mathcal{N}_s(\mathcal{M}_s(f_s))(g). \end{aligned}$$

Note that if $n = 2$, $\mathcal{U}^2_{w_2,v}(s) = \mathcal{M}_s$. Those two operators $\mathcal{M}_s$ and $\mathcal{N}_s$ will be implicitly described in the following Lemma.

LEMMA 2.2.1. *Assume that $n \geq 3$. Let $Q^n_{1,n-2}$ be a parabolic subgroups of G_n in the following form:*

$$Q^n_{1,n-2} = \left\{ \begin{pmatrix} t & 0 & * & & & \\ * & m & * & & * & \\ 0 & 0 & a & & & b \\ & & & t^{-1} & * & 0 \\ & 0 & & 0 & {}^t m^{-1} & 0 \\ & & c & * & * & d \end{pmatrix} \right\} = (GL(1) \times GL(n-2) \times Sp(1))N^{n,'}_{1,n-2}.$$

Then the $G_{n,v}$-intertwining operators $\mathcal{M}_s$ and $\mathcal{N}_s$ can be described as follows: For generic complex values of s,

(a) $\mathcal{M}_s$ *is a $G_{n,v}$-intertwining operator*

$$\mathcal{M}_s : \ I^n_{n-1,v}(s) \to ind^{Sp(n,F_v)}_{Q^n_{1,n-2}(F_v)}(|t|^{-s-\frac{n}{2}+3}|\det m|^{s+\frac{n+2}{2}}),$$

which is defined by the following integral

$$\mathcal{M}_s(f_s)(g) = \int_{F_v^3} f_s(w_2 n(x,y,z)g)dxdydz.$$

(b) $\mathcal{N}_s$ *is a $G_{n,v}$-intertwining operator from*

$$ind^{Sp(n,F_v)}_{Q^n_{1,n-2}(F_v)}(|t|^{-s-\frac{n}{2}+3}|\det m|^{s+\frac{n+2}{2}})$$

to

$$ind^{Sp(n,F_v)}_{P^n_{1,n-2}(F_v)}(|t|^{-s+\frac{n}{2}+1}|\det m|^{s+\frac{n}{2}})$$

which is defined as

$$\mathcal{N}_s(f_s)(g) = \int_{F_v^{n-2}} f_s(n(x_{n-2})g)dx_{n-2}.$$

The following lemma is another version of the Rallis Lemma [**PSRa1**] and the proof is also similar. For completion, we will give the proof below.

LEMMA 2.2.2. *Let $B' = TN'$ be the Borel subgroup of $Sp(n)$ contained in $Q^n_{1,n-2}$ and $\lambda_s(b) = |t_1|^{-s-\frac{n}{2}+3}|t_2 t_3 \cdots t_{n-1}|^{s+\frac{n+2}{2}}$ a character of B'. Let $w_\circ = \begin{pmatrix} 0 & I_n \\ -I_n & 0 \end{pmatrix}$, the longest element in the Weyl group of $Sp(n)$ and*

$$S = \{\phi \in I^n_{n-1,v}(s) \ : \ \phi \text{ is smooth with compact support in } P^n_{n-1} w_\circ P^n_{n-1}\}.$$

Then we have

(a) $\mathcal{M}_s$ *is an intertwining operator from $I^n_{n-1,v}(s)$ to $ind^{Sp(n,F_v)}_{B'(F_v)}(\lambda_s)$.*

(b) *The analytic properties of the family $\{\mathcal{M}_s(\phi)(w_\circ) \ : \ \phi \in S\}$ coincide with the analytic properties of the family*

$$\{\mathcal{M}_s(f_s) \ : \ f_s \text{ varies as holomorphic sections in } I^n_{n-1,v}(s)\}.$$

PROOF. (a) It is easy to compute, for $\underline{t} = diag(t_1, \cdots, t_n, t_1^{-1}, \cdots, t_n^{-1}) \in T$ and $\underline{n} \in N'$,

$$
\begin{aligned}
\mathcal{M}_s(f_s)(\underline{t}g) &= \int_{F_v^3} f_s(w_2 n(x, y, z)\underline{t}g)dxdydz \\
&= \int_{F_v^3} f_s(w_2 \underline{t} w_2^{-1} w_2 n(t_1^{-1}x, t_1^{-1}y, t_1^{-2}z)g)dxdydz \\
&= |t_1|^{-s-\frac{n}{2}+3}|t_2\cdots t_{n-1}|^{s+\frac{n+2}{2}}\mathcal{M}_s(f_s)(g)
\end{aligned}
$$

and $\mathcal{M}_s(f_s)(\underline{n}g) = \mathcal{M}_s(f_s)(g)$.

(b) For $s_\circ$, the Laurent series of $\mathcal{M}_s(f_s)$ at $s = s_\circ$ is defined as

$$
\mathcal{M}_s(f_s) = \sum_{k \geq a} l_k(f_{s_\circ})(s - s_\circ)^k.
$$

By the general theory of intertwining operators, there is the smallest integer a so that the map $f_{s_\circ} \mapsto l_a(f_{s_\circ})$ is a nontrivial intertwining operator from $I_{n-1,v}^n(s_\circ)$ to $ind_{B'(F_v)}^{Sp(n,F_v)}(\lambda_{s_\circ})$.

To prove (b) is equivalent to prove that there must be a function $\phi \in S$ such that $l_a(\phi)(w_\circ) \neq 0$.

Now, if for any $\phi \in S$, $l_a(\phi)(w_\circ) = 0$, then $l_a(\phi)(B'w_\circ P_{n-1}^n) \equiv 0$ and further $l_a(\phi) \equiv 0$ as an element in $ind_{B'(F_v)}^{Sp(n,F_v)}(\lambda_{s_\circ})$ since $B'w_\circ P_{n-1}^n$ is dense in $Sp(n, F_v)$. Therefore the kernel $ker(l_a)$ is a nonzero $Sp(n, F_v)$-subspace in $I_{n-1,v}^n(s_\circ)$.

Since for any s, $I_{n-1,v}^n(s)$ is dual to $I_{n-1,v}^n(-s)$ according to the non-degenerate $Sp(n, F_v)$-invariant pair

$$
\int_{P_{n-1}^n(F_v)\backslash Sp(n,F_v)} f_s(g)f_{-s}(g)d\dot{g},
$$

the dual subspace $X^\perp$ of $X = ker(l_a)$ is a $Sp(n, F_v)$-subspace in $I_{n-1,v}^n(-s_\circ)$. For a $f_{-s_\circ} \in X^\perp$ and any $\phi \in S$, we have

$$
\begin{aligned}
0 &= \int_{P_{n-1}^n(F_v)\backslash Sp(n,F_v)} \phi(g)f_{-s_\circ}(g)d\dot{g} \\
&= \int_{P_{n-1}^n(F_v)\backslash P_{n-1}^n w_\circ P_{n-1}^n(F_v)} \phi(g)f_{-s_\circ}(g)d\dot{g} \\
&= \int_{N_{n-1}^n(F_v)} \phi(w_\circ n)f_{-s_\circ}(w_\circ n)dn.
\end{aligned}
$$

It is not difficult to see that the space S is isomorphic to $C_c^\infty(N_{n-1}^n(F_v))$ via $\phi \mapsto \phi|_{N_{n-1}^n(F_v)}$. Hence $f_{-s_\circ}(w_\circ N_{n-1}^n(F_v)) \equiv 0$. This implies that $f_{-s_\circ}(P_{n-1}^n w_\circ N_{n-1}^n(F_v)) \equiv 0$ and then $f_{-s_\circ} \equiv 0$ as an element in $I_{n-1,v}^n(-s_\circ)$ by the density of $P_{n-1}^n w_\circ N_{n-1}^n$ in $Sp(n)$. Therefore $X = ker(l_a) = I_{n-1,v}^n(s)$, that is $l_a \equiv 0$. However, this contradicts the assumption that l_a is nonzero. Thus there must be a $\phi \in S$ so that $l_a(\phi)(w_\circ) \neq 0$.   □

LEMMA 2.2.3. *For any* $\phi \in S$, $\mathcal{M}_s(\phi)(w_\circ)$ *has the form :* $\zeta_v(s + \frac{n}{2} - 2)$ *times a holomorphic function in* s.

PROOF. Considering the Bruhat decomposition of $Sp(n, F_v)$, we have that the element $w_2 n(x, y, z^{-1}) w_\circ$ is equal to

$$
\begin{pmatrix}
-z & 0 & -xz & 1 & 0 & -yz \\
0 & I_{n-2} & 0 & 0 & 0 & 0 \\
0 & 0 & 1 - xyz & y & 0 & -y^2 z \\
0 & 0 & 0 & -z^{-1} & 0 & 0 \\
0 & 0 & 0 & 0 & I_{n-2} & 0 \\
0 & 0 & x^2 z & -x & 0 & 1 + xyz
\end{pmatrix}
\begin{pmatrix}
1 & 0 & yz & -z & 0 & -xz \\
0 & I_{n-2} & 0 & 0 & 0 & 0 \\
0 & 0 & 1 & -xz & 0 & 0 \\
0 & 0 & 0 & 1 & 0 & 0 \\
0 & 0 & 0 & 0 & I_{n-2} & 0 \\
0 & 0 & 0 & -yz & 0 & 1
\end{pmatrix} .
$$

Then we deduce in the way similar to that used in proof of Proposition 2.1.1 in the last subsection that for any $\phi \in S$,

$$
\begin{aligned}
&\mathcal{M}_s(\phi)(w_\circ) \\
&= \int_{F_v^3} \phi(w_2 n(x, y, z,) w_\circ) dx dy dz \\
&= \int_{F_v^3} |z|^{-s - \frac{n}{2} - 1} \phi(w_\circ n(yz^{-1}, -xz^{-1}, -z^{-1})) dx dy dz \\
&= \int_{F_v^\times} \int_{F_v^2} |z|^{-s - \frac{n}{2} + 2} \phi(w_\circ n(y, x, -z^{-1})) dx dy d^\times z \\
&= \int_{F_v^\times} \int_{F_v^2} |z|^{s + \frac{n}{2} - 2} \sum_{i,j} a_{i,j} \phi_i(z) \phi_j(x, y) dx dy d^\times z \\
&= \zeta_v(s + \frac{n}{2} - 2) \times \text{a holomorphic function in } s.
\end{aligned}
$$

$\square$

COROLLARY 2.2.1. *The modified intertwining operator* $\frac{1}{\zeta_v(s + \frac{n}{2} - 2)} \mathcal{M}_s(f_s)$ *is holomorphic, that is, for any holomorphic section* $f_s \in I_{n-1,v}^n(s)$, $\frac{1}{\zeta_v(s + \frac{n}{2} - 2)} \mathcal{M}_s(f_s)$ *is a holomorphic section in* $\text{ind}_{Q_{1,n-2}^n(F_v)}^{Sp(n, F_v)}(|t|^{-s - \frac{n}{2} + 3} |\det m|^{s + \frac{n+2}{2}})$.

For the case of $n = 2$, the analytic property of $\mathcal{U}_{w_2, v}^2(s)$ can be determined since $\mathcal{U}_{w_2, v}^2(s) = \mathcal{M}_s$.

COROLLARY 2.2.2. *The modified intertwining operator* $\frac{1}{\zeta_v(s-1)} \mathcal{U}_{w_2, v}^2(f_s)$ *is holomorphic, that is, for any holomorphic section* f_s *in* $I_{1,v}^2(s)$, $\frac{1}{\zeta_v(s-1)} \mathcal{U}_{w_2, v}^2(f_s)$ *is a holomorphic section in* $I_{1,v}^2(-s)$.

Now, we turn to consider the $Sp(n)$-intertwining operator $\mathcal{N}_s$ defined in Lemma 2.2.1. The composition of $\mathcal{M}_s$ and $\mathcal{N}_s$ gives our intertwining operator $\mathcal{U}_{w_2, v}^n(s)$. Note that the integrating variable x_{n-2} in the integral which defines $\mathcal{N}_s$ lives in the Levi subgroup M_{n-1}^n of the maximal parabolic subgroup P_{n-1}^n. Since the analytic properties of the family $\{\mathcal{N}_s(f_s)\}$ do not depend on the evaluation of $\mathcal{N}_s(f_s)$ at g, we deduce that the analytic properties of the family $\{\mathcal{N}_s(f_s)\}$ coincide with that of the restriction of $\{\mathcal{N}_s(f_s)\}$ to the subgroup $GL(n-1, F_v)$, where $GL(n-1)$ is canonically embedded into M_{n-1}^n via $g \mapsto (g, I_2)$ for $g \in GL(n-1)$. As a $GL(n-1)$-intertwining operator, $\mathcal{N}_s$ carries sections in $ind_{P_{1,n-2}^-(F_v)}^{GL(n-1,F_v)}(|t|^{-s-\frac{n}{2}+3}|\det m|^{s+\frac{n+2}{2}})$ to sections in $ind_{P_{1,n-2}^-(F_v)}^{GL(n-1,F_v)}(|t|^{-s+\frac{n}{2}+1}|\det m|^{s+\frac{n}{2}})$, for generic complex values of s, where $P_{1,n-2}^-$ and $P_{1,n-2}$ are maximal parabolic subgroups of $GL(n-1)$ in the following forms:

$$P_{1,n-2}^- = \left\{\begin{pmatrix} t & 0 \\ * & m \end{pmatrix}\right\} \text{ and } P_{1,n-2} = \left\{\begin{pmatrix} t & * \\ 0 & m \end{pmatrix}\right\} \text{ for } m \in GL(n-2).$$

We deduce that for holomorphic sections f_s in $ind_{P_{1,n-2}^-(F_v)}^{GL(n-1,F_v)}(|t|^{-s-\frac{n}{2}+3}|\det m|^{s+\frac{n+2}{2}})$, the analytic properties of the family $\{\mathcal{N}_s(f_s)\}$ coincide with that of $\{\mathcal{N}_s^-(f_s)\}$, where the $GL(n-1)$-intertwining operator $\mathcal{N}_s^-$ is defined as

$$\mathcal{N}_s^-(f_s)(g) = \int_{F_v^{n-2}} f_s(\overline{w}_o' n^-(x_{n-3}, z)g)dx_{n-3}dz$$

with $\overline{w}_o' = J_{n-1}$ the longest element in the Weyl group of $GL(n-1)$ and the element in the unipotent subgroup is of form: $n^-(x_{n-3}, z) = \begin{pmatrix} 1 & 0 & 0 \\ 0 & I_{n-3} & 0 \\ z & x_{n-3} & 1 \end{pmatrix}$. It is easy to prove that $\mathcal{N}_s^-$ carries sections in $ind_{P_{1,n-2}^-(F_v)}^{GL(n-1,F_v)}(|t|^{-s-\frac{n}{2}+3}|\det m|^{s+\frac{n+2}{2}})$ to sections in $ind_{P_{n-2,1}^-(F_v)}^{GL(n-1,F_v)}(|t|^{-s+\frac{n}{2}+1}|\det m|^{s+\frac{n}{2}})$, where $P_{n-2,1}^-$ is a maximal parabolic subgroup of $GL(n-1)$ of the following form:

$$P_{n-2,1}^- = \left\{\begin{pmatrix} m & 0 \\ * & t \end{pmatrix} : m \in GL(n-2)\right\}.$$

Using the same argument as in the determination of the holomorphy of $\mathcal{U}_{w_1}^n(s)$, we need to study the analytic properties of the family $\{\mathcal{N}_s^-(\phi)(\overline{w}_o')\}$ for sections ϕ in $ind_{P_{1,n-2}^-(F_v)}^{GL(n-1,F_v)}(|t|^{-s-\frac{n}{2}+3}|\det m|^{s+\frac{n+2}{2}})$ with compact support in $P_{1,n-2}^-\overline{w}_o' P_{1,n-2}^-$. The holomorphy of $\mathcal{N}_s^-(\phi)(\overline{w}_o')$ is determined in the following standard way:

$$\begin{aligned}
\mathcal{N}_s^-(\phi)(\overline{w}_o') &= \int_{F_v^{n-2}} \phi(\overline{w}_o' n^-(x_{n-3}, z)\overline{w}_o')dx_{n-3}dz \\
&= \int_{F_v^{n-2}} \phi(n(x_{n-3}, z))dx_{n-3}dz.
\end{aligned}$$

By means of the Iwasawa decomposition with respect to the lower Borel subgroup, the last integral is equal to

$$
= \int_{F_v^\times} \int_{F_v^{n-3}} \phi\left(\begin{pmatrix} z & & \\ 0 & J_{n-3} & \\ 1 & -x_{n-3}J_{n-3}z^{-1} & -z^{-1} \end{pmatrix} \overline{w}_\circ' \begin{pmatrix} 1 & & \\ 0 & I_{n-3} & \\ z^{-1} & x_{n-3}z^{-1} & 1 \end{pmatrix}\right) dx_{n-3} dz
$$

$$
= \int_{F_v^\times} \int_{F_v^{n-3}} |z|^{-2s-n+3} \phi(\overline{w}_\circ' \begin{pmatrix} 1 & & \\ 0 & I_{n-3} & \\ z^{-1} & x_{n-3}z^{-1} & 1 \end{pmatrix}) dx_{n-3} d^\times z
$$

$$
= \int_{F_v^\times} \int_{F_v^{n-3}} |z|^{2s} \phi(\overline{w}_\circ' \begin{pmatrix} 1 & & \\ 0 & I_{n-3} & \\ z & x_{n-3} & 1 \end{pmatrix}) dx_{n-3} d^\times z
$$

$$
= \zeta_v(2s) \times \text{ a holomorphic function in } s.
$$

This yields the following Lemma.

LEMMA 2.2.4. *The modified intertwining operator* $\frac{1}{\zeta_v(2s)}\mathcal{N}_s$ *is holomorphic in the following sense: for any holomorphic section f_s in the induced representation*

$$
\text{ind}_{Q_{1,n-2}^n(F_v)}^{Sp(n,F_v)}(|t|^{-s-\frac{n}{2}+3}|\det m|^{s+\frac{n+2}{2}}),
$$

$\frac{1}{\zeta_v(2s)}\mathcal{N}_s(f_s)$ *is a holomorphic section in* $\text{ind}_{P_{1,n-2}^n(F_v)}^{Sp(n,F_v)}(|t|^{-s+\frac{n}{2}+1}|\det m|^{s+\frac{n}{2}})$.

Combining the results about the intertwining operators $\mathcal{M}_s$ and $\mathcal{N}_s$, we obtain the holomorphy of the intertwining operator $\mathcal{U}_{w_2,v}^n(s)$.

PROPOSITION 2.2.1. *The modified intertwining operator* $\frac{1}{\zeta_v(s+\frac{n}{2}-2)\zeta_v(2s)}\mathcal{U}_{w_2,v}^n(s)$ *is holomorphic, that is, for a holomorphic section f_s in $I_{n-1,v}^n(s)$,*

$$
\frac{1}{\zeta_v(s+\frac{n}{2}-2)\zeta_v(2s)}\mathcal{U}_{w_2,v}^n(s)(f_s)
$$

is a holomorphic section in $\text{ind}_{P_{1,n-2}^n(F_v)}^{Sp(n,F_v)}(|t_1|^{-s+\frac{n+2}{2}}|\det m|^{s+\frac{n+2}{2}})$.

2.3. Intertwining Operator $\mathcal{U}_{w_\circ,v}^n(s)$: general sections. The local intertwining operator $\mathcal{U}_{w_\circ,v}^n(s)$ is the v-component of the global intertwining operator $\mathcal{U}_{w_\circ}^n(s)$ attached to the longest Weyl group element $w_\circ$ and the unipotent radical $N_{w_\circ} = N_{n-1}^n$ of the maximal parabolic subgroup P_{n-1}^n of G_n. More precisely, we have

$$
w_\circ = \begin{pmatrix} 0 & 0 & I_{n-1} & 0 \\ 0 & 1 & 0 & 0 \\ -I_{n-1} & 0 & 0 & 0 \\ 0 & 0 & 0 & 1 \end{pmatrix} \text{ and } N_{w_\circ} = \begin{pmatrix} I_{n-1} & X & W & Y \\ 0 & 1 & {}^tY & 0 \\ 0 & 0 & I_{n-1} & 0 \\ 0 & 0 & -{}^tX & 1 \end{pmatrix}.
$$

The (local) intertwining operator $\mathcal{U}_{w_\circ,v}^n(s)$ is defined by the following integral: for any section $f_s \in I_{n-1,v}^n(s)$,

$$
\mathcal{U}_{w_\circ,v}^n(s)(f_s)(g) = \int_{N_{w_\circ}(F_v)} f_s(w_\circ n g) dn, \tag{81}
$$

which takes the sections in $I_{n-1,v}^n(s)$ to the sections in $I_{n-1,v}^n(-s)$. In order to determine the holomorphy of this intertwining operator $\mathcal{U}_{w_o,v}^n(s)$, we are doing to decompose $\mathcal{U}_{w_o,v}^n(s)$ into three intertwining operators, the analytic properties of which are easier to be determined. Let us use the notations $N(X,W,Y) = N_{w_o}$, $N(W,Y) = N(0,W,Y)$, $N(X) = N(X,0,0)$, and similarly $N(W) = N(W,0)$ and $N(Y) = N(0,Y)$. Then one has $N(X,W,Y) = N(W,Y)N(X) = N(Y)N(W)N(X)$ as varieties. The first step is to decompose $\mathcal{U}_{w_o,v}^n(s)$ into the following two two intertwining operators $\mathcal{M}_{W,Y}(s)$ and $\mathcal{M}_X(s)$:

$$
\begin{aligned}
\mathcal{U}_{w_o,v}^n(s)(f_s)(g) &= \int_{N(X)(F_v)} \int_{N(W,Y)(F_v)} f_s(w_o n(w,y)n(x)g)dn(w,y)dn(x) \\
&= \int_{N(X)(F_v)} \mathcal{M}_{W,Y}(s)(f_s)(n(x)g)dn(x) \\
&= \mathcal{M}_X(s)[\mathcal{M}_{W,Y}(s)(f_s)](g). \quad (82)
\end{aligned}
$$

Those two operators $\mathcal{M}_{W,Y}(s)$ and $\mathcal{M}_X(s)$ are described in the following Lemma.

LEMMA 2.3.1. *Let* $Q_{n-1,1}^n$ *and* $P_{n-1,1}^n$ *be two parabolic subgroups of* $Sp(n)$ *of the following types:*

$$
Q_{n-1,1}^n = \{\begin{pmatrix} a & 0 & * & * \\ * & b & * & * \\ 0 & 0 & {}^t a^{-1} & * \\ 0 & 0 & 0 & b^{-1} \end{pmatrix}\} \text{ and } P_{n-1,1}^n = \{\begin{pmatrix} a & * & * & * \\ 0 & b & * & * \\ 0 & 0 & {}^t a^{-1} & 0 \\ 0 & 0 & * & b^{-1} \end{pmatrix}\} \text{ with } a \in GL(n-1).
$$

Then the $G_{n,v}$-*intertwining operators* $\mathcal{M}_{W,Y}(s)$ *and* $\mathcal{M}_X(s)$ *can be described in the following ways:*

(a) $\mathcal{M}_{W,Y}(s)$ *is a* $G_{n,v}$-*intertwining operator*

$$
\mathcal{M}_{W,Y}(s) \; : \; I_{n-1}^n(s) \to ind_{Q_{n-1,1}^n(F_v)}^{G_{n,v}}(|a(q)|^{-s+\frac{n}{2}}|b|^{n-1}),
$$

which is defined by the following integral

$$
\mathcal{M}_{W,Y}(s)(f_s)(g) = \int_{N(W,Y)(F_v)} f_s(w_o n(w,y)g)dn(w,y). \quad (83)
$$

(b) $\mathcal{M}_X(s)$ *is a* $G_{n,v}$-*intertwining operator*

$$
\mathcal{M}_X(s) \; : \; ind_{Q_{n-1,1}^n(F_v)}^{G_{n,v}}(|a(q)|^{-s+\frac{n}{2}}|b|^{n-1}) \to ind_{P_{n-1,1}^n(F_v)}^{G_{n,v}}(|a(p)|^{-s+\frac{n+2}{2}}|b|^0),
$$

which is defined by the integral below

$$
\mathcal{M}_X(s)(f_s)(g) = \int_{N(X)(F_v)} f_s(n(x)g)dn(x). \quad (84)
$$

We start with the intertwining operator $\mathcal{M}_X(s)$ first. Since integrating variable $n(x)$ in the integral defining $\mathcal{M}_X(s)$ belongs to the Levi subgroup M_{n-1}^n of the maximal parabolic subgroup P_{n-1}^n, the analytic properties of the family of

$$
\{\mathcal{M}_X(s)(f_s) \; : \; f_s \in ind_{Q_{n-1,1}^n(F_v)}^{G_{n,v}}(|a(q)|^{-s+\frac{n}{2}}|b|^{n-1})\}
$$

coincides with that of the restriction of $\{\mathcal{M}_X(s)(f_s)\}$ to the Levi subgroup M_n^n of the standard maximal parabolic subgroup P_n^n of Siegel type, i.e. $M_n^n = \{\begin{pmatrix} a & 0 \\ 0 & {}^t a^{-1} \end{pmatrix}\}$, which is isomorphic to $GL(n)$. By the restriction, we can regard $\mathcal{M}_X(s)$ as an intertwining operator from the representation

$$ind_{P_{n-1,1}^-}^{GL(n,F_v)}(|a(p)|^{-s+\frac{n}{2}}|b|^{n-1})$$

to the representation

$$ind_{P_{n-1,1}}^{GL(n,F_v)}(|a(p)|^{-s+\frac{n+2}{2}}|b|^0),$$

where $P_{n-1,1}^-$ and $P_{n-1,1}$ are maximal parabolic subgroups of $GL(n)$ of the following forms: $P_{n-1,1}^- = \{\begin{pmatrix} a & 0 \\ * & b \end{pmatrix}\}$ and $P_{n-1,1} = \{\begin{pmatrix} a & * \\ 0 & b \end{pmatrix}\}$ with $a \in GL(n-1)$.

With applying the argument for the intertwining operator $\mathcal{N}_s$ in the previous subsection to the present case, one will easily see that the analytic properties of the family of $\mathcal{M}_X(s)(f_s)$ with sections f_s varying in $ind_{P_{n-1,1}^-}^{GL(n,F_v)}(|a(p)|^{-s+\frac{n}{2}}|b|^{n-1})$ coincides with that of the family of $\mathcal{M}_X(s)(\phi)(1)$ for ϕ in $ind_{P_{n-1,1}^-}^{GL(n,F_v)}(|a(p)|^{-s+\frac{n}{2}}|b|^{n-1})$ with compact support inside the open Bruhat cell $P_{n-1,1}^- \overline{w}_\circ P_{n-1,1}^-$.

The possible poles of $\mathcal{M}_X(s)(\phi)(1)$ will be determined by the following standard computation: As before, we denote the longest Weyl group element of $GL(n)$ by $\overline{w}_\circ = J_n$ as in Lemma 2.2.1. Then we have

$$\mathcal{M}_X(s)(\phi)(1)$$

$$= \int_{N(X)(F_v)} f_s(n(x)g)dn(x)$$

$$= \int_{F_v^{n-1}} \phi(\begin{pmatrix} 1 & 0 & z \\ 0 & I_{n-2} & x_{n-2} \\ 0 & 0 & 1 \end{pmatrix})dzdx_{n-2}$$

$$= \int_{F_v^{n-1}} \phi(\begin{pmatrix} z & 0 & 0 \\ x_{n-2} & J_{n-2} & 0 \\ 1 & 0 & z^{-1} \end{pmatrix} J_n \begin{pmatrix} 1 & 0 & 0 \\ -z^{-1}x_{n-2} & I_{n-2} & 0 \\ z^{-1} & 0 & 1 \end{pmatrix})dzdx_{n-2}.$$

By the left homogeneity of ϕ and by changing the variable $-z^{-1}x_{n-2} \mapsto x_{n-2}$, we deduce that

$$\mathcal{M}_X(s)(\phi)(1)$$

$$= \int_{F_v^{n-1}} |z|^{-s+\frac{n}{2}-(n-1)}\phi(J_n \begin{pmatrix} 1 & 0 & 0 \\ -z^{-1}x_{n-2} & I_{n-2} & 0 \\ z^{-1} & 0 & 1 \end{pmatrix})dzdx_{n-2}$$

$$= \int_{F_v^{n-1}} |z|^{-s+\frac{n}{2}}\phi(J_n \begin{pmatrix} 1 & 0 & 0 \\ x_{n-2} & I_{n-2} & 0 \\ z^{-1} & 0 & 1 \end{pmatrix})d^\times z dx_{n-2}$$

$$= \zeta_v(s - \frac{n}{2}) \times \text{ a holomorphic function in } s.$$

Therefore we obtain the holomorphy of the intertwining operator $\mathcal{M}_X(s)$, which is stated as

LEMMA 2.3.2. *The modified intertwining operator $\frac{1}{\zeta_v(s-\frac{n}{2})}\mathcal{M}_X(s)$ is holomorphic, that is, for any holomorphic section f_s in $ind_{Q^n_{n-1,1}}^{G_{n,v}}(|a(p)|^{-s+\frac{n}{2}}|b|^{n-1})$,*

$$\frac{1}{\zeta_v(s-\frac{n}{2})}\mathcal{M}_X(s)(f_s)$$

is a holomorphic section $ind_{P^n_{n-1,1}}^{G_{n,v}}(|a(p)|^{-s+\frac{n+2}{2}}|b|^0)$.

We now turn to determine the holomorphy of the intertwining operator $\mathcal{M}_{W,Y}(s)$. As stated in Lemma 2.3.1, $\mathcal{M}_{W,Y}(s)$ is a $G_{n,v}$-intertwining operator taking sections in $I_{n-1}^n(s)$ to sections in $ind_{Q^n_{n-1,1}}^{G_{n,v}}(|a(p)|^{-s+\frac{n}{2}}|b|^{n-1})$. The holomorphy of $\mathcal{M}_{W,Y}(s)$ will be determined by decomposing it into two intertwining operators $\mathcal{M}_Y(s)$ and $\mathcal{M}_W(s)$ in the following way: For any section $f_s \in I_{n-1}^n(s)$,

$$\mathcal{M}_{W,Y}(s)(f_s)(g)$$
$$= \int_{N(W,Y)(F_v)} f_s(w_\circ n(W,Y)g)dn(W,Y)$$
$$= \int_{N(W)(F_v)}\int_{N(Y)(F_v)} f_s(w_\circ n(Y)w_\circ^{-1}w_\circ n(W)w_\circ^{-1}w_\circ g)dn(Y)dn(W)$$
$$= \int_{N(W)}\int_{N(Y)} f_s\left(\begin{pmatrix} I_{n-1} & 0 \\ {}^tY & 1 \\ & & I_{n-1} & -Y \\ & & 0 & 1 \end{pmatrix}\begin{pmatrix} I_{n-1} & 0 \\ 0 & 1 \\ W & 0 & I_{n-1} \\ 0 & 0 & 0 & 1 \end{pmatrix}w_\circ g\right)dn(Y)dn(W).$$

Conjugating by the Weyl group element w_1 as defined in subsection 3.1, one have

$$w_1\begin{pmatrix} I_{n-1} & 0 \\ {}^tY & 1 \\ & & I_{n-1} & -Y \\ & & 0 & 1 \end{pmatrix}w_1^{-1} = \begin{pmatrix} 1 & Y' \\ 0 & I_{n-1} \\ & & 1 & 0 \\ & & -{}^tY' & I_{n-1} \end{pmatrix} =: N'(Y')$$

and

$$w_1\begin{pmatrix} I_{n-1} & 0 \\ 0 & 1 \\ W & 0 & I_{n-1} \\ 0 & 0 & 0 & 1 \end{pmatrix}w_1^{-1} = \begin{pmatrix} 1 & 0 \\ 0 & I_{n-1} \\ 0 & 0 & 1 & 0 \\ 0 & W' & 0 & I_{n-1} \end{pmatrix}.$$

If we set $w^* = w_1 w_\circ w_1^{-1}$, then $w^{*-1}w_1 w_\circ = w_1$ and

$$(w^*)^{-1}\begin{pmatrix} 1 & 0 \\ 0 & I_{n-1} \\ 0 & 0 & 1 & 0 \\ 0 & W' & 0 & I_{n-1} \end{pmatrix}w^* = \begin{pmatrix} 1 & 0 & 0 & 0 \\ 0 & I_{n-1} & 0 & W' \\ 0 & 0 & 1 & 0 \\ 0 & 0 & 0 & I_{n-1} \end{pmatrix} =: N^*(W').$$

Hence we can deduce that

$$\mathcal{M}_{W,Y}(s)(f_s)(g)$$
$$= \int_{N^*(W')(F_v)} \int_{N'(Y')(F_v)} f_s(w_1 n'(Y') w^* n^*(W') w_1 g) dn'(Y') dn^*(W')$$
$$= \int_{N^*(W)(F_v)} \mathcal{M}_Y(s)(f_s)(w^* n^*(W) w_1 g) dn^*(W)$$
$$= \mathcal{M}_W(s)[\mathcal{M}_Y(s)(f_s)](w_1 g)$$
$$= \tau_{w_1}[\mathcal{M}_W(s) \circ \mathcal{M}_Y(s)(f_s)](g).$$

Since the left shift operator $\tau_{w_1}(f_s)(g) = f_s(w_1 g)$ is holomorphic, the analytic properties of $\mathcal{M}_{W,Y}(s)$ follows from that of the composition $\mathcal{M}_W(s) \circ \mathcal{M}_Y(s)$. On the other hand, it is easy to figure out that for any section f_s in $I^n_{n-1}(s)$, one has $\mathcal{M}_Y(s)(f_s) = \mathcal{U}^n_{w_1,v}(s)(f_s)$. By Proposition 2.1.1, the modified intertwining operator $\frac{1}{\zeta_v(s-\frac{n}{2}+2)}\mathcal{M}_Y(s)$ is holomorphic, that is, the operator $\frac{1}{\zeta_v(s-\frac{n}{2}+2)}\mathcal{M}_Y(s)$ takes holomorphic sections in $I^n_{n-1}(s)$ to holomorphic sections in the induced representation $ind^{Sp(n,F_v)}_{P^n_{1,n-1}(F_v)}(|t|^{n-1}|\det m|^{s+\frac{n}{2}})$, the parabolic subgroup $P^n_{1,n-1}$ of $G_{n,v}$ is described in (74). Therefore we only need to determine the holomorphy of the intertwining operator $\mathcal{M}_W(s)$.

LEMMA 2.3.3. *The Weyl group element w^* is* $\begin{pmatrix} 1 & & 0 & \\ & 0 & & I_{n-1} \\ & 1 & & \\ 0 & & & \\ & -I_{n-1} & & 0 \end{pmatrix}$. *The intertwining operator $\mathcal{M}_W(s)$ is defined by the following integral*

$$\mathcal{M}_W(s)(f_s)(g) = \int_{N^*(W)(F_v)} f_s(w^* n^*(W) g) dn^*(W) \qquad (85)$$

and maps, for generic complex values of s, from $ind^{Sp(n,F_v)}_{P^n_{1,n-1}(F_v)}(|t|^{n-1}|\det m|^{s+\frac{n}{2}})$ to $ind^{Sp(n,F_v)}_{P^n_{1,n-1}(F_v)}(|t|^{n-1}|\det m|^{-s+\frac{n}{2}})$.

Note that in the integral (82), the integrating variable $n^*(W)$ and the Weyl group element w^* belong to the symplectic group $Sp(n-1)$ of rank $n-1$, which is embedded in the Levi subgroup $M^n_1 = GL(1) \times Sp(n-1)$ of the maximal parabolic subgroup P^n_1 via $g \mapsto (1, g)$. The analytic properties of the family $\{\mathcal{M}_W(s)(f_s)\}$ for f_s in $ind^{Sp(n,F_v)}_{P^n_{1,n-1}(F_v)}(|t|^{n-1}|\det m|^{s+\frac{n}{2}})$ coincide with that of the restriction of $\{\mathcal{M}_W(s)(f_s)\}$ to the subgroup $Sp(n-1)$. Under this restriction, $\mathcal{M}_W(s)$ can be viewed as an $Sp(n-1)$-intertwining operator from $ind^{Sp(n-1,F_v)}_{P^{n-1}_{n-1}(F_v)}(|\det m|^{s+\frac{n}{2}})$ to $ind^{Sp(n-1,F_v)}_{P^{n-1}_{n-1}(F_v)}(|\det m|^{-s+\frac{n}{2}})$, where P^{n-1}_{n-1} is the standard maximal parabolic subgroup of Siegel-type. This reduces our operator $\mathcal{M}_W(s)$ to the case which was extensively studied by Piatetski-Shapiro and Rallis in [**PSRa1**]. According to the Appendix to section 4 in [**PSRa1**], one has the following Lemma.

LEMMA 2.3.4. *Let* $a_{n-1,v}(s) = \zeta_v(s - \frac{n+2}{2}) \prod_{j=1,j\equiv n(2)}^{n-2} \zeta_v(2s - j + 1)$. *then the modified intertwining operator* $\frac{1}{a_{n-1,v}(s)} \mathcal{M}_W(s)$ *is holomorphic.*

Combining Lemma 2.3.4 with Lemma 2.3.3 and 2.3.2, we finally obtain the holomorphy of the intertwining operator $\mathcal{U}_{w_o}^n(s)$.

PROPOSITION 2.3.1. *Let*

$$a_{o,v}(s) = \zeta_v(s - \frac{n}{2} + 2)\zeta_v(s - \frac{n}{2} + 1)\zeta_v(s - \frac{n}{2}) \prod_{j=1,j\equiv n(2)}^{n-2} \zeta_v(2s - j + 1).$$

Then the modified intertwining operator $\frac{1}{a_{o,v}(s)} \mathcal{U}_{w_o,v}^n(s)$ *is holomorphic.*

The method we have used to determine the holomorphy of intertwining operators $\mathcal{U}_{w_1}^n(s)$, $\mathcal{U}_{w_2}^n(s)$, and $\mathcal{U}_{w_o}^n(s)$ depends on two basic techniques: (1) Decompose an intertwining operator into several intertwining operators which are in some sense easier to deal with. (2) Embeds the section into some larger spaces (by restriction) where we can apply the method developed by Piatetski-Shapiro and Rallis in [**PSRa1**]. In general, those two techniques may create extra poles. For instance, for $n = 2$, one has $\mathcal{U}_{w_o,v}^2(s) \equiv \mathcal{U}_{w_2,v}^2(s)$ for all s. We have used different way to determine the *possible* poles for this intertwining operator and obtained different results as shown in Corollary 2.2.1 and Proposition 2.3.1. It is clear that the method used in establishing Proposition 2.3.1 creates two extra possible poles for the operator $\mathcal{U}_{w_o,v}^2(s)$. In order to make sure that our application of those two techniques to our special cases is *safe*, we will use the method of Gindinkin and Karpelevich, which was described globally by Piatetski-Shapiro and Rallis in [**PSRa**], to prove that the poles of our intertwining operators determined by our methods are actually achieved by the spherical section.

By eulerian property of those intertwining operators, we can state as follows the global versions of the local results we obtained in previous three subsections.

THEOREM 2.3.1 (Global Versions). *All of those three modified intertwining operators* $\frac{1}{\zeta(s-\frac{n}{2}+2)} \mathcal{U}_{w_1}^n(s)$, $\frac{1}{\zeta(s+\frac{n}{2}-2)\zeta(2s)} \mathcal{U}_{w_2}^n(s)$, *and* $\frac{1}{a_o(s)} \mathcal{U}_{w_o}^n(s)$ *are holomorphic, where we let* $a_o(s) := \zeta(s - \frac{n}{2} + 2)\zeta(s - \frac{n}{2} + 1)\zeta(s - \frac{n}{2}) \prod_{j=1,j\equiv n(2)}^{n-2} \zeta(2s - j + 1)$.

2.4. Intertwining Operators $\mathcal{U}_{w_1}^n(s)$, $\mathcal{U}_{w_2}^n(s)$, and $\mathcal{U}_{w_o}^n(s)$: spherical section. In this subsection, we will prove that the analytic properties of intertwining operators $\mathcal{U}_{w_1}^n(s)$, $\mathcal{U}_{w_2}^n(s)$, and $\mathcal{U}_{w_o}^n(s)$ described in previous three subsections can be realized by the spherical sections in each case. To this end, we need the formula of Gindinkin and Karpelevich. The formula of such type was first established by Gindinkin and Karpelevich for real groups and by Langlands for groups over $\mathbb{Q}_p$. The global version of such formula was described by Piatetski-Shapiro and Rallis in [**PSRa**].

We are going to recall the formula of Gindinkin and Karpelevich (global version) from [**PSRa**]. We only focus on our own case although the formula works generally. Let T_n be the standard split maximal torus of $G_n = Sp(n)$, the elements of which are diagonal matrix of form: $t = \mathrm{diag}(t_1, \cdots, t_n, t_1^{-1}, \cdots, t_n^{-1})$. Let B_n be the standard Borel subgroup of G_n containing T_n and its unipotent radical N^n is on upper triangular matrix. Let $X^*(T_n)$ be the group of characters of T_n and ε_i the character of T_n so that $\varepsilon_i(t) = t_i$ for $i = 1, 2, \cdots, n$. Then one has $X^*(T_n) = \{\sum_{i=1}^n n_i \varepsilon_i\}$. Let $\Phi_{G_n} = \Phi(G_n, T_n)$ be the set of roots of T_n in G_n, $\Phi_{G_n}^+$ the set of positive roots of Φ_{G_n} determined by N_n, and $\triangle_{G_n}$ the set of simple roots in $\Phi_{G_n}^+$. Then these root data can be described as follows:

$$\begin{aligned}
\Phi_{G_n} &= \{\pm(\varepsilon_i \pm \varepsilon_j), \pm 2\varepsilon_i \ : i < j, \ i, j = 1, 2, \cdots, n\}, \\
\Phi_{G_n}^+ &= \{(\varepsilon_i \pm \varepsilon_j), 2\varepsilon_i \ : i < j, \ i, j = 1, 2, \cdots, n\}, \\
\triangle_{G_n} &= \{\alpha_i = \varepsilon_i - \varepsilon_{i+1}, \alpha_n = 2\varepsilon_n \ : \ i = 1, 2, \cdots, n-1\}.
\end{aligned} \tag{86}$$

Let $X_*(T_n)$ be the set of the one-parameter subgroups of T_n. Then $X_*(T_n)$ pairs non-degenerately with $X^*(T_n)$ by $(\ ,\ )$, which is defined by

$$x(x^\vee(\xi)) = \xi^{(x, x^\vee)}, \text{ for } \xi \in F^\times, x \in X^*(T_n), \text{ and } x^\vee \in X_*(T_n). \tag{87}$$

Let $\chi = \sum_{i=1}^n s_i \varepsilon_i$ be a complex character in $X^*(T_n) \otimes \mathbb{C}$ and $\delta = \sum_{i=1}^n 2(n - i + 1)\varepsilon_i$ the modulus character of the Borel subgroup B_n. The (normalized) induced representation $Ind_{B_n(\mathbb{A})}^{G_n(\mathbb{A})}(\chi)$ of $G_n(\mathbb{A})$ is defined to be the space consisting of smooth functions $\phi : G_n(\mathbb{A}) \to \mathbb{C}$ satisfying the following condition: $f(bg) = \chi(b)\delta^{\frac{1}{2}}(b)f(g)$. To each Weyl group element $w \in W_{G_n}$ we can associate a unipotent subgroup N_w^n of N^n so that $N^n = N_n^w N_w^n$ with $N_n^w = w^{-1}N^n w \cap N^n$, and an intertwining operator $\mathcal{M}_w$ defined by the following integral

$$\mathcal{M}_w(f)(g) = \int_{N_w^n(\mathbb{A})} f(wng)dn, \tag{88}$$

which takes sections in $Ind_{B_n(\mathbb{A})}^{G_n(\mathbb{A})}(\chi)$ to sections in $Ind_{B_n(\mathbb{A})}^{G_n(\mathbb{A})}(w^{-1} \cdot \chi)$, where the Weyl group W_{G_n} acts on $X^*(T_n) \otimes \mathbb{C}$ by $w^{-1} \cdot \chi(t) = \chi(w^{-1}tw)$. It is evident that in each $Ind_{B_n(\mathbb{A})}^{G_n(\mathbb{A})}(\chi)$, the subspace of K_n-fixed vectors is of at most one dimension and there is a unique K_n-fixed vector ϕ_χ, normalized so that $\phi_\chi(1) = 1$. Since the intertwining operator $\mathcal{M}_w$ takes K_n-fixed vectors in $Ind_{B_n(\mathbb{A})}^{G_n(\mathbb{A})}(\chi)$ to K_n-fixed vectors in $Ind_{B_n(\mathbb{A})}^{G_n(\mathbb{A})}(w^{-1} \cdot \chi)$, there exists a function $c_w(\chi)$ so that $\mathcal{M}_w(\phi_\chi) = c_w(\chi) \cdot \phi_{w^{-1} \cdot \chi}$. The c-function can be written in terms of the following integral

$$c_w^n(\chi) = \int_{N_w^n(\mathbb{A})} \phi_\chi(wn)dn. \tag{89}$$

Let E be any half-space in $X^*(T_n) \otimes_{\mathbb{Z}} \mathbb{Q}$ containing $\Phi_{G_n}^+$ and $R_w = wE$ for $w \in W_{G_n}$ and $\Phi_{G_n}^-(R_w) = \{\alpha \in \Phi_{G_n} : \alpha < 0, \ w^{-1}\alpha > 0\}$. According to [**PSRa**], the method of Gindinkin and Karpelevich gives that the following formula

$$\int_{N(R_w)(\mathbb{A})} \phi_\chi(u)du = \prod_{\alpha \in -\Phi_{G_n}^-(R_w)} \frac{\zeta((\chi, \alpha^\vee))}{\zeta((\chi, \alpha^\vee) + 1)} \tag{90}$$

where $N(R_w) = \prod_{\alpha \in \Phi_G^-(R_w)} N_\alpha = w N_w w^{-1}$, N_α is the one-parameter subgroup associated to the root α, and the coroot $\alpha^\vee = \alpha$ if $\alpha = \varepsilon_i \pm \varepsilon_j$ and $i \neq j$; and $\alpha^\vee = \frac{1}{2}\alpha$ if $\alpha = 2\varepsilon_i$. Thus by formula (85), (86), the function $c_w(\chi)$ is determined by

$$c_w^n(\chi) = \prod_{\alpha \in -\Phi_{G_n}^-(R_w)} \frac{\zeta((\chi, \alpha^\vee))}{\zeta((\chi, \alpha^\vee) + 1)}. \tag{91}$$

Now we are going to use formula (86) to determine the holomorphy of our three intertwining operators $\mathcal{U}_{w_1}^n(s)$, $\mathcal{U}_{w_2}^n(s)$, and $\mathcal{U}_{w_o}^n(s)$ evaluated at the spherical sections. More precisely, our degenerate principal series representation $I_{n-1}^n(s)$ can be naturally embedded into the normalized induced representation $Ind_{B_n(\mathbb{A})}^{G_n(\mathbb{A})}(\chi_s)$, where χ_s is a character of T_n defined by $\chi_s = [\sum_{i=1}^{n-1}(s - \frac{n}{2} + i)\varepsilon_i] - \varepsilon_n$. Note that under this embedding, the image of the normalized K_n-fixed (spherical) section f_s° is the K_n-fixed vector ϕ_{χ_s}. Then we have to compute the following three c-functions:

$$c_{w_1}^n(\chi_s) = \int_{N_{w_1}^n(\mathbb{A})} f_s^\circ(w_1 n) dn; \tag{92}$$

$$c_{w_2}^n(\chi_s) = \int_{N_{w_2}^n(\mathbb{A})} f_s^\circ(w_2 n) dn; \tag{93}$$

$$c_{w_o}^n(\chi_s) = \int_{N_{w_o}^n(\mathbb{A})} f_s^\circ(w_o n) dn. \tag{94}$$

In order to use formulà (86), we have to check that $N(R_w) = w N_w^n w^{-1}$ for $w = w_1, w_2$, and w_o.

Since T_n is also a maximal split torus in the Levi factor M_{n-1}^n of the maximal parabolic subgroup P_{n-1}^n, one can consider the set of roots $\Phi_{M_{n-1}^n} = \Phi(M_{n-1}^n, T_n)$ for the reductive group M_{n-1}^n with respect to T_4 and the set of the positive roots is $\Phi_{M_{n-1}^n}^+ = \Phi_{G_n}^+ \cap \Phi_{M_{n-1}^n}$. Let Ω be the distinguished set of coset representatives for $W_{M_{n-1}^n} \backslash W_{Sp(n)}$ obtained by choosing the unique element of minimal length in each coset. Then by Casselman [**Cas**]. one has a description of Ω as follows:

$$\Omega = \{w \in W_{Sp(n)} : w^{-1}\Phi_{M_{n-1}^n}^+ \subset \Phi_{Sp(n)}^+\}. \tag{95}$$

It is easy to check that for $w \in \Omega$, $N(R_w) = w N_{n,w} w^{-1}$ holds. So, we have to modify our Weyl group elements w_1 and w_2 by an element $w' \in W_{M_{n-1}^n}$ so that our intertwining operators $\mathcal{U}_{w_1}^n(s)$ and $\mathcal{U}_{w_2}^n(s)$ are attached to the Weyl group elements in Ω, while w_o is already in Ω. We choose w' to be the following element in $W_{M_{n-1}^n}$

$$w' = \begin{pmatrix} 0 & I_{n-2} & 0 & & & \\ 1 & 0 & 0 & & 0 & \\ 0 & 0 & 1 & & & \\ & & & 0 & I_{n-2} & 0 \\ & 0 & & 1 & 0 & 0 \\ & & & 0 & 0 & 1 \end{pmatrix}.$$

Then it is easy to check that $w_i^* = w'w_i \in \Omega$ and $\mathcal{U}_{w_i}^n(s) = \mathcal{U}_{w_i^*}^n(s)$ for $i = 1, 2$.

Now, applying formula (87), the c-functions $c_w^n(\chi_s)$ for $w = w_1^*, w_2^*$, and $w_\circ$ can be computed in the following way:

$$
\begin{aligned}
c_w^n(\chi_s) &= \int_{N_w^n(\mathbb{A})} \phi_{\chi_s}(wn) dn \\
&= \int_{N_w^n(\mathbb{A})} \phi_{\chi_s}(wnw^{-1}) dn \\
&= \int_{wN_w^n(\mathbb{A})w^{-1}} \phi_{\chi_s}(n) dn \\
&= \prod_{\alpha \in -\Phi_{G_n}^-(R_w)} \frac{\zeta((\chi, \alpha^\vee))}{\zeta((\chi, \alpha^\vee) + 1)},
\end{aligned}
$$

since $wN_w^n(\mathbb{A})w^{-1} = N(R_w)(\mathbb{A})$. Note that

$$
\begin{aligned}
-\Phi_{G_n}^-(R_{w_1^*}) &= \{\varepsilon_i - \varepsilon_n : i = 1, 2, \cdots, n-1\}; \\
-\Phi_{G_n}^-(R_{w_2^*}) &= \{\varepsilon_{n-1} - \varepsilon_n, \varepsilon_{n-1} + \varepsilon_i : i = 1, 2, \cdots, n\}; \qquad (96) \\
-\Phi_{G_n}^-(R_{w_\circ}) &= \{\varepsilon_i \pm \varepsilon_n, \varepsilon_i + \varepsilon_j : i, j = 1, 2, \cdots, n-1, i \neq j\}.
\end{aligned}
$$

It is a straightforward computation that

$$
\begin{aligned}
c_{w_1^*}^n(s) &= c_{w_1}^n(s) = \frac{\zeta(s - \frac{n}{2} + 2)}{\zeta(s + \frac{n}{2} + 1)}; \\
c_{w_2^*}^n(s) &= c_{w_2}^n(s) = \frac{\zeta(2s)\zeta(s + \frac{n}{2} - 2)}{\zeta(2s + n - 2)\zeta(s + \frac{n}{2} + 1)}; \\
c_{w_\circ}^n(s) &= \frac{\zeta(s - \frac{n}{2} + 2)\zeta(s - \frac{n}{2} + 1)\zeta(s - \frac{n}{2})}{\zeta(s + \frac{n}{2} + 1)\zeta(s + \frac{n}{2})\zeta(s + \frac{n}{2} - 1)} \prod_{j=1, j \equiv n(2)}^{n-2} \frac{\zeta(2s - j + 1)}{\zeta(2s + j)}.
\end{aligned}
$$

Note that if $n = 2$, one has $c_{w_2}^2(s) = c_{w_\circ}^2(s) = \frac{\zeta(s-1)}{\zeta(s+2)}$. Therefore we obtain that

PROPOSITION 2.4.1. *For $Re(s) > 0$ and the normalized spherical section $f_s^\circ \in I_{n-1}^n(s)$, the poles of the intertwining operators $\mathcal{U}_{w_1}^n(s)$, $\mathcal{U}_{w_2}^n(s)$, and $\mathcal{U}_{w_\circ}^n(s)$ can be described as follows:*

(a) *$\mathcal{U}_{w_1}^n(s)(f_s^\circ)$ achieves the same poles as the function $\zeta(s - \frac{n}{2} + 2)$ does.*

(b) *$\mathcal{U}_{w_2}^n(s)(f_s^\circ)$ achieves the same poles as the function $\zeta(2s)\zeta(s + \frac{n}{2} - 2)$ does, for $n \geq 3$.*

(c) *$\mathcal{U}_{w_\circ}^n(s)(f_s^\circ)$ achieves the same poles as the function $\zeta(s - \frac{n}{2} + 2)\zeta(s - \frac{n}{2} + 1)\zeta(s - \frac{n}{2}) \prod_{j=1, j \equiv n(2)}^{n-2} \zeta(2s - j + 1)$ does, for $n \geq 3$.*

PROOF. Since $\mathcal{U}_w^n(s)(f_s^\circ)(g) = c_w(\chi_s) \cdot f_{w \cdot s}^\circ(g)$ for $w = w_1, w_2$, and $w_\circ$, where f_w° is the normalized spherical section in image space under $\mathcal{U}_w^n(s)$, results (a), (b), and (c) follow from the computations above and the nonvanishing of $f_{w \cdot s}^\circ$. $\qquad\square$

Combining the results in Proposition 2.4.1 and these in Theorem 2.3.1, we have the following theorem, which is important for us to study the poles of our Eisenstein series.

THEOREM 2.4.1 (Global). *For $Re(s) > 0$ and holomorphic sections f_s in $I_{n-1}^n(s)$, we have the following three statements:*

(a) $\mathcal{U}_{w_1}^n(s)(f_s)$ *and* $\zeta(s - \frac{n}{2} + 2)$ *share the same poles.*

(b) $\mathcal{U}_{w_2}^n(s)(f_s)$ *and* $\zeta(2s)\zeta(s + \frac{n}{2} - 2)$ *share the same poles, for $n \geq 3$.*

(c) $\mathcal{U}_{w_0}^n(s)(f_s)$ *and* $\zeta(s - \frac{n}{2} + 2)\zeta(s - \frac{n}{2} + 1)\zeta(s - \frac{n}{2}) \prod_{j=1, j \equiv n(2)}^{n-2} \zeta(2s - j + 1)$ *share the same poles, for $n \geq 3$.*

For the special case $n = 2$, we may state our results as

PROPOSITION 2.4.2. *For $Re(s) > 0$, the intertwining operator $\mathcal{U}_{w_2}^2(s)$ and the global zeta function $\zeta(s - 1)$ share the same poles.*

3. Poles of Eisenstein Series: $n \leq 3$

In this section, we are going to study, for $Re(s) \geq 0$, the poles of our (unnormalized) Eisenstein series $E_{n-1}^n(g, s; f_s)$ for the special case $n \leq 3$. The case of $n = 1$ is well known. We will use our inductive formula to determine the cases $n = 2$ and $n = 3$. Actually, the case $n = 3$ is one of the most technical parts of our determination of the poles of Eisenstein series $E_{n-1}^n(g, s; f_s)$. First of all, we recall from [**KuRa**] the results on the Siegel Eisenstein series.

THEOREM 3.0.2 (Kudla-Rallis [**KuRa**]). *For any holomorphic section f_s in $I_n^n(s)$, the (unnormalized) Siegel Eisenstein series $E_n^n(g, s; f_s)$ enjoys the following properties:*

(a) $E_n^n(g, s; f_s)$ *is holomorphic for $Re(s) \geq 0$ except for $s \in \{\frac{e'(n)}{2}, \cdots, \frac{n-1}{2}, \frac{n+1}{2}\}$.*

(b) *At $s_0 = \frac{e'(n)}{2}, \cdots, \frac{n-1}{2}, \frac{n+1}{2}$, the residue representation $Res_{s=s_0}(E_n^n(g, s; f_s))$ does not vanish.*

Here $e'(n) = 1$ if n is even and 2 if odd.

Actually, Kudla and Rallis' result on the family of Eisenstein series $E_n^n(g, s; f_s)$ is much more general. They also proved that the normalized Eisenstein series is holomorphic for $Re(s) \leq 0$.

For $n = 1$, one has the following properties that will be used later to determine the poles of $E_n^n(g, s; f_s)$ for $n = 3$.

COROLLARY 3.0.1. *Let any holomorphic section $f_s \in I_1^1(s)$. Then we have:*

(a) *The Eisenstein series $E_1^1(g, s; f_s)$ has a zero at $s = 0$, and*

(b) *The residue of $E_1^1(g, s; f_s)$ at $s = 1$ is*

$$Res_{s=1}(E_1^1(g, s; f_s)) = Res_{s=1}(\int_{\mathbb{A}} f_s(w_0 n(x)g)dx)$$

where $w_0 = \begin{pmatrix} 0 & 1 \\ -1 & 0 \end{pmatrix}$ *and* $n(x) = \begin{pmatrix} 1 & x \\ 0 & 1 \end{pmatrix}$.

PROOF. (b) is straightforward by considering the constant term along the standard Borel subgroup of $Sp(1)$. (a) is from Kudla and Rallis [**KuRa1**] or [**KuRa2**] since the normalized Eisenstein series $E_1^{1*}(g, s : f_s) := b^S(s, 1)E_1^1(g, s : f_s)$ is holomorphic at $s = 0$, while the normalizing factor $b^S(s, 1) = \zeta^S(s + 1)$ has a simple pole at $s = 0$. $\qquad\qquad\square$

For $n = 2$, the analytic properties of the (unnormalized) Eisenstein series $E_1^2(g, s; f_s)$ can be described as follows.

THEOREM 3.0.3. *For any holomorphic section f_s in $I_1^2(s)$, the (unnormalized) Eisenstein series $E_1^2(g, s; f_s)$ is holomorphic for $Re(s) \geq 0$ except for $s = 2$ where it achieves a simple pole.*

PROOF. According to formula in Theorem 1.0.2, the constant term of $E_1^2(g, s; f_s)$ along the maximal parabolic subgroup P_1^2 is equal to

$$E_{1,P_1^2}^2(m_1(t_1, g'), s; f_s) \tag{97}$$
$$= |t_1|^{s+2} f_s(g') + |t_1| E_1^1(g', s; i_1^* \circ \mathcal{U}_{w_1}^2(s)(f_s)) + |t_1|^{-s+2} \mathcal{U}_{w_2}^2(s)(f_s)(g').$$

Since $\mathcal{U}_{w_1}^{2,*}(s) = \frac{\zeta(s+2)}{\zeta(s+1)}\mathcal{U}_{w_1}^2(s)$ and $\mathcal{U}_{w_2}^{2,*}(s) = \frac{\zeta(s+2)}{\zeta(s-1)}\mathcal{U}_{w_2}^2(s)$ are holomorphic for $Re(s) > 0$, we can rewrite the above formula as

$$E_{1,P_1^2}^2(m_1(t_1, g'), s; f_s) = |t_1|^{s+2} f_s(g') \tag{98}$$
$$+ |t_1| E_1^1(g', s; i_1^* \circ \mathcal{U}_{w_1}^{2,*}(s)(f_s))\frac{\zeta(s+1)}{\zeta(s+2)}$$
$$+ |t_1|^{-s+2} \mathcal{U}_{w_2}^{2,*}(s)(f_s)(g')\frac{\zeta(s-1)}{\zeta(s+2)}.$$

The first term is always holomorphic as a function of s. For $Re(s) > 0$, the second term is holomorphic except for $s = 1$ and the third term is holomorphic except for $s = 1$ and $s = 2$. Hence the constant term achieves a simple pole at $s = 2$.

We are going to prove that the constant term is actually holomorphic at $s = 1$. Taking the residue at $s = 1$ of the constant term, we have

$$Res_{s=1}[E_{1,P_1^2}^2(m_1(t_1, g'), s; f_s)] = \frac{|t_1|\zeta(2)}{\zeta(3)} Res_{s=1}[E_1^1(g', s; i_1^* \circ \mathcal{U}_{w_1}^{2,*}(s)(f_s))]$$
$$= + \frac{|t_1|}{\zeta(3)} Res_{s=1}[\zeta(s-1)] \lim_{s \to 1}[\mathcal{U}_{w_2}^{2,*}(s)(f_s)(g')]].$$

By Corollary 3.0.1, the residue at $s = 1$ of $E_1^1(g', s; i_1^* \circ \mathcal{U}_{w_1}^{2,*}(s)(f_s))$ is equal to

$$Res_{s=1}[\int_{\mathbb{A}} \mathcal{U}_{w_1}^{2,*}(s)(f_s)(\begin{pmatrix} 1 & & & 0 \\ & 0 & 1 & \\ & 0 & 1 & \\ & -1 & & 0 \end{pmatrix}\begin{pmatrix} 1 & & & \\ & 1 & 1 & z \\ & & 1 & \\ & & & 1 \end{pmatrix}g')dz]. \qquad (99)$$

We denote the integral by $\mathcal{M}_z(s)(\mathcal{U}_{w_1}^{2,*}(s)(f_s))(g')$. Note that $\mathcal{U}_{w_1}^{2,*}(s)$ may be regarded as an $Sp(2)$-intertwining operator from $I_1^2(s)$ to $Ind_{B_2(\mathbb{A})}^{Sp(2,\mathbb{A})}(|t_1|^{-1}|t_2|^s)$. Then $\mathcal{M}_z(s)$ is an $Sp(2)$-intertwining operator from

$$Ind_{B_2(\mathbb{A})}^{Sp(2,\mathbb{A})}(|t_1|^{-1}|t_2|^s) \text{ to } Ind_{B_2(\mathbb{A})}^{Sp(2,\mathbb{A})}(|t_1|^{-1}|t_2|^{-s})$$

for generic values of s. After normalizing by $\frac{\zeta(s+1)}{\zeta(s)}$, i.e. $\mathcal{M}_z^*(s) = \frac{\zeta(s+1)}{\zeta(s)}\mathcal{M}_z(s)$, the modified operator $\mathcal{M}_z^*(s)$ is holomorphic for $Re(s) > 0$. Hence the residue at $s = 1$ of $E_1^1(g', s; i_1^* \circ \mathcal{U}_{w_1}^{2,*}(s)(f_s))$ is equal to $\frac{Res_{s=1}\zeta(s)}{\zeta(2)} \lim_{s\to1}[\mathcal{M}_z^*(s) \circ \mathcal{U}_{w_1}^{2,*}(s)(f_s)](g')$. In order to prove the holomorphy at $s = 1$ of the constant term $E_{1,P_1^2}^2(m_1(t_1, g'), s; f_s)$, that is, the vanishing at $s = 1$ of the residue, it suffices to prove the following identity of global intertwining operators:

$$\lim_{s\to1}[\mathcal{U}_{w_2}^{2,*}(s)(f_s)](g') = \lim_{s\to1}[\mathcal{M}_z^*(s) \circ \mathcal{U}_{w_1}^{2,*}(s)(f_s)](g'). \qquad (100)$$

Since the residue $Res_{s=1}\zeta(s)$ and the residue $Res_{s=1}\zeta(s - 1)$ are reciprocal. Because both sides of the above identity are eulerian, we shall first prove the corresponding local identity for each local place and then the global identity will follow the local one by the standard local-global argument.

For each local place v, the local intertwining operator $\mathcal{U}_{w_2,v}^{2,*}(s)$ maps from $I_{1,v}^2(s)$ to $Ind_{B_2(F_v)}^{Sp(2,F_v)}(|t_1|^{-s}|t_2|^{-1})$, and similarly $\mathcal{M}_{z,v}^*(s) \circ \mathcal{U}_{w_1,v}^{2,*}(s)$ takes sections in $I_{1,v}^2(s)$ to sections in $Ind_{B_2(F_v)}^{Sp(2,F_v)}(|t_1|^{-1}|t_2|^{-s})$. Moreover, both of them take the unique normalized spherical section in $I_{1,v}^2(s)$ to the unique normalized spherical section in the respective space of induced representation. Let us consider a typical intertwining operator $\mathcal{M}_{y,v}^*(s) = \frac{\zeta(s)}{\zeta(s-1)}\mathcal{M}_{y,v}(s)$, which is defined by the following integral

$$\mathcal{M}_{y,v}^*(s)(f_s)(g) = \frac{\zeta(s)}{\zeta(s-1)} \int_{\mathbb{A}} f_s(\begin{pmatrix} 1 & & & \\ & 1 & & \\ & & 1 & \\ & & & 1 \end{pmatrix}\begin{pmatrix} 1 & y & & \\ & 1 & & \\ & & 1 & \\ & & -y & 1 \end{pmatrix}g)dy$$

and maps from $Ind_{B_2(F_v)}^{Sp(2,F_v)}(|t_1|^{-1}|t_2|^{-s})$ to $Ind_{B_2(F_v)}^{Sp(2,F_v)}(|t_1|^{-s}|t_2|^{-1})$. Further it is easy to check that the intertwining operator $\mathcal{M}_{y,v}^*(s)$ is holomorphic for $Re(s) > 0$ and takes the unique normalized spherical section in $Ind_{B_2(F_v)}^{Sp(2,F_v)}(|t_1|^{-1}|t_2|^{-s})$ to that in $Ind_{B_2(F_v)}^{Sp(2,F_v)}(|t_1|^{-s}|t_2|^{-1})$. The reason we introduce the operator $\mathcal{M}_{y,v}^*(s)$ is that we have the following identity for $Re(s) > 0$:

$$\mathcal{U}^{2,*}_{w_2,v}(s) \equiv \mathcal{M}^*_{y,v}(s) \circ \mathcal{M}^*_{z,v}(s) \circ \mathcal{U}^{2,*}_{w_1,v}(s). \tag{101}$$

This identity can be verified by straightforward computations of the integrals which define those intertwining operators and the relevant normalizing factors. Note that those intertwining operators $\mathcal{M}^*_{y,v}(s)$, $\mathcal{M}^*_{z,v}(s)$, and $\mathcal{U}^{2,*}_{w_1,v}(s)$ are well-defined, nonzero, and holomorphic for $Re(s) > 0$. The above decomposition is valid for $Re(s) > 0$.

For $s = 1$, $\mathcal{M}^*_{y,v}(s)$ becomes an endomorphism of $Ind^{Sp(2,F_v)}_{B_2(F_v)}(|t_1|^{-1}|t_2|^{-1})$. By means of induction on stages, we have

$$Ind^{Sp(2,F_v)}_{B_2(F_v)}(|t_1|^{-1}|t_2|^{-1}) = Ind^{Sp(2,F_v)}_{P_2(F_v)}(|\det a(p)|^{-1} \otimes Ind^{GL(2,F_v)}_{B}(|\frac{t_1}{t_2}|^0))$$

and the intertwining operator $\mathcal{M}^*_{y,v}(1)$ can be interpreted as the canonical induction of the normalized standard intertwining operator from $Ind^{GL(2,F_v)}_{B}(|\frac{t_1}{t_2}|^0)$ to itself, which is the identity map. In other words, (97) at $s = 1$ has following refined form:

$$\mathcal{U}^{2,*}_{w_2,v}(1) \equiv \mathcal{M}^*_{z,v}(1) \circ \mathcal{U}^{2,*}_{w_1,v}(1). \tag{102}$$

Next we return to prove the global version, identity (96). Without loss of generality, we assume that the section f_s is eulerian, i.e. $f_s = \otimes_v f_{s,v} \in I^2_1(s)$. Then there exists a finite set S of places of the number field F, so that $f_s = (\otimes_{v \in S} f_{s,v}) \otimes (\otimes_{v \notin S} f^0_{s,v})$, where $f^0_{s,v}$ is the unique normalized spherical section in $I^2_{1,v}(s)$. Applying the intertwining operators to f_s, one has, for $g = (g_v) \in Sp(2, \mathbb{A})$,

$$\mathcal{U}^{2,*}_{w_2}(s)(f_s)(g) = [\prod_{v \in S} \mathcal{U}^{2,*}_{w_2,v}(s)(f_{s,v})(g_v)][\prod_{v \notin S} \mathcal{U}^{2,*}_{w_2,v}(s)(f^0_{s,v})(g_v)]$$

and

$$\begin{aligned}
\mathcal{M}^*_z(s) \circ \mathcal{U}^{2,*}_{w_1}(s)(f_s)(g) &= [\prod_{v \in S} \mathcal{M}^*_{z,v}(s) \circ \mathcal{U}^{2,*}_{w_1,v}(s)(f_{s,v})(g_v)] \\
&= [\prod_{v \notin S} \mathcal{M}^*_{z,v}(s) \circ \mathcal{U}^{2,*}_{w_1,v}(s)(f^0_{s,v})(g_v)].
\end{aligned}$$

Note that the products above are actually ones with finite number of local factors since $\mathcal{U}^{2,*}_{w_2,v}(s)(f^0_{s,v})$ and $\mathcal{M}^*_{z,v}(s) \circ \mathcal{U}^{2,*}_{w_1,v}(s)(f^0_{s,v})$ are the unique normalized spherical section in the respective induced space and for almost every local place v, g_v belongs to the maximal open compact subgroup $Sp(2, \mathcal{O}_v)$. Thus each of those products is holomorphic in s for $Re(s) > 0$ and the limit $\lim_{s \to 1}$ can be exchanged with those products $\prod$. Therefore we obtain that, for $g = (g_v) \in Sp(2, \mathbb{A})$,

$$
\lim_{s \to 1} \mathcal{M}_z^*(s) \circ \mathcal{U}_{w_1}^{2,*}(s)(f_s)(g)
$$

$$
= [\prod_{v \in S} \mathcal{M}_{z,v}^*(1) \circ \mathcal{U}_{w_1,v}^{2,*}(1)(f_{1,v})(g_v)][\prod_{v \notin S} \mathcal{M}_{z,v}^*(1) \circ \mathcal{U}_{w_1,v}^{2,*}(1)(f_{1,v}^0)(g_v)]
$$

$$
= [\prod_{v \in S} \mathcal{U}_{w_2,v}^{2,*}(1)(f_{1,v})(g_v)][\prod_{v \notin S} \mathcal{U}_{w_2,v}^{2,*}(1)(f_{1,v}^0)(g_v)]
$$

$$
= \lim_{s \to 1} \mathcal{U}_{w_2}^{2,*}(s)(f_s)(g).
$$

Since the subspace of factorizable sections is dense in $I_1^2(s)$, identity (96) follows. This proves the Theorem. $\square$

Finally, we come to deal with the case $n = 3$, which is crucial for us to determine the order of the poles for the Eisenstein series $E_{n-1}^n(g, s; f_s)$.

THEOREM 3.0.4. *For $Re(s) \geq 0$ and any holomorphic section f_s in $I_2^3(s)$, the (unnormalized) Eisenstein series $E_2^3(g, s; f_s)$ is holomorphic except for $s = \frac{1}{2}, \frac{3}{2}$, and $\frac{5}{2}$, and at $s = \frac{1}{2}, \frac{3}{2}$, and $\frac{5}{2}$, the Eisenstein series $E_2^3(g, s; f_s)$ achieves a simple pole.*

Among the ingredients of our proof of this Theorem are: (1) the reducibility of degenerate principal series representations of $Sp(n)$ for $n \leq 3$, (2) the results of Kudla and Rallis about the poles of Eisenstein series of Siegel type, and (3) more techniques on intertwining operators. The proof of this theorem will be completed separately for the case of $s \neq \frac{1}{2}$ and the case of $s = \frac{1}{2}$. Those two cases will be treated in following two subsections.

3.1. Proof of Theorem 3.0.4 for the case of $s \neq \frac{1}{2}$. For any section $f_s = [\otimes_{v \notin S} f_{s,v}^\circ] \otimes [\otimes_{v \in S} f_{s,v}]$ in $I_2^3(s)$, we denote that

$$
f_{s,r} = i_2^*(f_s),
$$

$$
f_{s,1} = [\otimes_{v \notin S} f_{s,v,1}^\circ] \otimes [\otimes_{v \in S} \frac{\zeta_v(s + \frac{5}{2})}{\zeta_v(s + \frac{1}{2})} i_2^* \circ \mathcal{U}_{w_1,v}^3(s)(f_{s,v})), \tag{103}
$$

$$
f_{s,2} = [\otimes_{v \notin S} f_{s,v,2}^\circ] \otimes [\otimes_{v \in S} \frac{\zeta_v(s + \frac{5}{2})\zeta_v(2s + 1)}{\zeta_v(s - \frac{1}{2})\zeta_v(2s)} i_2^* \circ \mathcal{U}_{w_2,v}^3(s)(f_{s,v})],
$$

where $f_{s,v,1}^\circ$ and $f_{s,v,2}^\circ$ are the normalized $K_{3,v}$- spherical sections in the corresponding local representation spaces. Then the sections $f_{s,r}$, $f_{s,1}$, and $f_{s,2}$ are holomorphic for $Re(s) > 0$ and $K_{3,v}$-finite. Applying the results in §1 and §2, and the inductive formula in Theorem 1.0.2 to the case $n = 3$, the constant term $E_{2,P_1^3}^3(g, s; f_s)$ can be expressed in terms of Eisenstein series of $Sp(2)$ and the global zeta functions as follows:

$$
\begin{aligned}
E^3_{2,P_1^3}(m(t_1, g'), s; f_s) \;=\;& |t_1|^{s+\frac{5}{2}} E^2_1(g', s + \frac{1}{2}; i^*_2(f_s)) \\
& + |t_1|^2 E^2_2(g', s; f_{s,1}) \frac{\zeta(s + \frac{1}{2})}{\zeta(s + \frac{5}{2})} \\
& + |t_1|^{-s+\frac{5}{2}} E^2_1(g', s - \frac{1}{2}; f_{s,2}) \frac{\zeta(s - \frac{1}{2})\zeta(2s)}{\zeta(s + \frac{5}{2})\zeta(2s + 1)}.
\end{aligned}
\tag{104}
$$

By means of Theorem 3.0.2 and 3.0.3, we are be able to conclude that the constant term $E^3_{2,P_1^3}(m(t_1, g'), s; f_s)$ is holomorphic for $Re(s) \geq 0$ except for $s \in X_3 = \{\frac{1}{2}, \frac{3}{2}, \frac{5}{2}\}$. In fact, if we call the three terms in order term I, term II, and term III, then we see easily that for $Re(s) \geq 0$, term I only has a pole at $s = \frac{3}{2}$, term II only has a pole at $s = \frac{1}{2}$ and $\frac{3}{2}$, and term III has a pole at $s \in X_3$. Hence the constant term is holomorphic at $s \notin X_3$. In addition, it is easy to see that the constant term $E^3_{2,P_1^3}(m(t_1, g'), s; f_s)$ achieves a simple pole at $s = \frac{3}{2}, \frac{5}{2}$. At $s = \frac{3}{2}$, each of those three terms achieves a simple pole and there are no cancelations among those three terms because the 'exponents' at t_1 of three terms are different, and at $s = \frac{5}{2}$, the first two terms are holomorphic while the third term achieves a simple pole. Therefore we have completed the proof of the Theorem for the case $s \neq \frac{1}{2}$. In the next subsection, we have to prove that the constant term $E^3_{2,P_1^3}(m(t_1, g'), s; f_s)$ also achieves only a simple pole at $s = \frac{1}{2}$!

3.2. Proof of Theorem 3.0.4 for the case of $s = \frac{1}{2}$. The difficulty to determine the order of the pole at $s = \frac{1}{2}$ of the constant term is that (1) term I is holomorphic and both term II and term III have a double pole, (2) both term II and term III have the same 'exponent' at t_1, which implies that there should be some cancelations between those two terms. We will work out such cancelations in this subsection.

We need some more intertwining operators of $Sp(3)$. The analytic properties of those intertwining operators will be critical to our determination of the order of the pole at $s = \frac{1}{2}$ of the constant term $E^3_{2,P_1^3}(m(t_1, g'), s; f_s)$.

For any section f_s in $I^3_2(s)$, we will define $Sp(3, \mathbb{A})$-intertwining operators $\mathcal{M}_u(s)$ and $\mathcal{M}_d(s)$ as follows:

$$
\mathcal{M}_u(s)(f_s)(g) \;:=\; \int_{\mathbb{A}^4} f_s(w_u n_u(w, x, y, z)g)dwdxdydz
\tag{105}
$$

$$
\mathcal{M}_d(s)(f_s)(g) \;:=\; \int_{\mathbb{A}^5} f_s(w_d n_d(v, w, x, y, z)g)dvdwdxdydz
\tag{106}
$$

Those integrals converge absolutely for the real part of s large and have meromorphic continuations to the whole complex plane of s. The Weyl group elements w_u and w_d

and the corresponding unipotent subgroups are described as follows:

$$w_u = \begin{pmatrix} 0 & & 1 & & 0 & \\ & 0 & & 1 & & \\ 1 & & 0 & 0 & & \\ & & 0 & 0 & & 1 \\ & -1 & & & 0 & \\ 0 & & 1 & & 0 & \end{pmatrix} \text{ and } n_u(w,x,y,z) = \begin{pmatrix} 1 & 0 & w & 0 & x & 0 \\ & 1 & y & x & z & 0 \\ & & 1 & 0 & 0 & 0 \\ & & & 1 & & \\ 0 & & & 0 & 1 & \\ & & -w & -y & 1 \end{pmatrix};$$

and

$$w_d = \begin{pmatrix} 0 & & 1 & & 0 & \\ & 0 & 1 & & 0 & \\ & 1 & 0 & 0 & & 0 \\ -1 & & 0 & 0 & & \\ & 0 & & & 0 & 1 \\ & 0 & & 0 & 1 & 0 \end{pmatrix} \text{ and } n_d(v,w,x,y,z) = \begin{pmatrix} 1 & v & w & z & x & 0 \\ & 1 & y & x & 0 & 0 \\ & & 1 & 0 & 0 & 0 \\ & & & 1 & & \\ 0 & & -v & 1 & & \\ & & -w' & -y & 1 \end{pmatrix}.$$

Then for generic complex values of s, $\mathcal{M}_u(s)$ and $\mathcal{M}_d(s)$ are $Sp(3)$-intertwining operators in the following sense.

$$\mathcal{M}_u(s) \; : \; I_2^3(s) \to I^3(-1, -s-\frac{1}{2}, s-\frac{1}{2}) := Ind_{B(\mathbb{A})}^{Sp(3,\mathbb{A})}(|t_1|^{-1}|t_2|^{-s-\frac{1}{2}}|t_3|^{s-\frac{1}{2}}),$$

$$\mathcal{M}_d(s) \; : \; I_2^3(s) \to I^3(-s-\frac{1}{2}, -1, s-\frac{1}{2}) := Ind_{B(\mathbb{A})}^{Sp(3,\mathbb{A})}(|t_1|^{-s-\frac{1}{2}}|t_2|^{-1}|t_3|^{s-\frac{1}{2}}).$$

where the induced representations from the Borel subgroup are normalized and smooth (We always assume that the archimedean components are K_∞- finite).

By induction on stages, we can rewrite those principal series representations as follows:

$$I^3(-1, -s-\frac{1}{2}, s-\frac{1}{2})$$

$$= Ind_{P_2^3(\mathbb{A})}^{Sp(3,\mathbb{A})}([\| \det |^{-\frac{2s+3}{4}} \otimes Ind_{B_2(\mathbb{A})}^{GL(2,\mathbb{A})}(|\frac{t_1}{t_2}|^{\frac{2s-1}{4}})] \otimes I_1^1(s-\frac{1}{2})), \qquad (107)$$

$$I^3(-s-\frac{1}{2}, -1, s-\frac{1}{2})$$

$$= Ind_{P_2^3(\mathbb{A})}^{Sp(3,\mathbb{A})}([\| \det |^{-\frac{2s+3}{4}} \otimes Ind_{B_2(\mathbb{A})}^{GL(2,\mathbb{A})}(|\frac{t_1}{t_2}|^{-\frac{2s-1}{4}})] \otimes I_1^1(s-\frac{1}{2})). \qquad (108)$$

The canonical intertwining operator $\mathcal{M}_z$ from $[\| \det |^{-\frac{2s+3}{4}} \otimes Ind_{B_2(\mathbb{A})}^{GL(2,\mathbb{A})}(|\frac{t_1}{t_2}|^{\frac{2s-1}{4}})]$ to $[\| \det |^{-\frac{2s+3}{4}} \otimes Ind_{B_2(\mathbb{A})}^{GL(2,\mathbb{A})}(|\frac{t_1}{t_2}|^{-\frac{2s-1}{4}})]$ and the identity map of $I_1^1(s-\frac{1}{2})$ together will induce a $Sp(3,\mathbb{A})$-intertwining operator denoted by $\mathcal{M}_{u,d}(s)$ from $I^3(-1, -s-\frac{1}{2}, s-\frac{1}{2})$ to $I^3(-s-\frac{1}{2}, -1, s-\frac{1}{2})$. More precisely, $\mathcal{M}_{u,d}(s)$ can be expressed by following integral: for any section f_s in $Ind_{P_2^3(\mathbb{A})}^{Sp(3,\mathbb{A})}([\| \det |^{-\frac{2s+3}{4}} \otimes Ind_{B_2(\mathbb{A})}^{GL(2,\mathbb{A})}(|\frac{t_1}{t_2}|^{\frac{2s-1}{4}})] \otimes I_1^1(s-\frac{1}{2}))$,

$$\mathcal{M}_{u,d}(s)(f_s)(g) := \int_{\mathbb{A}} f_s(w_{u,d} n_{u,d}(z)g)dz, \qquad (109)$$

where the Weyl group element $w_{1,u,d}$ and the corresponding unipotent subgroup is as follows:

$$w_{u,d} = \begin{pmatrix} 0 & 1 & & & & \\ 1 & 0 & & & 0 & \\ & & 1 & & & \\ & & & 0 & 1 & \\ & 0 & & 1 & 0 & \\ & & & & & 1 \end{pmatrix} \text{ and } n_{u,d}(z) = \begin{pmatrix} 1 & z & & & & \\ & 1 & & & 0 & \\ & & 1 & & & \\ & & & 1 & & \\ & 0 & & -z & 1 & \\ & & & & & 1 \end{pmatrix}.$$

The following lemma describes the relations among those intertwining operators just defined as above.

LEMMA 3.2.1. *We have following identities of* $Sp(3, \mathbb{A})$*-intertwining operators:*

(a) *For generic complex values of* s, $\mathcal{M}_d(s) \equiv \mathcal{M}_{u,d}(s) \circ \mathcal{M}_u(s)$.
(b) *The normalized intertwining operators enjoy the same property:*

$$\mathcal{M}_d^*(s) \equiv \mathcal{M}_{u,d}^*(s) \circ \mathcal{M}_u^*(s).$$

Those operators are modified in the following way so that the normalized operators are holomorphic for $Re(s) > 0$ *and take the unique normalized spherical sections to the unique normalized spherical sections in relevant spaces of induced representations:*

$$\mathcal{M}_u^*(s) = \frac{\zeta(2s+1)\zeta(s+\frac{3}{2})\zeta(s+\frac{5}{2})}{\zeta(2s)\zeta(s+\frac{1}{2})\zeta(s+\frac{1}{2})} \mathcal{M}_u(s),$$

$$\mathcal{M}_{u,d}^*(s) = \frac{\zeta(s+\frac{1}{2})}{\zeta(s-\frac{1}{2})} \mathcal{M}_{u,d}(s),$$

$$\mathcal{M}_d^*(s) = \frac{\zeta(2s+1)\zeta(s+\frac{3}{2})\zeta(s+\frac{5}{2})}{\zeta(2s)\zeta(s+\frac{1}{2})\zeta(s-\frac{1}{2})} \mathcal{M}_d(s).$$

PROOF. The identity in (a) holds since the integral defining the intertwining operator on one side coincides with that defining the intertwining operator on the other side when they are absolutely convergent (i.e. for $Re(s)$ large). The identity in (b) follows from (a) and the computations of those normalizing factors (c-functions). Our methods to determine the holomorphy of those intertwining operators are similar to those used in the previous section and we will omit the details here. □

The reason that we introduce those intertwining operators $\mathcal{M}_u(s)$, $\mathcal{M}_{u,d}(s)$ and $\mathcal{M}_d(s)$ is made clear by the following lemma.

LEMMA 3.2.2. *Let* $Sp(1) = Sp(Fe_3 \oplus Fe_3')$, *which is canonically embedded in* $Sp(3)$ *via* $g \mapsto m(1,1,g)$, *where* $m(t_1, t_2, g)$ *in* $GL(1) \times GL(1) \times Sp(1) \subset M_2^3$, *the Levi factor of the standard maximal parabolic subgroup* P_2^3 *of* $Sp(3)$. *Then, for any* section $f_s \in I_2^3(s)$ and $Re(s) > 0$, the following identities hold:

$$\mathcal{M}_u^*(s)(f_s)(g) = i_1^* \circ \mathcal{U}^{2,*}(s) \circ i_2^* \circ \mathcal{U}_{w_1}^{3,*}(s)(f_s)(g),$$

$$\mathcal{M}_d^*(s)(f_s)(g) = i_1^* \circ \mathcal{U}_{w_1}^{2,*}(s - \frac{1}{2}) \circ i_2^* \circ \mathcal{U}_{w_2}^{3,*}(s)(f_s)(g)$$

where $\mathcal{U}^{2,*}(s)$ is the $Sp(2)$-intertwining operator from $I_2^2(s)$ to $Ind_{P_1^2}^{Sp(2,\mathbb{A})}(|t_2|^{-s-\frac{1}{2}} \otimes I_1^1(s - \frac{1}{2})$ for $Re(s) > 0$, which is defined in [**KuRa**] by the following integral:

$$\mathcal{U}^{2,*}(s) = \frac{\zeta(2s+1)\zeta(s+\frac{3}{2})}{\zeta(2s)\zeta(s+\frac{1}{2})} \int_{\mathbb{A}^2} f_s(w_2^2 n(x,y)g\,dn, \tag{110}$$

where the Weyl group element and the unipotent subgroup are

$$w_2^2 = \begin{pmatrix} 0 & & 1 & \\ & 1 & & 0 \\ -1 & & 0 & \\ & 0 & & 1 \end{pmatrix} \text{ and } n(x,y) = \begin{pmatrix} 1 & x & y & \\ & 1 & & 0 \\ & & 1 & \\ & & -x & 1 \end{pmatrix}.$$

PROOF. The proof goes exactly the same as that of Lemma 3.2.1. We omit it here. $\qquad\square$

With all those lemmas stated above, we are able to determine the order of the pole at $s = \frac{1}{2}$ of the constant term $E_{2,P_1^3}^3(m(t_1,g'),s;f_s)$. Again by the inductive formula (100) the constant term $E_{2,P_1^3}^3(m(t_1,g'),s;f_s)$ have the following expression:

$$\begin{aligned} E_{2,P_1^3}^3(m(t_1,g'),s;f_s) = & |t_1|^{s+\frac{5}{2}} E_1^2(g', s+\frac{1}{2}; i^*(f_s)) \\ & + |t_1|^2 E_2^2(g', s; f_{s,1}) \frac{\zeta(s+\frac{1}{2})}{\zeta(s+\frac{5}{2})} \\ & + |t_1|^{-s+\frac{5}{2}} E_1^2(g', s-\frac{1}{2}; f_{s,2}) \frac{\zeta(s-\frac{1}{2})\zeta(2s)}{\zeta(s+\frac{5}{2})\zeta(2s+1)}. \end{aligned}$$

Note that the first term $|t_1|^{s+\frac{5}{2}} E_1^2(g', s+\frac{1}{2}; i^*(f_s))$ has at most a simple pole at $s = \frac{3}{2}$ by Theorem 3.0.3, while both the second and the third terms have the same 'exponent' for t_1, the center $GL(1)$ of the Levi factor M_1^3, and achieve a double pole at $s = \frac{1}{2}$ since the value representation of the Eisenstein series $E_1^2(g, s; f_s)$ as $s = 0$ does not vanish according to Kudla and Rallis' argument for the 'first term' of regularized Siegel-Weil formula in [**KuRa1**].

In order to make the cancelation of the poles in both terms, we will consider the P_1^2-constant term of

$$|t_1|^2 E_2^2(g', s; f_{s,1}) \frac{\zeta(s+\frac{1}{2})}{\zeta(s+\frac{5}{2})} + |t_1|^{-s+\frac{5}{2}} E_1^2(g', s-\frac{1}{2}; f_{s,2}) \frac{\zeta(s-\frac{1}{2})\zeta(2s)}{\zeta(s+\frac{5}{2})\zeta(2s+1)}. \tag{111}$$

Since we concern the leading term in the Laurent expansion at $s = \frac{1}{2}$ of the above expansion as a function in s, the factor $|t_1|^*$ can be omitted. Then the P_1^2-constant term evaluated at $g' = m_1(t_2, g'')$ of (107) equals

$$
\frac{\zeta(s+\frac{1}{2})}{\zeta(s+\frac{5}{2})}[|t_2|^{s+\frac{3}{2}} E_1^1(g'', s + \frac{1}{2}; f_{s,1,0}) + |t_2|^{\frac{3}{2}-s} E_1^1(g'', s - \frac{1}{2}; i_1^* \circ \mathcal{U}^2(s)(f_{s,1}))]
$$

$$
+ \frac{\zeta(s-\frac{1}{2})\zeta(2s)}{\zeta(s+\frac{5}{2})\zeta(2s+1)}[|t_2|^{s+\frac{3}{2}} f_{s,2}(m(1,g'')) + |t_2|^{\frac{5}{2}-s} \mathcal{U}_{w_2}^2(s-\frac{1}{2})(f_{s,2})(m(1,g''))]
$$

$$
+ \frac{\zeta(s-\frac{1}{2})\zeta(2s)}{\zeta(s+\frac{5}{2})\zeta(2s+1)} |t_2| E_1^1(g'', s - \frac{1}{2}; i_1^* \circ \mathcal{U}_{w_1}^2(s-\frac{1}{2})(f_{s,2})).
$$

It suffices to prove the following two **statements**:

(a) The sum of two terms with the exponent 1 for t_2 is

$$
E_1(s) := \frac{\zeta(s+\frac{1}{2})}{\zeta(s+\frac{5}{2})} E_1^1(g'', s - \frac{1}{2}; i_1^* \circ \mathcal{U}^2(s)(f_{s,1}))
$$

$$
+ \frac{\zeta(s-\frac{1}{2})\zeta(2s)}{\zeta(s+\frac{5}{2})\zeta(2s+1)} E_1^1(g'', s - \frac{1}{2}; i_1^* \circ \mathcal{U}_{w_1}^2(s-\frac{1}{2})(f_{s,2})),
$$

which has at most a simple pole at $s = \frac{1}{2}$.

(b) The sum of three terms with the exponent 2 for t_2 is

$$
E_2(s) := \frac{\zeta(s+\frac{1}{2})}{\zeta(s+\frac{5}{2})} E_1^1(g'', s + \frac{1}{2}; f_{s,1,0})
$$

$$
+ \frac{\zeta(s-\frac{1}{2})\zeta(2s)}{\zeta(s+\frac{5}{2})\zeta(2s+1)} (f_{s,2}(m(1,g'')) + \mathcal{U}_{w_2}^2(s-\frac{1}{2})(f_{s,2})(m(1,g''))),
$$

which has an at most simple pole at $s = \frac{1}{2}$.

The proof of Statement (a): According to Kudla and Rallis [**KuRa**], for any holomorphic section $f_{s,1}$ in $I_2^2(s)$ with $Re(s) > 0$, one has

$$
i_1^* \circ \mathcal{U}^2(s)(f_{s,1}) = \frac{\zeta(2s)\zeta(s+\frac{1}{2})}{\zeta(2s+1)\zeta(s+\frac{3}{2})} \cdot f_{s,1,2'} \tag{112}
$$

where the section $f_{s,1,2'}$ is defined by

$$
f_{s,1,2'} = (\otimes_{v \notin S} f_{s,1,v,2'}^{\circ}) \otimes (\otimes_{v \in S} \frac{\zeta_v(2s+1)\zeta_v(s+\frac{3}{2})}{\zeta_v(2s)\zeta_v(s+\frac{1}{2})} i_1^* \circ \mathcal{U}_v^2(s)(f_{s,1,v})), \tag{113}
$$

which is a holomorphic and K-finite section in $I_1^1(s-\frac{1}{2})$. By means of (99), we have

$$
i_1^* \circ \mathcal{U}_{w_1}^2(s-\frac{1}{2})(f_{s,2}) = \frac{\zeta(s+\frac{1}{2})}{\zeta(s+\frac{3}{2})} \cdot f_{s,2,1}. \tag{114}
$$

Let us set $\Gamma_1(s) := \frac{\zeta(2s)\zeta(s+\frac{1}{2})}{\zeta(s+\frac{5}{2})\zeta(s+\frac{3}{2})\zeta(2s+1)}$. Note that the function $\Gamma_1(s)$ has a double pole at $s = \frac{1}{2}$. Substituting those data into $E_1(s)$, we deduce that

$$E_1(s) = \Gamma_1(s)[\zeta(s+\frac{1}{2})E_1^1(g'', s - \frac{1}{2}; f_{s,1,2'}) + \zeta(s-\frac{1}{2})E_1^1(g'', s - \frac{1}{2}; f_{s,2,1})]. \tag{115}$$

Following Corollary 3.0.1, $E_1^1(g, s; f_s)$ vanishes at $s = 0$. In order to prove that $E_1(s)$ has at most a simple pole at $s = \frac{1}{2}$, we need more information about those two Eisenstein series $E_1^1(g'', s - \frac{1}{2}; f_{s,1,2'})$ and $E_1^1(g'', s - \frac{1}{2}; f_{s,2,1})$.

Actually, $f_{s,1,2'} = \mathcal{M}_u^*(s)(f_s)$ and $f_{s,2,1} = \mathcal{M}_d^*(s)(f_s)$. They are holomorphic and is in $I^3(-1, -s-\frac{1}{2}, s-\frac{1}{2})$ and $I^3(-s-\frac{1}{2}, -1, s-\frac{1}{2})$, respectively, see (103) and (104). Moreover, one has $\mathcal{M}_{u,d}^*(f_{s,1,2'}) = f_{s,2,1}$ by Lemma 3.2.1 and 3.2.2. Since those two Eisenstein series live on the factor $Sp(1)$ of the Levi part M_2^3 of the maximal parabolic subgroup P_2^3 and the intertwining operator $\mathcal{M}_{u,d}^*(s)$ is essentially $\mathcal{M}_z^*(s)$ on the factor $GL(2)$ of the Levi part M_2^3 of P_2^3, we consider the restriction of those two sections $f_{s,1,2'}$ and $f_{s,2,1}$ to the Levi part $M_2^3 = GL(2) \times Sp(1)$. According to our general assumption that those induced modules at the real archimedean place are Harish-Chandra modules (i.e. K-finite), we can separate the variables of the restriction of those two sections. In other words, for $(g_1, g_2) \in GL(2) \times Sp(1)$, we have

$$f_{s,1,2'}((g_1, g_2)) = \sum_i \phi_{s,i} \otimes \phi'_{s,i}((g_1, g_2)) \tag{116}$$

and

$$f_{s,2,1}((g_1, g_2)) = \sum_i \mathcal{M}_z^*(s)(\phi_{s,i}) \otimes \phi'_{s,i}((g_1, g_2)) \tag{117}$$

where the summation is finite and $\phi_{s,i}$ is in $[|\det|^{-\frac{2s+3}{4}} \otimes Ind_{B_2(\mathbb{A})}^{GL(2,\mathbb{A})}(|\frac{t_1}{t_2}|^{\frac{2s-1}{4}})]$ and $\mathcal{M}_z^*(s)(\phi_{s,i})$ is in $[|\det|^{-\frac{2s+3}{4}} \otimes Ind_{B_2(\mathbb{A})}^{GL(2,\mathbb{A})}(|\frac{t_1}{t_2}|^{-\frac{2s-1}{4}})]$, and $\phi'_{s,i}$ is in $I_1^1(s-\frac{1}{2})$.

Without loss of generality, we may assume that $f_{s,1,2'}((g_1, g_2)) = \phi_s \otimes \phi'_s((g_1, g_2))$ and both ϕ_s and ϕ'_s are holomorphic in s. Then our Eisenstein series can be written more specifically as follows:

$$E_1^1(g'', s - \frac{1}{2}; f_{s,1,2'}) = \phi_s(1) \cdot E_1^1(g'', s - \frac{1}{2}; \phi'_s) \tag{118}$$

$$E_1^1(g'', s - \frac{1}{2}; f_{s,2,1}) = \mathcal{M}_z^*(\phi_s)(1) \cdot E_1^1(g'', s - \frac{1}{2}; \phi'_s). \tag{119}$$

Plugging the above into $E_1(s)$, we obtain the following:

$$E_1(s) = \Gamma_1(s)E_1^1(g'', s - \frac{1}{2}; \phi'_s)[\zeta(s+\frac{1}{2})\phi_s(1) + \zeta(s-\frac{1}{2})\mathcal{M}_z^*(s)(\phi_s)(1)]. \tag{120}$$

It is easy to see that $E_1(s)$ has at most a simple pole at $s = \frac{1}{2}$ since $\Gamma_1(s)E_1^1(g'', s - \frac{1}{2}; \phi'_s)$ has at most a simple pole at $s = \frac{1}{2}$ and $[\zeta(s+\frac{1}{2})\phi_s(1) + \zeta(s-\frac{1}{2})\mathcal{M}_z^*(s)(\phi_s)(1)]$

can be proved to be holomorphic at $s = \frac{1}{2}$. The last claim is equivalent to the identity:
$\phi_{\frac{1}{2}}(g_1) = \mathcal{M}_z^*(\frac{1}{2})(\phi_{\frac{1}{2}})(g_1)$. Note that $Res_{s=\frac{1}{2}}\zeta(s + \frac{1}{2}) = -Res_{s=\frac{1}{2}}\zeta(s - \frac{1}{2})$.

In fact, at $s = \frac{1}{2}$, both $\phi_{\frac{1}{2}}$ and $\mathcal{M}_z^*(\frac{1}{2})(\phi_{\frac{1}{2}})$ are in $[|\det|^{-1} \otimes Ind_{B_2(\mathbb{A})}^{GL(2,\mathbb{A})}(|\frac{t_1}{t_2}|^0)]$. At each local place v, $[|\det|^{-1} \otimes Ind_{B_2}^{GL(2,F_v)}(|\frac{t_1}{t_2}|^0)]$ is irreducible and the normalized intertwining operator $\mathcal{M}_z^*(\frac{1}{2})$ is the identity map of $[|\det|^{-1} \otimes Ind_{B_2}^{GL(2,F_v)}(|\frac{t_1}{t_2}|^0)]$. Thus we have $\phi_{\frac{1}{2},v}(g_{1,v}) = \mathcal{M}_{z,v}^*(\frac{1}{2})(\phi_{\frac{1}{2},v})(g_{1,v})$. Without loss of generality, we may assume that $\phi_{\frac{1}{2}}$ is factorizable, i.e. $\phi_{\frac{1}{2}} = [\otimes_{v \in S}\phi_{\frac{1}{2},v}] \otimes [\otimes_{v \notin S}\phi_{\frac{1}{2},v}^0]$. Then

$$
\begin{aligned}
\mathcal{M}_z^*(\frac{1}{2})(\phi_{\frac{1}{2}})(g_1) &= \lim_{s \to \frac{1}{2}}[\prod_{v \in S} \mathcal{M}_{z,v}^*(s)(\phi_{s,v})(g_{1,v})][\prod_{v \notin S} \mathcal{M}_{z,v}^*(s)(\phi_{s,v}^0)(g_{1,v})] \\
&= [\prod_{v \in S} \mathcal{M}_{z,v}^*(\frac{1}{2})(\phi_{\frac{1}{2},v})(g_{1,v})][\prod_{v \notin S} \mathcal{M}_{z,v}^*(\frac{1}{2})(\phi_{\frac{1}{2},v}^0)(g_{1,v})] \\
&= [\prod_{v \in S} \phi_{\frac{1}{2},v}(g_{1,v})][\prod_{v \notin S} \phi_{\frac{1}{2},v}^0(g_{1,v})] \\
&= \phi_{\frac{1}{2}}(g_1).
\end{aligned}
$$

This proves the global identity: $\phi_{\frac{1}{2}}(g_1) = \mathcal{M}_z^*(\frac{1}{2})(\phi_{\frac{1}{2}})(g_1)$, and therefore proves that $E_1(s)$ has at most a simple pole at $s = \frac{1}{2}$, i.e. Statement (a).

The proof of Statement (b): Now we have to prove that $E_2(s)$, which is

$$
\begin{aligned}
E_2(s) = &\frac{\zeta(s + \frac{1}{2})}{\zeta(s + \frac{5}{2})} E_1^1(g'', s + \frac{1}{2}; f_{s,1,0}) \\
&+ \frac{\zeta(s - \frac{1}{2})\zeta(2s)}{\zeta(s + \frac{5}{2})\zeta(2s + 1)}[f_{s,2}(m(1, g'')) + \mathcal{U}_{w_2}^2(s - \frac{1}{2})(f_{s,2})(m(1, g''))],
\end{aligned}
$$

has at most a simple pole at $s = \frac{1}{2}$. To this end, we need a lemma, which will be proved in §5 of this chapter.

LEMMA 3.2.3. *Let v be any local place of the totally real number field F.*

(a) The normalized $Sp(2, F_v)$-intertwining operator $i_2^* \circ \mathcal{U}_{w_2,v}^{3,*}(\frac{1}{2})$ from $I_{2,v}^3(\frac{1}{2})$ to $I_{1,v}^2(0)$ is not surjective.

(b) Let $\mathcal{M}_1 := \mathcal{M}_{w_0,v}^*(1) \circ \mathcal{U}_{w_1,v}^{3,*}(\frac{1}{2})$ and $\mathcal{M}_2 := \mathcal{U}_{w_2,v}^{3,*}(\frac{1}{2})$. Then $\mathcal{M}_1 \equiv \mathcal{M}_2$ as $Sp(3, F_v)$-intertwining operators from $I_{2,v}^3(\frac{1}{2})$ to $Ind_{P_1^3}^{Sp(3,F_v)}(|t_1|^{-1} \otimes I_1^2(0))$, where the intertwining operator $\mathcal{M}_{w_0,v}^*(s + \frac{1}{2})$ is defined by following integral

$$
\mathcal{M}_{w_0}^*(s + \frac{1}{2})(f_{s,1})(g) = \frac{\zeta(s + \frac{3}{2})}{\zeta(s + \frac{1}{2})} \int_{\mathbb{A}} f_s\left(\begin{pmatrix} I_2 & 0 & 0 & 0 \\ 0 & 0 & 0 & 1 \\ 0 & 0 & I_2 & 0 \\ 0 & -1 & 0 & 0 \end{pmatrix} \begin{pmatrix} I_2 & 0 & 0 & 0 \\ 0 & 1 & 0 & x \\ 0 & 0 & I_2 & 0 \\ 0 & 0 & 0 & 1 \end{pmatrix} g\right) dx,
$$

for sections $f_s \in Ind_{P_1^3}^{Sp(3,F_v)}(|t_1|^{-1} \otimes I_{2,v}^2(\frac{1}{2}))$.

From this Lemma, we have a global results about sections f_s.

PROPOSITION 3.2.1. *For any section f_s in $I_2^3(s)$ holomorphic at $s = \frac{1}{2}$, we have*

$$\lim_{s \to \frac{1}{2}} f_{s,2} = \lim_{s \to \frac{1}{2}} \mathcal{U}_{w_2}^2(s)(f_{s,2}) = \lim_{s \to \frac{1}{2}} \mathcal{M}_{w_0}^*(s + \frac{1}{2})(f_{s,1}) \tag{121}$$

as sections in $Ind_{P_1^3}^{Sp(3,\mathbb{A})}(|t_1|^{-1} \otimes I_1^2(0))$.

PROOF. First we shall prove the identity: $\lim_{s \to \frac{1}{2}} f_{s,2} = \lim_{s \to \frac{1}{2}} \mathcal{U}_{w_2}^2(s)(f_{s,2})$. The idea to prove this global identity is that we prove the local version of this identity for each local place v first and then the global identity will follow from the standard local-global argument.

According to [**Jan**] and [**Cop**], for each local place v, the local $Sp(2, F_v)$-module $I_{1,v}^2(0)$ is a direct sum of two irreducible submodules denoted by V_0 and V_1, that is, $I_{1,v}^2(0) = V_0 \oplus V_1$. We assume that the submodule V_0 is generated by $K_{2,v}$-spherical functions in $I_1^2(0)$. By the Lemma above, $i_2^* \circ \mathcal{U}_{w_2,v}^{3,*}(\frac{1}{2})$ in not a surjective $Sp(2)$-intertwining operator from $I_{2,v}^3(\frac{1}{2})$ to $I_1^2(0)$ and takes the normalized $K_{3,v}$-spherical section to the normalized $K_{2,v}$-spherical section. This implies that

$$i_2^* \circ \mathcal{U}_{w_2,v}^{3,*}(\frac{1}{2})(I_{2,v}^3(\frac{1}{2})) = V_0. \tag{122}$$

In other words, for any section $f_{s,v}$ in $I_{2,v}^3(s)$ which is holomorphic at $s = \frac{1}{2}$, we always have $f_{s,2,v} \in V_0$.

By the standard normalization, $\mathcal{U}_{w_2,v}^{2,*}(s - \frac{1}{2}) = \frac{\zeta_v(s + \frac{3}{2})}{\zeta_v(s - \frac{3}{2})} \mathcal{U}_{w_2,v}^2(s - \frac{1}{2})$ for each local place v, and $\mathcal{U}_{w_2}^{2,*}(s - \frac{1}{2}) = \frac{\zeta(s + \frac{3}{2})}{\zeta(s - \frac{3}{2})} \mathcal{U}_{w_2}^2(s - \frac{1}{2})$ globally. At $s = \frac{1}{2}$, the normalized intertwining operator $\mathcal{U}_{w_2,v}^{2,*}(0)$ is the identity map when restricted to submodule V_0. In other words, we have, for each local place v,

$$\mathcal{U}_{w_2,v}^{2,*}(0)(f_{\frac{1}{2},2,v}) = f_{\frac{1}{2},2,v} \tag{123}$$

Note that this is the local version of the identity we are going to prove.

For the global identity, without loss of generality, we may assume that the section $f_{s,2}$ is factorizable, i.e., $f_{s,2} = [\otimes_{v \in S} f_{s,2,v}] \otimes [\otimes_{v \notin S} f_{s,2,v}^0]$. Then we have

$$\mathcal{U}_{w_2}^{2,*}(s - \frac{1}{2})(f_{s,2})(g) = [\prod_{v \in S} \mathcal{U}_{w_2,v}^{2,*}(s - \frac{1}{2})(f_{s,2,v})(g_v)][\prod_{v \notin S} \mathcal{U}_{w_2,v}^{2,*}(s - \frac{1}{2})(f_{s,2,v}^0)(g_v)].$$

Note that the second product is the one of finitely many terms since g_v is in $Sp(3, \mathcal{O}_v)$ for almost all v and $\mathcal{U}_{w_2,v}^{2,*}(s - \frac{1}{2})(f_{s,2,v}^0)$ is the unique normalized spherical section in

$Ind_{P_1^3}^{Sp(3,F_v)}(|t_1|^{-1} \otimes I_{1,v}^2(0))$. It follows by taking the limit and by Lemma 3.2.3 that

$$\lim_{s \to \frac{1}{2}}[\frac{\zeta(s+\frac{3}{2})}{\zeta(s-\frac{3}{2})}\mathcal{U}_{w_2}^2(s-\frac{1}{2})(f_{s,2})(g)]$$

$$= [\prod_{v \in S}\mathcal{U}_{w_2,v}^{2,*}(0)(f_{\frac{1}{2},2,v})(g_v)][\prod_{v \notin S}\mathcal{U}_{w_2,v}^{2,*}(0)(f_{\frac{1}{2},2,v}^0)(g_v)]$$

$$= [\prod_{v \in S}f_{\frac{1}{2},2,v}(g_v)][\prod_{v \notin S}f_{\frac{1}{2},2,v}^0(g_v)]$$

$$= f_{\frac{1}{2},2}(g).$$

Since $\lim_{s \to \frac{1}{2}}\frac{\zeta(s+\frac{3}{2})}{\zeta(s-\frac{3}{2})} = 1$, we finally obtain the global identity: $\mathcal{U}_{w_2}^2(0)(f_{\frac{1}{2},2}) = f_{\frac{1}{2},2}$.

For the identity: $\mathcal{M}_{w_0}^*(1)(f_{\frac{1}{2},1}) = f_{\frac{1}{2},2}$, we notice that $\mathcal{M}_{w_0,v}^*(1)(f_{\frac{1}{2},1,v}) \equiv \mathcal{M}_1(f_{\frac{1}{2},v})$ and $f_{\frac{1}{2},2,v} \equiv \mathcal{M}_2(f_{\frac{1}{2},v})$. By Lemma 3.2.3, we have the local version of the identity: $\mathcal{M}_{w_0,v}^*(1)(f_{\frac{1}{2},1,v}) = f_{\frac{1}{2},2,v}$. By the same local-global argument as above, the global identity follows. The Proposition is proved. $\quad\square$

We consider the Laurent expansion of $E_2(s)$ at $s = \frac{1}{2}$ as a function in s:

$$E_2(s) = \frac{B_2}{s - \frac{1}{2}} + \frac{B_1}{s - \frac{1}{2}} + \cdots . \tag{124}$$

Then the leading coefficient B_2 can be easily expressed as

$$\frac{Res_{s=\frac{1}{2}}[\zeta(s+\frac{1}{2})]}{\zeta(3)}Res_{s=\frac{1}{2}}[E_1^1(g'',s+\frac{1}{2};f_{s,1,0})]$$

$$+ \frac{Res_{s=\frac{1}{2}}[\zeta(s-\frac{1}{2})]Res_{s=\frac{1}{2}}[\zeta(2s)]}{\zeta(3)\zeta(2)}\lim_{s \to \frac{1}{2}}[f_{s,2} + \mathcal{U}_{w_2}^2(s-\frac{1}{2})(f_{s,2})](m(1,g'')).$$

Since the restriction to $Sp(1)$ of $f_{s,2}$ and $\mathcal{U}_{w_2}^2(s-\frac{1}{2})(f_{s,2})$ are constant, one has $f_{s,2}(m(1,g'')) = f_{s,2}(1)$ and $\mathcal{U}_{w_2}^2(s-\frac{1}{2})(f_{s,2})(m(1,g'')) = \mathcal{U}_{w_2}^2(s-\frac{1}{2})(f_{s,2})(1)$. Note that $Res_{s=\frac{1}{2}}[\zeta(2s)] = \frac{1}{2}Res_{s=1}[\zeta(s)]$ and $Res_{s=\frac{1}{2}}[\zeta(s+\frac{1}{2})] = Res_{s=1}[\zeta(s)]$. Thus, by the identity in Proposition 3.2.1, we have that

$$B_2 = \frac{Res_{s=1}(\zeta(s))}{\zeta(3)}Res_{s=\frac{1}{2}}[E_1^1(g'',s+\frac{1}{2};f_{s,1,r}) + \frac{\zeta(s-\frac{1}{2})}{\zeta(2)}f_{\frac{1}{2},2}(1)].$$

It suffices to prove that

$$Res_{s=\frac{1}{2}}[E_1^1(g'',s+\frac{1}{2};f_{s,1,r}) + \frac{\zeta(s-\frac{1}{2})}{\zeta(2)}f_{\frac{1}{2},2}(1)] = 0. \tag{125}$$

Taking again the constant term along the Borel B_1 of $Sp(1)$, we have that

$$E_{1,B_1}^1(g'',s+\frac{1}{2};f_{s,1,r}) = f_{s,1,r}(g'') + \frac{\zeta(s+\frac{1}{2})}{\zeta(s+\frac{3}{2})}\mathcal{M}_{w_0}^*(s+\frac{1}{2})(f_{s,1,r}))(g'')$$

where $\mathcal{M}_{w_0}^*(s + \frac{1}{2})$ is defined as in Lemma 3.2.3. Note that $\mathcal{M}_{w_0}^*(s + \frac{1}{2})(f_{s,1,r})(g'')$ is constant. It is reduced to prove that

$$Res_{s=\frac{1}{2}}[\zeta(s + \frac{1}{2})\mathcal{M}_{w_0}^*(s + \frac{1}{2})(f_{s,1,r})(1) + \zeta(s - \frac{1}{2})f_{s,2}(1)]$$

is zero. In other words, we have to prove that $\mathcal{M}_{w_0}^*(1)(f_{\frac{1}{2},1}) = f_{\frac{1}{2},2}$. But this is exactly the identity we have proved in Proposition 3.2.1. We are done. Theorem 3.0.4 is completely proved.

4. Poles of Eisenstein Series

It comes to determine the poles of Eisenstein series $E_{n-1}^n(g, s; f_s)$ for general n. We recall that as in Proposition 1.0.2, i_{n-1}^* is the restriction operator from $I_{n-1}^n(s)$ to $I_{n-2}^{n-1}(s + \frac{1}{2})$, and for generic complex values of s, $i_{n-1}^* \circ \mathcal{U}_{w_1}^n(s)$ is the operator from $I_{n-1}^n(s)$ to $I_{n-1}^{n-1}(s)$, $i_{n-1}^* \circ \mathcal{U}_{w_2}^n(s)$ is the one from $I_{n-1}^n(s)$ to $I_{n-2}^{n-1}(s - \frac{1}{2})$, and the intertwining operator $\mathcal{U}_{w_0}^n(s)$ maps from $I_{n-1}^n(s)$ to $I_{n-1}^n(-s)$. According to Theorem 2.4.1, the modified intertwining operators $\frac{1}{c_{w_1}^n(s)} \cdot i_{n-1}^* \circ \mathcal{U}_{w_1}^n(s)$, $\frac{1}{c_{w_2}^n(s)} \cdot i_{n-1}^* \circ \mathcal{U}_{w_2}^n(s)$, and $\frac{1}{c_{w_0}^n(s)} \cdot \mathcal{U}_{w_0}^n(s)$ are holomorphic for $Re(s) > 0$, with $c_{w_1}^n(s)$, $c_{w_2}^n(s)$, and $c_{w_0}^n(s)$ as in §2.4. Let $f_s = [\otimes_{v \in S} f_{s,v}] \otimes [\otimes_{v \notin S} f_{s,v}^\circ]$ be any holomorphic section in $I_{n-1}^n(s)$. we set that

$$
\begin{aligned}
f_{s,\circ} &:= \frac{1}{c_{w_\circ}^n(s)} \cdot \mathcal{U}_{w_\circ}^n(s)(f_s) \\
&= [\otimes_{v \in S} \frac{1}{c_{w_\circ,v}^n(s)} \cdot \mathcal{U}_{w_\circ,v}^n(s)(f_{s,v})] \otimes [\otimes_{v \notin S} f_{s,v,\circ}^\circ]; \\
f_{s,1} &:= \frac{1}{c_{w_1}^n(s)} \cdot i_{n-1}^* \circ \mathcal{U}_{w_1}^n(s)(f_s) \\
&= [\otimes_{v \in S} \frac{1}{c_{w_1,v}^n(s)} \cdot \mathcal{U}_{w_1,v}^n(s)(f_{s,v})] \otimes [\otimes_{v \notin S} f_{s,v,1}^\circ]; \\
f_{s,2} &:= \frac{1}{c_{w_2}^n(s)} \cdot i_{n-1}^* \circ \mathcal{U}_{w_2}^n(s)(f_s) \\
&= [\otimes_{v \in S} \frac{1}{c_{w_2,v}^n(s)} \cdot \mathcal{U}_{w_2,v}^n(s)(f_{s,v})] \otimes [\otimes_{v \notin S} f_{s,v,2}^\circ].
\end{aligned}
\tag{126}
$$

Then the sections $f_{s,\circ}$, $f_{s,1}$, and $f_{s,2}$ are holomorphic for $Re(s) > 0$ and K_n-finite in the corresponding image spaces of $I_{n-1}^n(s)$ under the intertwining operators $\mathcal{U}_{w_\circ}^n(s)$, $\mathcal{U}_{w_1}^n(s)$, and $\mathcal{U}_{w_2}^n(s)$, respectively.

Now we can determine the poles of our Eisenstein series $E_{n-1}^n(g, s; f_s)$ for $Re(s) > 0$ and $n \geq 4$. First we need following Lemma, which will be proved in the next section.

Let $\mathcal{U}_{w_2}^{n,*}(s) := \frac{1}{c_{w_2}^n(s)}\mathcal{U}_{w_2}^n(s)$ be the normalized intertwining operator associated to $\mathcal{U}_{w_2}^n(s)$. Then $\mathcal{U}_{w_2}^{n,*}(s)$ is holomorphic for $Re(s) > -\frac{n}{2} + 1$, which is greater than or equals -1 when $n \geq 4$. Hence, for any holomorphic section f_s in $I^n_{n-1}(s)$, $f_{s,2} = i^*_{n-1} \circ \mathcal{U}^{n,*}_{w_2}(s)(f_s)$ is holomorphic for $Re(s) > -1$.

LEMMA 4.0.4. *For any even integer $n \geq 4$, the Eisenstein series $E^{n-1}_{n-2}(g, s - \frac{1}{2}; f_s)$ vanishes at $s = \frac{1}{2}$ for any section $f_s = f_{s,2}$ in the image in $I^{n-1}_{n-2}(0)$ of $I^n_{n-1}(\frac{1}{2})$ under the map $i^*_{n-1} \circ \mathcal{U}^{n,*}_{w_2}(\frac{1}{2})$.*

Based on the results about the location and the order of the poles of the Eisenstein series $E^n_{n-1}(g, s; f_s)$ for $n \leq 3$ and the above Lemma, the location and the order of possible poles of the Eisenstein series $E^n_{n-1}(g, s; f_s)$ for general n can be determined as follows.

THEOREM 4.0.1. *Assume that $n \geq 4$. For any holomorphic section f_s in $I^n_{n-1}(s)$, the (unnormalized) Eisenstein series $E^n_{n-1}(g, s; f_s)$ enjoys the following properties:*

(a) *$E^n_{n-1}(g, s; f_s)$ is holomorphic for $Re(s) > 0$ except for*

$$s \in X^+_n := \{\frac{e(n)}{2}, \cdots, \frac{n-2}{2}, \frac{n}{2}, \frac{n+2}{2}\},$$

where $e(n) = 1$ if n is odd and $e(n) = 2$ if n is even.

(b) *$E^n_{n-1}(g, s; f_s)$ achieves a simple pole at $s = \frac{n+2}{2}$, $\frac{n}{2}$ and a double pole at $s = \frac{n-2}{2}$, and $E^n_{n-1}(g, s; f_s)$ achieves at most a double pole at $s \in \{\frac{e(n)}{2}, \cdots, \frac{n-4}{2}\}$.*

PROOF. First we prove the case $n = 4$. According to the constant term principal in the general theory of Eisenstein series, the analytic properties of $E^4_3(g, s; f_s)$ can be determined by its constant term $E^4_{3,P^4_1}(g, s; f_s)$ along the unipotent radical N^4_1 of the maximal parabolic subgroup P^4_1. By means of Theorem 1.0.2 and (122) above for the case $n = 4$, we can express the constant term $E^4_{3,P^4_1}(g, s; f_s)$ in terms of Eisenstein series of $Sp(3)$ and the c-functions $c^3_{w_1}(s)$ and $c^3_{w_2}(s)$ as follows:

$$
\begin{aligned}
E^4_{3,P^4_1}(m_1(t_1, g'), s; f_s) &= |t_1|^{s+3} E^3_2(g', s + \frac{1}{2}; i^*_3(f_s)) \\
&+ |t_1|^3 E^3_3(g', s; f_{s,1}) \frac{\zeta(s)}{\zeta(s+3)} \\
&+ |t_1|^{-s+3} E^3_2(g', s - \frac{1}{2}; f_{s,2}) \frac{\zeta(2s)\zeta(s)}{\zeta(2s+2)\zeta(s+3)}.
\end{aligned}
$$

According to the Theorem of Kudla and Rallis in [**KuRa**] the (unnormalized) Eisenstein series $E^3_3(g, s; f_{s,1})$ is holomorphic for $Re(s) > 0$ except for $s = 1, 2$, and at $s = 1, 2$, $E^3_3(g, s; f_{s,1})$ achieves a simple pole (by taking the normalized spherical section). On the other hand, by means of Theorem 3.0.4, we conclude that the (unnormalized) Eisenstein series $E^3_2(g', s + \frac{1}{2}; i^*_3(f_s))$ is holomorphic for $Re(s) > 0$ except for $s = 1, 2$, and $E^3_2(g', s + \frac{1}{2}; i^*_3(f_s))$ achieves a simple pole at $s = 1, 2$ (by taking the normalized spherical section); and the (unnormalized) Eisenstein series

$E_2^3(g', s - \frac{1}{2}; f_{s,2})$ is holomorphic for $Re(s) > 0$ with exception of $s = 1, 2, 3$, where $E_2^3(g', s - \frac{1}{2}; f_{s,2})$ achieves a simple pole by the same reason. According to the above Lemma, the third term $|t_1|^{-s+3} E_2^3(g', s - \frac{1}{2}; f_{s,2}) \frac{\zeta(2s)\zeta(s)}{\zeta(2s+2)\zeta(s+3)}$ is holomorphic at $s = \frac{1}{2}$. Hence the constant term $E_{3,P_1^4}^4(m_1(t_1, g'), s; f_s)$ is holomorphic for $Re(s) > 0$ except for $s = 1, 2, 3$.

The existence of the poles of $E_{3,P_1^4}^4(m_1(t_1, g'), s; f_s)$ at $s = 1, 2, 3$ can be determined in the following way: We number the three terms in the inductive formula by their order and call them term I, term II, and term III. Since for $s = 1, 2, 3$, those three terms have different 'exponents' at the center $GL(1) \times I_6$ of M_1^4, there are no cancelations among those three terms. Thus the poles of $E_{3,P_1^4}^4(m_1(t_1, g'), s; f_s)$ is determined by each term. Note that if f_s is the normalized K_4-spherical section in $I_3^4(s)$, then all three sections $i_3^*(f_s)$, $f_{s,1}$, and $f_{s,2}$ are simultaneously the normalized K_3-spherical sections in the corresponding degenerate principal series representations $I_2^3(s + \frac{1}{2})$, $I_3^3(s)$, and $I_2^3(s - \frac{1}{2})$, respectively. At $s = 3$, only term III achieves a simple pole, so does $E_{3,P_1^4}^4(m_1(t_1, g'), s; f_s)$. At $s = 2$, all three terms achieve a simple pole, so does $E_{3,P_1^4}^4(m_1(t_1, g'), s; f_s)$ since no cancelations will happen. Finally, at $s = 1$, term I achieves a simple pole, and both terms II and III achieve a double pole. We conclude that $E_{3,P_1^4}^4(m_1(t_1, g'), s; f_s)$ achieves a double pole at $s = 1$ just because there are no cancelations between term II and term III. This proves the theorem for the case $n = 4$.

Next we will prove the theorem for the case of general n. Inductively, we assume that the theorem hold for the case of $n - 1$. The constant term of Eisenstein series $E_{n-1}^n(g, s; f_s)$ along the maximal parabolic subgroup P_1^n can be expressed, by Theorem 1.0.2 and 4.3.17, as follows:

$$E_{n-1,P_1^n}^n(m_1(t_1, g'), s; f_s)$$

$$= |t_1|^{s + \frac{n+2}{2}} E_{n-2}^{n-1}\left(g', s + \frac{1}{2}; i_3^*(f_s)\right) + |t_1|^{n-1} E_{n-1}^{n-1}(g', s; f_{s,1}) \frac{\zeta\left(s - \frac{n-4}{2}\right)}{\zeta\left(s + \frac{n+2}{2}\right)}$$

$$+ |t_1|^{-s + \frac{n+2}{2}} E_{n-2}^{n-1}\left(g', s - \frac{1}{2}; f_{s,2}\right) \frac{\zeta(2s)\zeta\left(s + \frac{n-4}{2}\right)}{\zeta(2s + n - 2)\zeta\left(s + \frac{n+2}{2}\right)}.$$

By the assumption of induction, we have

(a) $E_{n-2}^{n-1}(g', s + \frac{1}{2}; i_3^*(f_s))$ achieves a simple pole at $s = \frac{n}{2}$ and $s = \frac{n-2}{2}$, a double pole at $s = \frac{n-4}{2}$, and has at most double pole at $s = \frac{n-6}{2}, \frac{n-8}{2}, \cdots, \frac{e(n-1)-1}{2}$.

(b) $E_{n-2}^{n-1}(g', s - \frac{1}{2}; f_{s,2})$ achieves a simple pole at $s = \frac{n+2}{2}$ and $s = \frac{n}{2}$, a double pole at $s = \frac{n-2}{2}$, and has at most double pole at $s = \frac{n-4}{2}, \frac{n-6}{2}, \cdots, \frac{e(n-1)+1}{2}$. Since $\zeta(2s)$ has a simple pole at $s = 0, \frac{1}{2}$ and $\zeta(s + \frac{n-4}{2})$ has a simple pole at $s = -\frac{n-4}{2}, -\frac{n-6}{2}$, and $\frac{e(n-1)+1}{2} \geq 1$, it is easy to see that the term $E_{n-2}^{n-1}(g', s - \frac{1}{2}; f_{s,2}) \frac{\zeta(2s)\zeta(s + \frac{n-4}{2})}{\zeta(2s+n-2)\zeta(s + \frac{n+2}{2})}$ achieves a simple pole at $s = \frac{n+2}{2}$ and $s = \frac{n}{2}$, a double pole at $s = \frac{n-2}{2}$, and has at most double pole at $s = \frac{n-4}{2}, \frac{n-6}{2}, \cdots, \frac{e(n-1)+1}{2}$. Note that when n is even, $\frac{1}{2}$ is not in the set X_n^+. However, the Lemma above implies the third term in the above inductive formula is holomorphic at $s = \frac{1}{2}$ and so is the constant term $E_{n-1,P_1^n}^n(m_1(t_1, g'), s; f_s)$.

On the other hand, by Theorem 3.0.2 or [**KuRa**], the term $E_{n-1}^{n-1}(g', s; f_{s,1}) \frac{\zeta(s - \frac{n-4}{2})}{\zeta(s + \frac{n+2}{2})}$ has a simple pole at $s = \frac{n}{2}$ and at $s = \frac{n-6}{2}, \frac{n-8}{2}, \cdots, \frac{e'(n-1)}{2}$ with $e'(n) = 1$ if n is even and $e'(n) = 2$ if n odd, and has a double pole at $s = \frac{n-2}{2}, \frac{n-4}{2}$. Therefore, by comparing the 'exponents' of $|t_1|$ in each term, we obtain that, at $s = \frac{n+2}{2}$ and $s = \frac{n}{2}$, the constant term $E_{n-1,P_1^n}^n(m_1(t_1, g'), s; f_s)$ achieves a simple pole and, at $s = \frac{n-2}{2}$, $E_{n-1,P_1^n}^n(m_1(t_1, g'), s; f_s)$ achieves a double pole since there are no cancelations among those three terms. At the value of s other than those three: $\frac{n+2}{2}, \frac{n}{2}$ and $\frac{n-2}{2}$, the constant term $E_{n-1,P_1^n}^n(m_1(t_1, g'), s; f_s)$ achieves at most a double pole. This proves the theorem. $\qquad\square$

Now we can prove our main Theorem of this Chapter. For a given section f_s, there is a finite set S of places which including all archimedean ones, the normalizing factor $d_{n-1}^{n,S}(s)$ is defined to be

$$d_{n-1}^{n,S}(s) = \prod_{v \notin S}[\zeta_v(s + \frac{n}{2} + 1)\zeta_v(s + \frac{n}{2})\zeta_v(s + \frac{n}{2} - 1) \prod_{j=1, j \equiv n(2)}^{n-2} \zeta_v(2s + j)]$$

and the normalized Eisenstein series $E_{n-1}^{n,*}(g, s; f_s)$ is defined by

$$E_{n-1}^{n,*}(g, s; f_s) = d_{n-1}^{n,S}(s)E_{n-1}^n(g, s; f_s).$$

THEOREM 4.0.2 (Main). *Assume that F is a totally real number field and $n \geq 3$. For any holomorphic section $f_s \in I_{n-1}^n(s)$, the normalized Eisenstein series enjoys the following properties:*

(a) *The set of possible poles (of order at most two) of the normalized Eisenstein series $E_{n-1}^{n,*}(g, s; f_s)$ is*

$$\{-\frac{n+2}{2}, -\frac{n}{2}, -\frac{n-2}{2}, \cdots, \hat{0}, \cdots, \frac{n-2}{2}, \frac{n}{2}, \frac{n+2}{2}\};$$

(b) *The (normalized) Eisenstein series achieves a simple pole at $s = \frac{n+2}{2}, \frac{n}{2}$ and a double pole at $s = \frac{n-2}{2}$.*

PROOF. We assume first that $n \geq 4$. Since the normalizing factor $d_{n-1}^{n,S}(s)$ is holomorphic for $Re(s) > 0$ ($n \geq 4$), the normalization does not change the holomorphy of the Eisenstein series $E_{n-1}^n(g, s; f_s)$. Thus the analytic properties of $E_{n-1}^{n,*}(g, s; f_s)$ for $Re(s) > 0$ follow from that of $E_{n-1}^n(g, s; f_s)$ as stated in Theorem 4.0.1.

Next we consider the situation for $Re(s) < 0$. By the general theory of Eisenstein series [**Lan2**], [**Art**], and [**MoWa**] one has a functional equation as follows:

$$E_{n-1}^n(g, s; f_s) = E_{n-1}^n(g, -s; \mathcal{U}_{w_o}^n(s)(f_s)).$$

By Theorem 2.4.1, we have, for $Re(s) < 0$,

$$
\begin{aligned}
E_{n-1}^{m,*}(g, s; f_s) &= d_{n-1}^{m,S}(s) E_{n-1}^n(g, s; f_s) \\
&= d_{n-1}^{m,S}(s) E_{n-1}^n(g, -s; \mathcal{U}_{w_o}^n(s)(f_s)) \\
&= \frac{d_{n-1}^{m,S}(s) c_{w_o}^n(s)}{d_{n-1}^{m,S}(-s)} E_{n-1}^{n,*}(g, -s; f_{s,o}).
\end{aligned}
$$

Since $f_{s,o}$ in K_n-finite and holomorphic as a section in $I_{n-1}^n(-s)$, it follows from the first part of the proof that $E_{n-1}^{n,*}(g, -s; f_{s,o})$ has at most double poles at $s \in \{-\frac{n+2}{2}, -\frac{n}{2}, -\frac{n-2}{2}, \cdots, -\frac{e(n)}{2}\}$. It is easy to check that the factor $\frac{d_{n-1}^{m,S}(s) c_{w_o}^n(s)}{d_{n-1}^{m,S}(-s)}$ is holomorphic for $Re(s) < 0$. In fact, one has, from §2.4

$$
\begin{aligned}
&\frac{d_{n-1}^{m,S}(s) c_{w_o}^n(s)}{d_{n-1}^{m,S}(-s)} \\
&= \prod_{v \in S} \frac{\zeta_v(-s + \frac{n}{2} + 1)\zeta_v(-s + \frac{n}{2})\zeta_v(-s + \frac{n}{2} - 1)\prod_{j=1, j \equiv n(2)}^{n-2} \zeta_v(-2s + j)}{\zeta_v(s + \frac{n}{2} + 1)\zeta_v(s + \frac{n}{2})\zeta_v(s + \frac{n}{2} - 1)\prod_{j=1, j \equiv n(2)}^{n-2} \zeta_v(2s + j)}
\end{aligned}
$$

because $c_{w_o}^n(s) = \frac{d_{n-1}^n(-s)}{d_{n-1}^n(s)}$. Therefore $\frac{d_{n-1}^{m,S}(s) c_{w_o}^n(s)}{d_{n-1}^{m,S}(-s)} E_{n-1}^{n,*}(g, -s; f_{s,o})$ has at most double poles at $s \in \{-\frac{n+2}{2}, -\frac{n}{2}, -\frac{n-2}{2}, \cdots, -\frac{e(n)}{2}\}$, and so does $E_{n-1}^{n,*}(g, s; f_s)$. This proves the theorem for $n \geq 4$.

Finally, for $n = 3$, the normalizing factor

$$
d_2^{3,S}(s) = \prod_{v \notin S}[\zeta_v(s + \frac{5}{2})\zeta_v(s + \frac{3}{2})\zeta_v(s + \frac{1}{2})\zeta_v(2s + 1)]
$$

is holomorphic for $Re(s) > 0$ except for $s = \frac{1}{2}$. According to §3, the Eisenstein series $E_2^3(g, s; f_s)$ is holomorphic for $Re(s) >$ except for $s = \frac{1}{2}, \frac{3}{2}$, and $\frac{5}{2}$. Thus the normalized Eisenstein series $E_2^{3,*}(g, s; f_s)$ has simple poles at $s = \frac{3}{2}$ and $s = \frac{5}{2}$, and has a double pole at $s = \frac{1}{2}$! The poles of $E_2^{3,*}(g, s; f_s)$ for $Re(s) < 0$ can be determined by the same argument as above. The theorem is proved. $\square$

The exact order of the poles at $s \in \{\frac{e(n)}{2}, \cdots, \frac{n-4}{2}\}$ will be determined in our later work.

5. Proof of Two Lemmas

This section is devoted to the proof of Lemma 3.2.3 and Lemma 4.0.1, which are very important to our determination of the poles of the Eisenstein series. The technical part of the proofs of those Lemmas are essentially the estimates of the dimensions of certain spaces of intertwining operators between degenerate principal series representations of $Sp(n)$, which follows from Bruhat theory on the estimate of the dimension of certain space of quasi-invariant distributions on $Sp(n)$ for both archimedean and non-archimedean cases.

let F_v be the local field of the totally real number field F corresponding to the local place v. Denote $G_n := Sp(n, F_v)$. As usual, denote by $C_c^\infty(G_n)$ the space of all smooth functions over G_n with compact support. Note that the space $C_c^\infty(G_n)$ has a canonically defined topology, called Schwartz topology [**War**]. Let E be any vector space over the complex number field $\mathbb{C}$. A distribution of $C_c^\infty(G_n)$ with value E is by definition a continuous linear functional of $C_c^\infty(G_n)$ with value in E. The space of all E-valued distributions is denoted by $\mathcal{L}(C_c^\infty(G_n); E)$. Following Lemma 1.0.1, one has

$$Sp(n, F_v) = [P_{n-1}^n P_1^n] \cup [P_{n-1}^n w_1 P_1^n] \cup [P_{n-1}^n w_2 P_1^n] = \mathcal{O}_0 \cup \mathcal{O}_1 \cup \mathcal{O}_2. \tag{127}$$

Let $\Omega_2 := \mathcal{O}_2$, $\Omega_1 := \Omega_2 \cup \mathcal{O}_1$, and $\Omega_0 := G_n$. They are open subset of G_n and $\mathcal{O}_i$ is a closed boundary orbit of Ω_i for $i = 0, 1, 2$. Let $P_{(i)}^n := P_{n-1}^n \cap [w_i^{-1} P_1^n w_i]$.

The following are the spaces of certain $P_{n-1}^n \times P_1^n$-quasi-invariant distributions, which we are going to study. Let (σ, E_σ) be the representation of G_{n-1} on the space $I_{n-2}^{n-1}(0)$. Then we define: for $(p_1, p_2) \in P_{n-1}^n \times P_1^n$,

$$\mathcal{U}^n(p_1, p_2) := |a(p_1)t_1(p_2)|^{-\frac{n+1}{2}} \sigma(g_{n-1}(p_2^{-1})). \tag{128}$$

This can be viewed as a representation of $P_{n-1}^n \times P_1^n$ on the space $E_\sigma := \mathbb{C} \otimes I_{n-2}^{n-1}(0)$. Then we define some spaces of E_σ-valued, $P_{n-1}^n \times P_1^n$-quasi-invariant distributions:

$$\mathbf{T}^n := \{T \in \mathcal{L}(C_c^\infty(G_n); E_\sigma) : (p_1, p_2) \circ T = U(p_1, p_2) \cdot T\}. \tag{129}$$

$$\mathbf{T}^n(\mathcal{O}_i) := \{T \in \mathcal{L}(C_c^\infty(\Omega_i); E_\sigma) : (p_1, p_2) \circ T = U(p_1, p_2) \cdot T\}. \tag{130}$$

By Bruhat theory as in Chapter 5 of [**War**] for archimedean cases and Chapter 2 of [**Sil**] for nonarchimedean cases, one has following inequality:

$$\dim \mathbf{T}^n \le \dim \mathbf{T}^n(\mathcal{O}_0) + \dim \mathbf{T}^n(\mathcal{O}_1) + \dim \mathbf{T}^n(\mathcal{O}_2). \tag{131}$$

Further, by the 'Fundamental Estimate' of Bruhat theory, one has following estimate:

$$\dim \mathbf{T}^n(\mathcal{O}_i) \le \sum_{m \ge 0} i^n(\mathcal{O}_i; m), \tag{132}$$

where $i^n(\mathcal{O}_i; m)$ is the dimension of the space of all $P_{(i)}^n$-intertwining maps from the representation

$$\Phi_{(i)}(p) := |a(p)|^{\frac{1}{2}} \cdot |t_1(w_i p w_i^{-1})|^{\frac{n-1}{2}} \sigma(g_{n-1}(w_i p w_i^{-1})) \tag{133}$$

to the representation

$$A_{(i)}^m(p) := [\delta_{P_{n-1}^n}(p)\delta_{P_1^n}(w_i p w_i^{-1})]^{-\frac{1}{2}} \delta_{(i)}(p)\Lambda_m(p), \tag{134}$$

where $\delta_{...}$ is the usual modular character of the group $\cdots$ as in previous Chapters. More precisely, we have

$$P_{(0)}^n = \begin{pmatrix} a & * & * & * & & * & * \\ & B & * & * & & * & \\ & & \alpha & * & & * & \beta \\ & & & a^{-1} & & & \\ & & & * & & {}^tB^{-1} & \\ & \gamma & & * & & * & \delta \end{pmatrix}$$

and for $p \in P_{(0)}^n$,

$$\begin{aligned} \Phi_{(0)}(p) &= |a|^{\frac{n}{2}} |\det B|^{\frac{1}{2}} \sigma(g_{n-1}(p)) \\ A_{(0)}^m(p) &= |a|^{\frac{n-2}{2}} |\det B|^{\frac{n}{2}} \Lambda_m(p); \end{aligned} \tag{135}$$

$$P_{(1)}^n = \begin{pmatrix} A & 0 & * & & * \\ 0 & b & * & & * \\ & & {}^tA^{-1} & & 0 \\ & & 0 & & b^{-1} \end{pmatrix}$$

and for $p \in P_{(1)}^n$,

$$\begin{aligned} \Phi_{(1)}(p) &= |b|^{\frac{n-1}{2}} |\det A|^{\frac{1}{2}} \sigma(g_{n-1}(w_1 p w_1^{-1})) \\ A_{(1)}^m(p) &= |b|^{1} |\det A|^{\frac{n}{2}} \Lambda_m(p); \end{aligned} \tag{136}$$

and

$$P_{(2)}^n = \begin{pmatrix} a & 0 & 0 & 0 & & 0 & 0 \\ * & B & * & 0 & & * & * \\ 0 & 0 & \alpha & 0 & & * & \beta \\ & & & a^{-1} & & * & 0 \\ & & & 0 & & {}^tB^{-1} & 0 \\ & \gamma & & 0 & & * & \delta \end{pmatrix}$$

and for $p \in P_{(2)}^n$,

$$\begin{aligned} \Phi_{(2)}(p) &= |a|^{-\frac{n-2}{2}} |\det B|^{\frac{1}{2}} \sigma(g_{n-1}(w_2 p w_2^{-1})) \\ A_{(2)}(p) &= |a|^{-\frac{n-2}{2}} |\det B|^{\frac{n+2}{2}}. \end{aligned} \tag{137}$$

Note that in the case of nonarchimedean field, all of the distributions have transversal order zero. In other words, one have in this case:

$$\dim \mathbf{T}^n(\mathcal{O}_i) \leq i^n(\mathcal{O}_i; 0). \tag{138}$$

To compute the dimension $i^n(\mathcal{O}_i; m)$ for $m \geq 0$ and $i = 0, 1, 2$, one need certain information about the representation Λ_m of $P_{(i)}^n$, which is by definition the m-th symmetric power of Λ_1, and Λ_1 is the representation of $P_{(i)}^n$ induced from the adjoint representation on the quotient space

$$\mathcal{V}_{(i)} := \frac{\mathfrak{sp}(n)}{\mathfrak{p}_{n-1}^n + w_i^{-1}\mathfrak{p}_1^n w_i}, \tag{139}$$

where $\mathfrak{g}$ denotes the complexification of the real Lie algebra of the corresponding real group G.

Let $M_{(i)}^n$ be the reductive part of the subgroup $P_{(0)}^n$. Then we have

$$\begin{aligned}
M_{(0)}^n &= GL(1) \times GL(n-2) \times Sp(1), \\
M_{(1)}^n &= GL(n-1) \times GL(1), \\
M_{(2)}^n &= GL(1) \times GL(n-2) \times Sp(1).
\end{aligned} \tag{140}$$

By restricted to $M_{(i)}^n$, the representation $\mathcal{V}_{(i)}$ for $i = 0, 1$ can be described as follows:

$$\begin{aligned}
\mathcal{V}_{(0)} &= \mathbb{C} \oplus \mathbb{C}^{n-2} \oplus \mathbb{C}^2, \\
\mathcal{V}_{(1)} &= \mathbb{C}^{n-1}.
\end{aligned} \tag{141}$$

More precisely, for $i = 0$, we let (a, B, s) be an element in $M_{(0)}^n$ and (w, z, x) a vector in $\mathbb{C} \oplus \mathbb{C}^{n-2} \oplus \mathbb{C}^2$, then, as an $M_{(0)}^n$-module, the action of $M_{(0)}^n$ on $\mathbb{C} \oplus \mathbb{C}^{n-2} \oplus \mathbb{C}^2$ is

$$(a, B, s) \circ (w, z, x) = (a^{-2}w, a^{-1}zB^{-1}, sxa^{-1}). \tag{142}$$

For $i = 1$, we let (A, b) be an element in $GL(n-1) \times GL(1)$ and y a vector in $\mathbb{C}^{n-1}$, then we have

$$(A, b) \circ y = {}^tA^{-1}yb^{-1}. \tag{143}$$

Then, as $M_{(i)}^n$-modules, the m-th symmetric power of $\mathcal{V}_{(i)}$ for $i = 0, 1$ can be described as

$$\begin{aligned}
\Lambda_m(\mathcal{V}_{(0)}) &= \oplus_{p+q+r=m} \Lambda_p(\mathbb{C}) \otimes \Lambda_q(\mathbb{C}^{n-2}) \otimes \Lambda_r(\mathbb{C}^2), \\
\Lambda_m(\mathcal{V}_{(1)}) &= b^{-m}\Lambda_m(\mathbb{C}^{n-1,*}).
\end{aligned} \tag{144}$$

It follows then that $i^n(\mathcal{O}_0; m) \neq 0$ implies the following equation $-\frac{n}{2} = -\frac{n-2}{2} - 2p - q - r$. But this is impossible. We thus deduce that $i^n(\mathcal{O}_0; m) = 0$ for all $m \geq 0$. Similarly, $i^n(\mathcal{O}_1; m) \neq 0$ implies that $\frac{n-1}{2} = 1 - m$, which has no solution for $n \geq 4$ and $m \geq 0$. But for $n = 3$, it has a solution $m = 0$. In other words, we have

$$i^n(\mathcal{O}_1; m) = \begin{cases} 0, & \text{if } n \geq 4, m \geq 0, \\ 1, & \text{if } n = 3, m = 0. \end{cases} \tag{145}$$

Finally, it is sure that $i^n(\mathcal{O}_2; m) = 0$ for $m \geq 1$ and $i^n(\mathcal{O}_2; 0) = 1$ since $\mathcal{O}_2$ is the open orbit. Therefore we obtain the following Lemma.

LEMMA 5.0.5.   (a) *For any place v of F and $n \geq 4$, the space*

$$Hom_{G_n}(I_{n-1}^n(\tfrac{1}{2}), Ind_{P_1^n}^{G_n}(|t_1|^{-\frac{n-1}{2}} \otimes I_{n-2}^{n-1}(0)))$$

is of one dimension.

(b) *For any place v, the subspace $\mathbf{T}_{0,1}^3$ of $\mathbf{T}^3$ consisting of distributions in $\mathbf{T}^3$ with support off the open orbit $\mathcal{O}_2$ is of at most one dimension.*

PROOF. Part (b) follows from the inequality (141). For part (a), we need to figure out the correspondence between intertwining operators and distributions. In fact we have following isomorphism as vector space:

$$Hom_{G_n}(I_{n-1}^n(\tfrac{1}{2}), Ind_{P_1^n}^{G_n}(|t_1|^{-\frac{n-1}{2}} \otimes I_{n-2}^{n-1}(0))) \quad \rightarrow \quad \mathbf{T}^n \qquad (146)$$
$$\mathcal{M} \quad \mapsto \quad T_\mathcal{M}$$

where $T_\mathcal{M} := (\mathcal{M} \circ pr)(\varphi)(e)$ for $\varphi \in C_c^\infty(G_n)$, which is a distribution in $\mathcal{L}(C_c^\infty(G_n); E_\sigma)$. Here pr is the canonical projection from $C_c^\infty(G_n)$ onto the induced module $I_{n-1}^n(s)$ given by

$$pr(\varphi)(g) := \int_{P_{n-1}^n} \varphi(pg)|a(p)|^{-\frac{n+3}{2}} dp, \qquad (147)$$

with the right Haar measure dp on P_{n-1}^n. The quasi-invariance for $P_{n-1}^n \times P_1^n$ of the distribution $T_\mathcal{M}$ can be checked as follows: for any $(p_1, p_2) \in P_{n-1}^n \times P_1^n$ and $\varphi \in C_c^\infty(G_n)$,

$$
\begin{aligned}
(p_1, p_2) \circ T_\mathcal{M}(\varphi) &= T_\mathcal{M}((p_1, p_2)^{-1} \circ \varphi) \\
&= \mathcal{M}(\int_{P_{n-1}^n} (p_1, p_2)^{-1} \circ \varphi(p)|a(p)|^{-\frac{n+3}{2}} dp)(e) \\
&= \mathcal{M}(\int_{P_{n-1}^n} \varphi(p_1 p p_2^{-1})|a(p)|^{-\frac{n+3}{2}} dp)(e) \\
&= |a(p_1)|^{-\frac{n+1}{2}} \mathcal{M}(p_2^{-1} \circ pr(\varphi))(e) \\
&= |a(p_1)|^{-\frac{n+1}{2}} \mathcal{M}(pr(\varphi))(p_2^{-1}) \\
&= |a(p_1)t_1(p_2)|^{-\frac{n+1}{2}} \sigma(g_{n-1}(p_2^{-1})) \mathcal{M}(pr(\varphi))(e) \\
&= |a(p_1)t_1(p_2)|^{-\frac{n+1}{2}} \sigma(g_{n-1}(p_2^{-1})) T_\mathcal{M}(\varphi).
\end{aligned}
$$

By Chapter 5 of [**War**], both vector spaces have the same dimension, and because the map, $\mathcal{M} \mapsto T_\mathcal{M}$, is injective, we conclude that the map, $\mathcal{M} \mapsto T_\mathcal{M}$, gives the isomorphism as required. This proves part (a).   □

It is important to mention that, in the archimedean case, we have restricted ourselves to $(\mathfrak{g}_\infty, K_{n,\infty})$-modules. Theoretically, Bruhat theory gives an estimate of the dimension of the intertwining operators from one smooth induced representation to another smooth induced representation of group G_n. According to the works of Casselman [**Cas2**] and Wallach [**Wal1**], those smooth degenerate principal series representations of G_n are of moderate growth and become the canonical extension of the corresponding $K_{n,\infty}$-finite degenerate principal series representations with respect to $(\mathfrak{g}_{n,\infty}, K_{n,\infty})$, and also any $(\mathfrak{g}_{n,\infty}, K_{n,\infty})$-intertwining map can be extended to be a unique G_n-intertwining map. Therefore part (a) of Lemma 5.0.5 really means, in the archimedean case, that there exists essentially one $(\mathfrak{g}_{n,\infty}, K_{n,\infty})$-intertwining operator from $(\mathfrak{g}_{n,\infty}, K_{n,\infty})$-module $I_{n-1}^n(\frac{1}{2})$ to $(\mathfrak{g}_{n,\infty}, K_{n,\infty})$-module $Ind_{P_1^n}^{G_n}(|t_1|^{-\frac{n-1}{2}} \otimes I_{n-2}^{n-1}(0))$.

COROLLARY 5.0.1. *For any place v and $n \geq 4$, the image $i_{n-1}^* \circ \mathcal{U}_{w_2,v}^{n,*}(\frac{1}{2})(I_{n-1,v}^n(\frac{1}{2}))$ in $I_{n-2,v}^{n-1}(0)$ of the representation $I_{n-1,v}^n(\frac{1}{2})$ under the M_1^n-intertwining operator $i_{n-1}^* \circ \mathcal{U}_{w_2,v}^{n,*}(\frac{1}{2})$ is irreducible.*

PROOF. Suppose the image $i_{n-1}^* \circ \mathcal{U}_{w_2,v}^{n,*}(\frac{1}{2})(I_{n-1,v}^n(\frac{1}{2}))$ is not irreducible, i.e., it is a reducible submodule in $I_{n-2,v}^{n-1}(0)$. Since $I_{n-2,v}^{n-1}(0)$ is completely reducible as a G_{n-1}-module, the image can be written as a direct sum of two submodules, say, $V_0 \oplus V_1$. Then one can construct two G_{n-1}-intertwining operators Λ_0 and Λ_1 with following properties:

(a) $Im(\Lambda_0) = V_0$ and $ker(\Lambda_0)$ contains V_1, and
(b) $Im(\Lambda_1) = V_1$ and $ker(\Lambda_1)$ contains V_0.

Composing the intertwining operator $i_{n-1}^* \circ \mathcal{U}_{w_2,v}^{n,*}(\frac{1}{2})$ with Λ_0 and Λ_1, respectively, we obtain two different intertwining maps $\Lambda_0 \circ i_{n-1}^* \circ \mathcal{U}_{w_2,v}^{n,*}(\frac{1}{2})$ and $\Lambda_1 \circ i_{n-1}^* \circ \mathcal{U}_{w_2,v}^{n,*}(\frac{1}{2})$ from $I_{n-1,v}^n(\frac{1}{2})$ to $|t_1|^{\frac{n+1}{2}} \otimes I_{n-2,v}^{n-1}(0)$. By Frobenius Reciprocity Law [Cas1], [Cas2], and [Car], those two intertwining maps give two G_n-intertwining maps from $I_{n-1,v}^n(\frac{1}{2})$ to $Ind_{P_1^n}^{G_n}(|t_1|^{-\frac{n-1}{2}} \otimes I_{n-2,v}^{n-1}(0))$. According to part (a) of the previous Lemma, those two different intertwining maps must be proportional to each other. But this is impossible since they have unisomorphic image V_0, V_1, respectively (one is spherical, but the other is not). Therefore the image $i_{n-1}^* \circ \mathcal{U}_{w_2,v}^{n,*}(\frac{1}{2})(I_{n-1,v}^n(\frac{1}{2}))$ is irreducible. Because the map $i_{n-1}^* \circ \mathcal{U}_{w_2,v}^{n,*}(\frac{1}{2})$ takes the normalized spherical section to the normalized spherical section, the image $i_{n-1}^* \circ \mathcal{U}_{w_2,v}^{n,*}(\frac{1}{2})(I_{n-1,v}^n(\frac{1}{2}))$ must be generated by the spherical section. This proves the Corollary.   □

Proof of Lemma 4.0.1: According to Corollary 5.0.1, we obtain the following global consequence that for any even integer $n \geq 4$, the image $i_{n-1}^* \circ \mathcal{U}_{w_2}^{n,*}(\frac{1}{2})(I_{n-1}^n(\frac{1}{2}))$ in $I_{n-2}^{n-1}(0)$ of the $I_{n-1}^n(\frac{1}{2})$ under the $Sp(n, \mathbb{A})$-intertwining map $i_{n-1}^* \circ \mathcal{U}_{w_2}^{n,*}(\frac{1}{2})$ is an irreducible submodule, which is generated by the spherical section $f_{\frac{1}{2},2}^\circ$. In order to prove that the Eisenstein series $E_{n-2}^{n-1}(g, 0; f_{\frac{1}{2},2})$ vanishes for such sections $f_{\frac{1}{2},2}$ in the image $i_{n-1}^* \circ \mathcal{U}_{w_2}^{n,*}(\frac{1}{2})(I_{n-1}^n(\frac{1}{2}))$, it suffices to prove the vanishing of $E_{n-2}^{n-1}(g, 0; f_{\frac{1}{2},2}^\circ)$ for the normalized spherical section $f_{\frac{1}{2},2}^\circ$. According to the general theory of Eisenstein series [MoWa], [Art] and [Lan2], one has following functional equation:

$$E_{n-2}^{n-1}(g, s; f_s) = c_{w_\circ}^{n-1}(s) E_{n-2}^{n-1}(g, -s; \mathcal{U}_{w_\circ}^{n,*}(s)(f_s)). \tag{148}$$

Note that the Eisenstein series $E_{n-2}^{n-1}(g, s; f_s)$ is holomorphic at $s = 0$. According to §2.4, one has

$$c_{w_o}^{n-1}(s) = \frac{\zeta(s - \frac{n-1}{2} + 2)\zeta(s - \frac{n-1}{2} + 1)\zeta(s - \frac{n-1}{2})}{\zeta(s + \frac{n-1}{2} + 1)\zeta(s + \frac{n-1}{2})\zeta(s + \frac{n-1}{2} - 1)} \prod_{j=1, j \equiv n-1(2)}^{n-3} \frac{\zeta(2s - j + 1)}{\zeta(2s + j)}.$$

At $s = 0$, we have $c_{w_o}^{n-1}(0) = (-1)^{n-1} = -1$ since n is even. Thus we get following functional equation:

$$E_{n-2}^{n-1}(g, 0; f_{\frac{1}{2}, 2}^\circ) = -E_{n-2}^{n-1}(g, 0; f_{\frac{1}{2}, 2}^\circ). \tag{149}$$

Therefore we have that $E_{n-2}^{n-1}(g, 0; f_{\frac{1}{2}, 2}^\circ) = 0$. Lemma 4.0.1 is proved.

Proof of Lemma 3.2.3:

We are going to prove part (b) of Lemma 3.2.3 first. We have constructed two intertwining operators $\mathcal{M}_2 = \mathcal{U}_{w_2, v}^{3,*}(\frac{1}{2})$ and

$$\mathcal{M}_1(f_s)(g) = \frac{\zeta(s + \frac{3}{2})\zeta(s + \frac{5}{2})}{\zeta(s + \frac{1}{2})^2} \int_{F_v^3} f_s(w_1 n_1 w_0 n_0 g) dn_1 dn_0 \tag{150}$$

where w_0 and n_0 are as in Lemma 3.2.3. As in the case of arbitrary n, those two intertwining operators $\mathcal{M}_1$ and $\mathcal{M}_2$ correspond to two distributions $T_{\mathcal{M}_1}$ and $T_{\mathcal{M}_2}$ in the space $\mathbf{T}^3$, all $P_2^3 \times P_1^3$-quasi-invariant distributions on G_3. Because the intertwining operator $\mathcal{U}_{w_2, v}^3(s)(f_s)$ has a simple pole at $s = \frac{1}{2}$ for any general holomorphic section f_s in $I_{2,v}^3(s)$, the distribution $T_{\mathcal{M}_2}$ corresponding to the normalized intertwining operator $\mathcal{U}_{w_2, v}^{3,*}(\frac{1}{2})$ must have support off the open orbit $P_2^3 w_2 P_1^3$. In other words, the distribution $T_{\mathcal{M}_2}$ belongs to the subspace $\mathbf{T}_{0,1}^3$, which is defined in part (b) of Lemma 5.0.5. Since the intertwining operator $\mathcal{M}_1$ is given by a convergent integral and its integrating variables $w_1 n_1 w_0 n_0$ lives in the 'middle' orbit $P_2^3 w_1 P_1^3$, the distribution $T_{\mathcal{M}_1}$ will supported in the closed subset $\overline{P_2^3 w_1 P_1^3} = G_3 - P_2^3 w_2 P_1^3$. Thus we also have that $T_{\mathcal{M}_1}$ belongs to $\mathbf{T}_{0,1}^3$. By means of part (b) of Lemma 5.0.5, the subspace $\mathbf{T}_{0,1}^3$ is of at most one dimension. There exists a nonzero constant c, so that $T_{\mathcal{M}_1} = c \cdot T_{\mathcal{M}_2}$. Equivalently, we have that $\mathcal{M}_1(f)(e) = c \cdot \mathcal{M}_2(f)(e)$ for f in $I_{2,v}^3(\frac{1}{2})$. Then for any $g \in G_3$, we have

$$\mathcal{M}_1(f)(g) = \mathcal{M}_1(g \circ f)(e) = c \cdot \mathcal{M}_2(g \circ f)(e) = c \cdot \mathcal{M}_2(f)(g).$$

In other words, $\mathcal{M}_1(f) = c \cdot \mathcal{M}_2(f)$. Because both $\mathcal{M}_1$ and $\mathcal{M}_2$ take the normalized spherical section f° in $I_{2,v}^3(\frac{1}{2})$ to the normalized spherical section in $Ind_{P_1^3}^{Sp(3,F_v)}(|t_1|^{-1} \otimes I_1^2(0))$, the nonzero constant c must be 1, that is, $\mathcal{M}_1 = \mathcal{M}_2$. This completes the proof of part (b) of Lemma 3.2.3.

From the above proof, the distribution $T_{\mathcal{M}_2}$ corresponding to the intertwining operator $\mathcal{M}_2 = \mathcal{U}_{w_2, v}^{3,*}(\frac{1}{2})$ belongs to the subspace $\mathbf{T}_{0,1}^3$. By the similar argument used in the proof of Corollary 5.0.1, the property that the subspace $\mathbf{T}_{0,1}^3$ is of one dimension implies the irreducibility of the image $i_2^* \circ \mathcal{U}_{w_2, v}^{3,*}(\frac{1}{2})(I_{2,v}^3(\frac{1}{2}))$. Since $I_{1,v}^2(0)$ is a direct sum of two submodules, part (a) of Lemma 3.2.3 follows. Therefore Lemma 3.2.3 is proved.

CHAPTER 4

Residue Representations of Eisenstein Series

The residue representations in our case are essentially the quotient representations of the relevant degenerate principal series representations. In this meaning, the information on the irreducible quotient representations of degenerate principal series representations are crucial in our study. Over the p-adic field, theory is now known from the recent work of C. Jantzen (for symplectic groups and odd-orthogonal groups). In the opposite, over the real archimedean case, it is far away from having satisfying results of degenerate principal series induced from arbitrary maximal parabolic subgroups. We shall determine the irreducible quotient representation of $I_{3,\infty}^4(s)$ by using Bruhat theory and an argument suggested by S. Rallis on the existence of certain quasi-invariant distributions. This is done in §2. We shall finally get two first term identities which interpret the residue representations of our Eisenstein series.

1. Eisenstein Series of $Sp(4)$

1.1. Poles of Eisenstein Series. In this first section we shall recall the results on the location and the order of the poles of some family of Eisenstein series of $Sp(4)$ associated to some maximal parabolic subgroups. More precisely, we consider three families of degenerate principal series representations of $Sp(4, \mathbb{A})$, which are $I_4^4(s)$, $I_3^4(s)$, and $I_1^4(s)$. The three families of Eisenstein series attached to these three families of degenerate principal series representations are $E_4^4(g, s; \phi_s)$, $E_3^4(g, s; f_s)$, and $E_1^4(g, s; \varphi_s)$, respectively.

For the family of Eisenstein series $E_4^4(g, s; \phi_s)$ of Siegel type, the location and the order of poles of $E_4^4(g, s; \phi_s)$ follow from the general results of Kudla and Rallis [**KuRa**]. we state them as follows:

PROPOSITION 1.1.1 (Kudla-Rallis). *For any holomorphic section $\phi_s \in I_4^4(s)$, the Eisenstein series $E_4^4(g, s; \phi_s)$ is holomorphic for $Re(s) \geq 0$ except for $s = \frac{1}{2}, \frac{3}{2}$, and $\frac{5}{2}$, where the Eisenstein series $E_4^4(g, s; \phi_s)$ achieves a pole of order one.*

For the family of Eisenstein series $E_3^4(g, s; f_s)$, the location and the order of their poles are determined in Chapter II. For convenience, we restate them here.

PROPOSITION 1.1.2 (Theorem 4.0.1). *For any holomorphic section $f_s \in I_3^4(s)$, the Eisenstein series $E_3^4(g, s; f_s)$ is holomorphic for $Re(s) \geq 0$ except for $s = 1, 2$, and 3, where the Eisenstein series $E_3^4(g, s; f_s)$ achieves a pole of order one at $s = 2, 3$ and a pole of order two at $s = 1$.*

For the family of Eisenstein series $E_1^4(g, s; \varphi_s)$, the location and the order of their poles are well known and can be described as follows.

PROPOSITION 1.1.3. *For any holomorphic section $\varphi_s \in I_1^4(s)$, the Eisenstein series $E_1^4(g, s; \varphi_s)$ is holomorphic for $\mathrm{Re}(s) \geq 0$ except for $s = 4$, where the Eisenstein series $E_1^4(g, s; \varphi_s)$ achieves a pole of order one.*

REMARK 1.1.1. *The goal here is to study the residue representations of $E_3^4(g, s; f_s)$ at $s = 1, 2$ (at $s = 3$ the residue is the trivial one). To do so, one need the specific knowledge about the irreducible quotient representations of the degenerate principal series representations $I_{r,v}^4(s)$ of $Sp(4, F_v)$ over the local field F_v.*

1.2. Degenerate Principal Series of $Sp(4, F_v)$. Let F_v be the complete local field at the place v of the totally real number field F. We consider the degenerate principal series representation $I_{r,v}^4(s)$ of $Sp(4, F_v)$, which is the v-component of the global degenerate principal series representation $I_r^4(s)$ of $Sp(4, \mathbb{A})$ as introduced in Chapter III. More precisely, $I_{r,v}^4(s)$ is the space of all smooth (K-finite if $v = \infty$) functions $f_s : G_{4v} \mapsto \mathbb{C}$ with compact support modulo P_r^4 and satisfying the condition:
$f_s(pg) = |a(p)|_r^{s + \frac{9-r}{2}} f_s(g)$. The meaning of $a(p)$ is clear, see section 3.1. Note that when $v = \infty$, $F_v = \mathbb{R}$ and $I_{r,v}^4(s)$ is only a $(\mathfrak{g}, K)$-module. The results we need are these stated in the following Theorem.

THEOREM 1.2.1. *Let F_v be the complete local field at the place v of a totally real number field F. Then, for $\mathrm{Re}(s) > 0$,*

(a) *$I_{1,v}^4(s)$ is irreducible except for $s = 4$,*

(b) *$I_{3,v}^4(s)$ has a unique (up to isomorphism) irreducible quotient representation.*

PROOF. Part (1) is known from [**Jan**] for $v < \infty$ and [**Cop**] for $v = \infty$. Part (2), when $v < \infty$, is obtain in [**Jia**] by using Hecke operator method. In his recent work [**Jan1**], Jantzen obtain the precise information on the irreducible constituents of general cases. When $v = \infty$, Part (2) will be proved in the next section of this Chapter.   □

COROLLARY 1.2.1 (Global). *$I_3^4(s) = \otimes_v' I_{3,v}^4(s)$ has a unique (up to isomorphism) irreducible quotient representation.*

PROOF. It suffices to prove that $I_3^4(-s)$ has a unique irreducible subrepresentation. Let X be an irreducible subrepresentation of $I_3^4(-s)$. Then $X = \otimes_v' X_v$. For a fixed place v, one has a map

$$\begin{array}{rcl}
\sigma_v \ : \ X_v & \to & X \\
x_v & \mapsto & x_v \otimes x^v
\end{array}$$

where $x^v = \otimes_{w \neq v}' x_w^0$ with x_w^0 the fixed vector in X_w. It is clear that σ_v is a G_{4v}-intertwining map. Another naturally map is the restriction to the v-component G_{4v}

of $G_4(\mathbb{A})$, that is,

$$\tau_v \ : \ I_3^4(-s) \ \to \ I_{3,v}^4(-s)$$
$$f_s \ \mapsto \ f_s|_{G_{4v}}.$$

The composition of σ_v and τ_v gives the natural inclusion

$$\tau_v \circ \sigma_v : X_v \to I_{3,v}^4(-s).$$

Now if $I_3^4(-s)$ has tow irreducible subrepresentations X and Y, then $X = \otimes_v' X_v$ and $Y = \otimes_v' Y_v$. For any local place v, following the argument above, X_v and Y_v are two irreducible subrepresentations in $I_{3,v}^4(-s)$. Since the unique quotient representation of $I_{3,v}^4(s)$ is spherical, the unique subrepresentation of $I_{3,v}^4(-s)$ is generated by the spherical section. Hence we conclude that $X_v = Y_v$. In other words, $X = Y$ as subrepresentations of $I_3^4(-s)$. This proves the uniqueness. $\square$

2. Intertwining Operators and Quasi-invariant Distributions

In this section, we are going to prove Part (2) of Theorem 1.2.1 for $v = \infty$. We assume that the underlying field $F_v = \mathbb{R}$.

2.1. The Uniqueness of the Irreducible Quotient Representation. In the real archimedean case, we have to consider also the smooth degenerate principal series representation $I_{3,\infty}^{4,sm}(s)$, which is the space of smooth $\mathbb{C}$-valued functions $f_s(\cdot)$ on G_4 with compact support modulo P_3^4 and satisfying following conditions: for $p \in P_3^4$ and $x \in G_4$, $f_s(px) = |a(p)|^{s+3} f_s(x)$, where $a(p)$ is the determinant of the $GL(3)$-factor of the Levi part of p. Let K_4 be the standard maximal compact subgroup of G_4, which is $U(4)$, and $I_{3,\infty}^4(s)$ be the subspace $I_3^{4,sm}(s)_{K_4}$ of $I_{3,\infty}^{4,sm}(s)$ consisting of all K_4-finite sections in $I_{3,\infty}^{4,sm}(s)$. Then $I_{3,\infty}^4(s)$ have a natural structure of $(\mathfrak{g}_4, K_4)$-module of finite length (Harish-Chandra module).

According to Casselman [**Cas**] and Wallach [**Wal**], $I_{3,\infty}^{4,sm}(s)$ is a smooth representation of G of moderate growth and the canonical extension of the Harish-Chandra module $I_{3,\infty}^4(s)$. Thus a $(\mathfrak{g}, K)$-intertwining operator from $I_{3,\infty}^4(s)$ to $I_{3,\infty}^4(-s)$ can be extended to be a unique G-intertwining operator from $I_{3,\infty}^{4,sm}(s)$ to $I_{3,\infty}^{4,sm}(-s)$. We shall prove in this special case that the space $Hom_G(I_{3,\infty}^{4,sm}(s), I_{3,\infty}^{4,sm}(-s))$ is of one dimension for any positive real number s by using Bruhat's estimate of the intertwining numbers and an argument about the extensions of some typical quasi-invariant distributions. In other words, there is essentially one $(\mathfrak{g}, K)$-intertwining operator from $I_{3,\infty}^4(s)$ to $I_{3,\infty}^4(-s)$. Based on this result, we can prove that, for any positive real number s,

(a) the Harish-Chandra module $I_{3,\infty}^4(s)$ is generated by the spherical section f_s^0 as a $(\mathfrak{g}, K)$-module and

(b) $I_{3,\infty}^4(s)$ has a unique irreducible quotient representation.

We shall first establish an archimedean version of a theorem of Waldspurger, which says that for any irreducible admissible representation π of a p-adic symplectic group, there exists an element δ in the similitude symplectic group such that the contragredient representation π^c of π is isomorphic to the twisted representation π^δ. In order to the archimedean analogue of the above result, we need the basic fact about the parameterization of Harish-Chandra modules in terms of the K-conjugacy classes of character data and the set of associated lowest K-types, from Vogan [**Vog**] and Przebinda [**Prz**].

We recall from Vogan [**Vog**] the relevant notations and results. Let θ be the Cartan involution of G which takes g to ${}^t g^{-1}$. On the Lie algebra level, one has a Cartan decomposition associated to θ: $\mathfrak{sp}(4,\mathfrak{F})_0 = \mathfrak{t}_0 + \mathfrak{p}_0$. Let $H = TA$ be a θ-stable Cartan subgroup with its compact part $T = H \cap K$ and its vector part $A = H \cap \exp(\mathfrak{p}_0)$. The the centralizer $C_G(A)$ of A in G has the Langlands decomposition MA. Let Γ be an ordinary character of H and $\gamma \in \mathfrak{h}^*$, where $\mathfrak{h}$ is the complexification of the real Lie algebra $\mathfrak{h}_0$ of the Cartan subgroup H. A character data is a triple (H,Γ,γ) defined in Definition 6.6.1 of [**Vog**]. Let us denote by $[H,\Gamma,\gamma]$ the K-conjugacy class of the character data (H,Γ,γ) if (H,Γ,γ) preserves conjugacy under K and by $Lk(\pi)$ the set of lowest K-types of π, an irreducible $(\mathfrak{g},K)$-module π. Then one has following basic results on Harish-Chandra modules.

THEOREM 2.1.1 ([**Vog**] and [**Prz**]). (a) *Each equivalence class of irreducible* $(\mathfrak{g},K)$-*module* π *is uniquely determined by a* K-*conjugacy class* $[H,\Gamma,\gamma]$ *and the set* $Lk(\pi)$ *of the lowest* K-*types of* π.

(b) *If we write* $\pi = \pi_{(\mathfrak{g},K)}[H,\Gamma,\gamma](Lk)$ *following (a), the contragredient representation* π^c *of* π *will be parameterized by* $\pi_{(\mathfrak{g},K)}[H,\Gamma^{-1},-\gamma](Lk^c)$, *i.e.*

$$(\pi_{(\mathfrak{g},K)}[H,\Gamma,\gamma](Lk))^c = \pi_{(\mathfrak{g},K)}[H,\Gamma^{-1},-\gamma](Lk^c).$$

(c) *If any automorphism* κ *of* G *stabilizes* K *and* H, *them one has*

$$\kappa(\pi_{(\mathfrak{g},K)}[H,\Gamma,\gamma](Lk)) = \pi_{(\mathfrak{g},K)}[H,\kappa(\Gamma),\kappa(\gamma)](\kappa(Lk)).$$

We also need specific information about the representatives of the conjugacy classes of Cartan subgroups of G, which was stated generally and completely in Sugiura [**Sug**].

PROPOSITION 2.1.1 ([**Sug**]). *In the real (split) symplectic group* $G = Sp(4,F)$, *the following is a list of representatives of conjugacy classes of Cartan subgroups in* G:

$$\begin{array}{ccc} A_1^4, & T_1 \times A_1^2, & T_1 \times T_1, \\ T_2 \times A_1^3, & T_2 \times T_1 \times A_1, & T_2^2 \times A_1^2, \\ T_2^2 \times T_1, & T_2^3 \times A_1, & T_2^4. \end{array}$$

where $A_1 = diag\{t, 1, 1, 1; t^{-1}, 1, 1, 1\}$ and T_1 and T_2 are matrix of form: $\begin{pmatrix} a & b \\ -b & a \end{pmatrix}$, *with $a^2 + b^2 = 1$, but T_1 and T_2 are embedded in G differently, for instance,*

$$T_2 \times T_1 \times A_1 = \begin{pmatrix} a & & & b & & \\ & T_1 & & & & \\ & & t & & & \\ -b & & & a & & \\ & & & & {}^tT_1^{-1} & \\ & & & & & t^{-1} \end{pmatrix}.$$

It is easy to see that all those standard Cartan subgroup are θ-stable and also stable under the adjoint action of elements $\delta^{\pm}$ of form: $\begin{pmatrix} & \pm I_4 \\ I_4 & \end{pmatrix}$. Note that δ^- is in K, but δ^+ is in the group of symplectic similitudes. By easy calculation of matrices, the restriction to K of the adjoint of δ^+ is equal to the complex conjugation (or the restriction to K of the Cartan involution). Thus K is also stable under the adjoint action of both $\delta^{\pm}$. Since δ^- belongs to K, the twisted representation π^{δ^-} is equivalent to π for any irreducible $(\mathfrak{g}, K)$-module π, where the twist is defined as follows: for any $g \in G$ and $v \in \pi$, $\pi^{\delta}(g)v = \pi(\delta g \delta^{-1})v$, which works for any element δ in $GL(8, F)$.

Next we are going to check that the twisted representation π^{δ^+} is equivalent to the contragredient representation π^c for any irreducible Harish-Chandra module π. Since $(\pi_{(\mathfrak{g},K)}[H, \Gamma, \gamma](Lk))^{\delta^+} = \pi_{(\mathfrak{g},K)}[H, \Gamma^{\delta^+}, \gamma^{\delta^+}](Lk^{\delta^+})$ and the restriction to K of the adjoint action of δ^+ equals the map taking both transpose and inverse, we have $Lk^{\delta^+} = Lk^c$, the set of lowest K-types of π^c. It suffices to show that $\Gamma^{\delta^+} = \Gamma^{-1}$ and $\gamma^{\delta^+} = -\gamma$. Notice that the restriction to each standard Cartan subgroup H as above of the adjoint action of δ^+ is equal to the map: $h \mapsto h^{-1}$. Hence one has $\Gamma^{\delta^+} = \Gamma^{-1}$. Similarly, on the Lie algebra level, the adjoint action of δ^+ gives the coadjoint action on $\mathfrak{h}^*$ which takes γ to $-\gamma$, i.e. $\gamma^{\delta^+} = -\gamma$. Therefore we obtain the following theorem.

THEOREM 2.1.2. *Let $\delta = \delta^+ \cdot \delta^-$. Then the twisted representation π^{δ} is equivalent to the contragredient representation π^c for any irreducible Harish-Chandra module π.*

Note that δ^- can be any element of K. Let δ the element as in the Theorem. Then it is the diagonal matrix $diag\{I_4; -I_4\}$, which is in the group of symplectic similitudes. The p-adic version of this result was obtained by Waldspurger [**MVW**]. It is easy to check that $I_{3,\infty}^{4,sm}(s)^{\delta}$ is equivalent to $I_{3,\infty}^{4,sm}(s)$ and then $I_{3,\infty}^4(s)^{\delta}$ is equivalent to $I_{3,\infty}^4(s)$. Further, we have the following Proposition, the p-adic version of which is obtain by the author in [**Jia**] by computation of certain Hecke operators associated to a parahori subgroup.

PROPOSITION 2.1.2. *Suppose that* $\dim Hom_{(\mathfrak{g},K)}(I_{3,\infty}^4(s_0), I_{3,\infty}^4(-s_0)) = 1$ *for* $s = s_0$.

(a) *The degenerate principal series representation $I_{3,\infty}^4(s_0)$ has a unique irreducible quotient representation.*

(b) *The spherical section $f_{s_0}^0$ generates $I_{3,\infty}^4(s_0)$ as a $(\mathfrak{g}, K)$-module.*

PROOF. We are going to prove part (b) first. Suppose that the spherical section $f_{s_0}^0$ does not generate $I_{3,\infty}^4(s_0)$ as a $(\mathfrak{g}_4, K_4)$-module. Then its unique irreducible quotient representation, say X, has no K-fixed vector. By the Theorem above, we have that

$$X \equiv (X^\delta)^\delta \equiv (X^c)^\delta.$$

Since $(X^c)^\delta$ is an irreducible subrepresentation of $I_{3,\infty}^4(-s_0)^\delta$ and then an irreducible subrepresentation of $I_{3,\infty}^4(-s_0)$, we can view the irreducible representation X as a subrepresentation of $I_{3,\infty}^4(-s_0)$. In this way, we obtain a nontrivial $(\mathfrak{g}_4, K_4)$-intertwining operator Λ which maps from $I_{3,\infty}^4(s_0)$ to its image X in $I_{3,\infty}^4(-s_0)$. Since we assume that the space $Hom_{(\mathfrak{g},K)}(I_{3,\infty}^4(s_0), I_{3,\infty}^4(-s_0))$ is of one dimension, it is easy to show that the normalized standard intertwining operator $\mathcal{M}_{w_0}^*$ (integral) associated to the longest Weyl group element w_0 generates the space $Hom_{(\mathfrak{g},K)}(I_{3,\infty}^4(s_0), I_{3,\infty}^4(-s_0))$ and takes the nonzero spherical section to the nonzero spherical section. Thus there exists a nonzero constant c such that $\Lambda = c \cdot \mathcal{M}_{w_0}^*$, which implies that $\Lambda(f_{s_0}^0) \neq 0$ in X. This contradicts the nonexistence of K-fixed vector in X. Part (b) is proved.

Now suppose that $I_{3,\infty}^4(s_0)$ has two irreducible quotient representations X and Y. From the proof of part (b), X and Y can be viewed as irreducible subrepresentations of $I_{3,\infty}^4(-s_0)$. So we have two intertwining operators $\mathcal{M}_X$ and $\mathcal{M}_Y$, taking from $I_{3,\infty}^4(s_0)$ to X, Y, resp., in $I_{3,\infty}^4(-s_0)$. Since $Hom_{(\mathfrak{g},K)}(I_{3,\infty}^4(s_0), I_{3,\infty}^4(-s_0))$ is of one dimension, those two intertwining operators must be proportional. This implies $X \equiv Y$. Part (a) follows. $\qquad\square$

The assumption that $\dim Hom_{(\mathfrak{g},K)}(I_{3,\infty}^4(s_0), I_{3,\infty}^4(-s_0)) = 1$ will be proved later to hold for all positive real values of s. Thus we have following unconditional results

THEOREM 2.1.3. *Let s be any positive real number.*

(a) *The degenerate principal series representation $I_{3,\infty}^4(s)$ has a unique irreducible quotient representation.*

(b) *The spherical section f_s^0 generates $I_{3,\infty}^4(s)$ as a $(\mathfrak{g}, K)$-module.*

2.2. Estimates of Intertwining Numbers. We shall estimate the dimension of the space $Hom_G(I_{3,\infty}^{4,sm}(s), I_{3,\infty}^{4,sm}(-s))$ by using Bruhat's Theory, which we recall from Chapter 5 in Warner's book [**War**]. Note that the induced representation $I_{3,\infty}^{4,sm}(s)$ considered here is equivalent to that in Warner's Book.

The distributions we are going to discuss are $\mathbb{C}$-distributions on G, i.e., continuous linear functional of $C_c^\infty(G)$. The space of all such distributions is denoted by $C_c^\infty(G)'$. Let H_1 and H_2 be any two subgroups of G. We let $H_1 \times H_2$ act on G by $(h_1, h_2) \circ g := h_1 g h_2^{-1}$, where $(h_1, h_2) \in H_1 \times H_2$ and $g \in G$. Then for distribution T, we define

$T^{(h_1,h_2)}$ as usual: for $\varphi \in C_c^\infty(G)$,

$$T^{(h_1,h_2)}(\varphi) := T((h_1, h_2)^{-1} \circ \varphi) \quad \text{and} \quad (h_1, h_2) \circ \varphi(g) := \varphi((h_1, h_2)^{-1} \circ g).$$

Let S be any closed subset of G. A distribution T is said to have support in S if T vanishes on $G - S$, that is, $T(\varphi) = 0$ for any $\varphi \in C_c^\infty(G - S)$.

As in Chapter 5 of [**War**], we let

$$\mathbf{T}_s := \{T \in C_c^\infty(G)' \ : \ T^{(p_1,p_2)} = |a(p_1 p_2)|^{s-3} \cdot T \text{ for } p_1, p_2 \in P_3^4\}$$

By Theorem 5.3.2.1 in [**War**], the dimension of $Hom_G(I_{3,\infty}^{4,sm}(s), I_{3,\infty}^{4,sm}(-s))$ is equal to the dimension of the space $\mathbf{T}_s$. As in the nonarchimedean case, we have generalized Bruhat decomposition: $G = \cup_{i,j} P_3^4 w_{(i,j)} P_3^4$, where $i, j = 0, 1, 2, 3$ and $i \leq j \leq i+1$, and $w_{(i,j)}$ are Weyl group elements as follows:

$$w_{(i,i)} = \begin{pmatrix} I_i & & & & 0 & \\ & 0 & & & & -I_{3-i} \\ & & 1 & & & & 0 \\ 0 & & & I_i & & \\ & I_{3-i} & & & 0 & \\ & & 0 & & & & 1 \end{pmatrix}$$

and

$$w_{(i,i+1)} = \begin{pmatrix} I_i & & & 0 & & \\ & 0 & & 1 & 0 & \\ & & 0 & & & I_{(2-i)} \\ & 1 & & 0 & & & 0 \\ 0 & & I_i & & \\ & 0 & & & 0 & & 1 \\ & I_{(2-i)} & & & 0 & \\ & & 0 & & 1 & & 0 \end{pmatrix}.$$

Set $O_{(i,j)} := P_3^4 w_{(i,j)} P_3^4$, and $\Omega_{(i,i)} := \Omega_{(i-1,i)} \cup O_{(i,i)}$ and $\Omega_{(i,i+1)} := \Omega_{(i,i)} \cup O_{(i,i+1)}$. Hence $\Omega_{(i,j)}$ are open subsets of G and linearly ordered as $i + j$ is increasing.

Set $\mathbf{T}_{(i,j)} := \{T \in \mathbf{T}_s \ : \ \text{supp}(T) \subset \overline{O_{(i,j)}}\}$ and we also set, for $p_1, p_2 \in P_3^4$,

$$\mathbf{T}_s[O_{(i,j)}] \ := \ \{T|_{\Omega_{(i,j)}} \ : \ T \in \mathbf{T}_{(i,j)}\}$$

$$\mathbf{T}_s(O_{(i,j)}) \ := \ \{T \in C_c^\infty(\Omega_{(i,j)})' \ : \ T^{(p_1,p_2)} = |a(p_1 p_2)|^{s-3} T, \ \text{supp}(T) \subset O_{(i,j)}\}.$$

Then we have that $\mathbf{T}_s[O_{(i,j)}] \subset \mathbf{T}_s(O_{(i,j)})$ and following inequalities:

$$\dim \mathbf{T}_s \leq \sum_{(i,j)} \dim \mathbf{T}_s[O_{(i,j)}] \leq \sum_{(i,j)} \dim \mathbf{T}_s(O_{(i,j)}).$$

The fundamental estimate of Bruhat's Theory gives a computable boundary for the dimension of the space $\mathbf{T}_s(O_{(i,j)})$. In our special case, we have following Proposition.

PROPOSITION 2.2.1. *For positive real values of s, we have following estimates:*

(a) $\dim \mathbf{T}_s(O_{(i,j)}) \leq i(s; O_{(i,j)}; 0) \leq 1$ *for all* (i,j). *(There is no distributions with positive transversal order).*

(b) $\sum_{(i,j)} \dim \mathbf{T}_s(O_{(i,j)})$

$$\leq \begin{cases} \dim \mathbf{T}_s(O_{(0,0)}), & \text{if } s \neq \frac{1}{2}, 1, 2; \\ \dim \mathbf{T}_s(O_{(0,0)}) + \dim \mathbf{T}_s(O_{(2,2)}), & \text{if } s = \frac{1}{2}; \\ \dim \mathbf{T}_s(O_{(0,0)}) + \dim \mathbf{T}_s(O_{(1,1)}) + \dim \mathbf{T}_s(O_{(1,2)}), & \text{if } s = 1; \\ \dim \mathbf{T}_s(O_{(0,0)}) + \dim \mathbf{T}_s(O_{(0,1)}), & \text{if } s = 2. \end{cases}$$

PROOF. We are going to prove (a). (b) will follow from (a) directly.

Let $P_{(i,j)} := P_3^4 \cap [w_{(i,j)}^{-1} P_3^4 w_{(i,j)}]$ and $\delta_{(i,j)}$ the modulus character defined by $\delta_{(i,j)}(p) := |\det Ad(p)|_{\mathfrak{n}_{(i,j)}}|$, where $\mathfrak{n}_{(i,j)}$ is the unipotent part of $P_{(i,j)}$. Then, by the fundamental estimate of Bruhat's Theory, Theorem 5.3.2.3 in [**War**], we have the following

$$\dim \mathbf{T}_s(O_{(i,j)}) \leq \sum_{n=0}^{\infty} i(s; O_{(i,j)}; n)$$

where $i(s; O_{(i,j)}; n)$ is the dimension of the space of all $P_{(i,j)}$-intertwining operators from the representation

$$\Psi_{(i,j)}^*(p) := |a(p \cdot w_{(i,j)} p w_{(i,j)}^{-1})|^{s-3}$$

to the representation

$$A_{(i,j)}^{*,m}(p) := |a(p \cdot w_{(i,j)} p w_{(i,j)}^{-1})|^{-6} \delta_{(i,j)}(p) \Lambda_m(p)$$

with $p \in P_{(i,j)}$. Following Warner's notation [**War**], $\Lambda_1(p)$ is the representation of the group $P_{(i,j)}$ on the quotient space $\mathfrak{V}_{(i,j)} := \mathfrak{sp}(4, \mathbb{C})/(\mathfrak{p}_3^4 + \mathfrak{ad}(\mathfrak{w}_{(i,j)}^{-1})\mathfrak{p}_3^4)$ induced from the adjoint representation of $P_{(i,j)}$ on $\mathfrak{sp}(4, \mathbb{C})$, where $\mathfrak{p}_3^4$ is the complexification of the real Lie algebra of P_3^4, and Λ_m is the m-th symmetric power of Λ_1.

In the case of $O_{(i,i)}$, we have

$$P_{(i,i)} = \begin{pmatrix} a & * & * & * & * & * \\ & b & 0 & * & 0 & 0 \\ & & x & * & 0 & y \\ & & & {}^t a^{-1} & & \\ & & & * & {}^t b^{-1} & \\ & & w & * & 0 & z \end{pmatrix}$$

Hence, following direct computations, we obtain that

$$\Psi_{(i,i)}^*(p) = |\det a|^{2s-6} \text{ and } A_{(i,i)}^{*,m}(p) = |\det a|^{-i-3} \Lambda_m(p).$$

Similarly, in the case of $O_{(i,i+1)}$, we have

$$P_{(i,i+1)} = \begin{pmatrix} a & * & * & * & * & * & * & * \\ & t & * & 0 & * & * & 0 & * \\ & & b & 0 & * & 0 & 0 & * \\ & & & x & * & * & * & * \\ & & & & {}^t a^{-1} & & & \\ & & & & * & t^{-1} & & \\ & & & & * & * & {}^t b^{-1} & \\ & & & & * & 0 & 0 & x^{-1} \end{pmatrix}$$

and also

$$\Psi^*_{(i,i+1)}(p) = |\det a|^{2s-6}|tx|^{s-3} \text{ and } A^{*,m}_{(i,i+1)}(p) = |\det a|^{-i-3}|tx|^{-i-1}\Lambda_m(p).$$

In the case of $O_{(i,i)}$, the restriction to $M_{(i,i)} = GL(i) \times GL(3-i) \times Sp(1)$ of the representation $\mathfrak{V}_{(i,i)}$ of the group $P_{(i,i)}$ is equivalent to $[Sym^2((\mathbb{C}^*)^i) \otimes \mathbb{C}] \oplus [(\mathbb{C}^*)^i \otimes \mathbb{C}^2]$. Note that the subgroup $GL(3-i)$ acts trivially. Hence we obtain that

$$\Lambda_m|_{M_{(i,i)}} \equiv \oplus_{p+q=m}|\det a|^{-2m} \otimes \mathfrak{V}'_{(i,i)}.$$

In other words, for $i \geq 1$ ($O_{(i,i)}$ is not the open orbit), the number $i(s; O_{(i,i)}; m)$ is not zero if $2s-6 = -i-3-2m$ has a solution for s real positive. However, $s = \frac{3-i}{2} - m \leq 0$ if $m, i \geq 1$. Thus, for any real positive number s, $i(s; O_{(i,i)}; m)$ is always zero if $m \geq 1$ (There is no distributions with positive transversal order). On the other hand, for $m = 0$ and $i = 1, 2, 3$, we have the equation $2s - 6 = -i - 3$. Hence we obtain that

$$i(s; O_{(i,i)}; 0) = \begin{cases} 0, & \text{if } s \neq \frac{3-i}{2}; \\ 1, & \text{if } s = \frac{3-i}{2}. \end{cases}$$

In the case of $O_{(i,i+1)}$, one can also check that the restriction to $M_{(i,i+1)} = GL(i) \times GL(1) \times GL(2-i) \times GL(1)$ of the representation $\mathfrak{V}_{(i,i+1)}$ of the group $P_{(i,i+1)}$ is isomorphic to

$$[Sym^2((\mathbb{C}^*)^i)] \oplus [\mathbb{C}_{t^{-1}} \otimes (\mathbb{C}^*)^i \otimes \mathbb{C}] \oplus [\mathbb{C} \otimes (\mathbb{C}^*)^i \otimes \mathbb{C}_{x^{-1}}] \oplus [\mathbb{C}_{t^{-1}} \otimes \mathbb{C}_{x^{-1}}].$$

So we have

$$\Lambda_m|_{M_{(i,i+1)}} = \oplus_{p+q+u+v=m}(\det a)^{-2p-q-u} \cdot t^{-u-v} \cdot x^{-q-v} \otimes \mathfrak{V}'_{(i,i+1)}.$$

If $i(s; O_{(i,i+1)}; m)$ is not zero, then the positive real number s must satisfy following equations:

$$\begin{cases} 2s - 6 &= -i - 3 - 2p - q - u, \\ s - 3 &= -i - 1 - q - u, \\ s - 3 &= -i - 1 - u - v. \end{cases}$$

In other words, s must be integral and also be equal to $\frac{7-3i-2m}{4}$. It is easy to that such positive integer s does not exist. Therefore $i(s; O_{(i,i+1)}; m)$ is zero if $m \geq 1$. On the other hand, for $m = 0$, the nonvanishing of $i(s; O_{(i,i+1)}; m)$ implies that $s = \frac{3-i}{2} = 2-i$ when $i = 1, 2$, or $s = 2$ when $i = 0$. Thus we obtain the following

$$i(s; O_{(i,i+1)}; 0) = \begin{cases} 0, & \text{if } s \neq 2 - i; \\ 1, & \text{if } s = 2 - i. \end{cases}$$

This proves part (a). Part (b) is implied in the proof of part (a). $\qquad\qquad\square$

COROLLARY 2.2.1. *The dimension of the space* $Hom_G(I_{3,\infty}^{4,sm}(s), I_{3,\infty}^{4,sm}(-s))$ *is one for all positive real number* s *except for* $s = \frac{1}{2}, 1, 2$. *For those three special values of* s, *we have following estimates:*

$$\dim Hom_G(I_{3,\infty}^{4,sm}(s), I_{3,\infty}^{4,sm}(-s)) \leq \begin{cases} 2, & \text{if } s = \frac{1}{2}, 2; \\ 3, & \text{if } s = 1. \end{cases}$$

Based on the estimate of the dimension of $Hom_G(I_{3,\infty}^{4,sm}(s), I_{3,\infty}^{4,sm}(-s))$, we shall prove here that the dimension of the space $Hom_G(I_{3,\infty}^{4,sm}(s), I_{3,\infty}^{4,sm}(-s))$ is one for any positive real value of s greater than zero by means of a simple argument on the extension of certain quasi-invariant distributions. We will do it next for $s = 2, 1$, and $\frac{1}{2}$, separately.

2.3. Dimension of $Hom_G(I_{3,\infty}^{4,sm}(2), I_{3,\infty}^{4,sm}(-2))$. . By Bruhat's theory used in the previous subsection, we have

$$\begin{aligned} \dim \mathbf{T}_2 \quad & \leq \dim \mathbf{T}_2[O_{(0,0)}] + \dim \mathbf{T}_2[O_{(0,1)}] \\ & \leq \dim \mathbf{T}_2(O_{(0,0)}) + \dim \mathbf{T}_2(O_{(0,1)}) \\ & \leq 1 + 1, \end{aligned}$$

where $\mathbf{T}_2[O_{(i,j)}]$ and $\mathbf{T}_2(O_{(i,j)})$ are defined as in §2.2. We are first going to construct nonzero distributions in $\mathbf{T}_2(O_{(0,0)})$ and $\mathbf{T}_2(O_{(0,1)})$, respectively. This means that the spaces $\mathbf{T}_2(O_{(0,0)})$ and $\mathbf{T}_2(O_{(0,1)})$ are of one dimension. Then we shall prove that $\mathbf{T}_2[O_{(0,1)}] = \mathbf{T}_2(O_{(0,1)})$, but $\mathbf{T}_2[O_{(0,0)}] = 0$. In other words, the distribution in $\mathbf{T}_2(O_{(0,1)})$ has an extension in $\mathbf{T}_2$ (which is unique up to a scalar), But the distribution in $\mathbf{T}_2(O_{(0,0)})$ has no extension in $\mathbf{T}_2$. This implies that the dimension of the space $\mathbf{T}_2$ is one and then the dimension of the space $Hom_G(I_{3,\infty}^{4,sm}(2), I_{3,\infty}^{4,sm}(-2))$ is one.

Let pr be the canonical projection from $C_c^\infty(G)$ to $I_{3,\infty}^{4,sm}(s)$ defined by following integral:

$$pr \;:\; \varphi \mapsto \int_{P_3^4} \varphi(pg) |a(p)|^{-(s+3)} dp$$

where dp is a right Haar measure on P_3^4. By Lemma 5.1.1.4 in [**War**], the projection pr is surjective as a intertwining operator of right regular representations.

The proof of $\dim \mathbf{T}_2[O_{(0,1)}] = 1$: As usual, to the orbit $O_{(0,1)}$, one has an intertwining operator $\mathcal{M}_{w_{(0,1)}}(s)$, which is given by following integral:

$$\mathcal{M}_{w_{(0,1)}}(s)(f_s)(g) := \int_{N^4_{w_{(0,1)}}} f_s(w_{(0,1)} \begin{pmatrix} 1 & x & 0 & Y & 0 \\ & I_2 & X & {}^tY & W & 0 \\ & & 1 & 0 & 0 & 0 \\ & & & 1 & & \\ & & & & I_2 & \\ & & & -x & -{}^tX & 1 \end{pmatrix} g)dn,$$

where the Weyl group element $w_{(0,1)}$ is in form:

$$\begin{pmatrix} 0 & & 1 & 0 & & \\ & 0 & & & I_2 & \\ 1 & & 0 & & & 0 \\ 0 & & & 0 & & 1 \\ & -I_2 & & & 0 & \\ & & 0 & 1 & & 0 \end{pmatrix}.$$

This integral converges absolutely for $Re(s)$ large, and define a holomorphic function in s over a right half plane and continues to a meromorphic function in s over the whole complex plane. At the point where the integral is holomorphic (i.e generic point of the integral), $\mathcal{M}_{w_{(0,1)}}(s)$ is an intertwining operator from the degenerate principal series representation $I^{4,sm}_{3,\infty}(s)$ to the (normalized) smooth induced module $Ind^{Sp(4)}_{P_3^4}(\| \det |^{-\frac{2s+2}{3}} \otimes Ind^{GL(3)}_{P_{1,2}}(|\frac{a^2}{\det A}|^{\frac{2s-1}{6}})] \otimes Ind^{Sp(1)}_{B_1}(|\alpha|^{s-1}))$. We introduce an intertwining operator $\mathcal{M}_{\overline{w}_\circ}(s)$ defined by

$$\mathcal{M}_{\overline{w}_\circ}(s)(f_s)(g) := \int_{N_{\overline{w}_\circ}} f_s(\overline{w}_\circ \begin{pmatrix} n(x,y) & 0 & & \\ & 1 & & z \\ & & -{}^tn(x,y) & \\ & & 0 & 1 \end{pmatrix} g)dn,$$

where we set $n(x,y) = \begin{pmatrix} 1 & & x \\ & 1 & y \\ & & 1 \end{pmatrix}$, and

$$\overline{w}_\circ := \begin{pmatrix} J_3 & & \\ & 0 & & 1 \\ & & J_3 & \\ & -1 & & 0 \end{pmatrix}$$

with $J_3 = \begin{pmatrix} & & 1 \\ & 1 & \\ 1 & & \end{pmatrix}$. Denote $\mathcal{M}_{(0,1)}(s) := \mathcal{M}_{\overline{w}_\circ}(s) \circ \mathcal{M}_{w_{(0,1)}}(s)$. Then $\mathcal{M}_{(0,1)}(s)$ is well defined for $Re(s)$ large and has a meromorphic continuation to the whole s-plane.

In particular, if $c_{(0,1)}(s) := \frac{\zeta_v(2s-1)\zeta_v^2(s-1)\zeta_v(s)}{\zeta_v(2s+2)\zeta_v(s+1)\zeta_v(s+2)\zeta_v(s+3)}$, the normalized intertwining operator $\mathcal{M}_{(0,1)}^*(s) := c_{(0,1)}^{-1}(s) \cdot \mathcal{M}_{(0,1)}(s)$ is holomorphic for $Re(s)$ positive. In other words, $\mathcal{M}_{(0,1)}(s)$ is holomorphic for $Re(s)$ positive except for $s = \frac{1}{2}, 1$. Hence for $s \neq \frac{1}{2}, 1$, $\mathcal{M}_{(0,1)}(s)$ is an intertwining operator from $I_{3,\infty}^{4,sm}(s)$ to the smooth induced representation $Ind_{P_3^4}^{Sp(4)}([|\det|^{-\frac{2s+2}{3}} \otimes Ind_{P_{1,2}}^{GL(3)}(|\frac{a}{\det A^{\frac{1}{2}}}|^{-\frac{2s-1}{3}})] \otimes Ind_{B_1}^{Sp(1)}(|\alpha|^{-s+1}))$. It is important to notice that at $s = 2$, $\mathcal{M}_{(0,1)}(2)$ is actually an intertwining operator form $I_{3,\infty}^{4,sm}(2)$ to $I_{3,\infty}^{4,sm}(-2)$.

Composed with the canonical projection pr and the δ-distribution at the identity e, the intertwining operator $\mathcal{M}_{(0,1)}(2)$ gives a distribution on G with required quasi-invariant properties. More precisely, we let, for $\varphi \in C_c^\infty(G)$,

$$\tilde{\mathcal{M}}_{(0,1)}(2)(\varphi) := \mathcal{M}_{(0,1)}(2)(pr(\varphi))(e).$$

Then $\tilde{\mathcal{M}}_{(0,1)}(2)$ is a distribution on G with support off the open orbit $O_{(0,0)}$ since the integrating variable in the integral defining the distribution $\tilde{\mathcal{M}}_{(0,1)}(2)$ is included in the orbit $O_{(0,1)}$. Furthermore, the distribution $\tilde{\mathcal{M}}_{(0,1)}(2)$ enjoys following quasi-invariant properties: for $(p_1, p_2) \in P_3^4$,

$$(p_1, p_2) \circ \tilde{\mathcal{M}}_{(0,1)}(2) = |a(p_1 p_2)|^{-1} \tilde{\mathcal{M}}_{(0,1)}(2).$$

In other words, $\tilde{\mathcal{M}}_{(0,1)}(2)|_{\Omega_{(0,1)}}$ is a nonzero distribution in $\mathbf{T}_2[O_{(0,1)}]$. This proves that

$$\dim \mathbf{T}_2[O_{(0,1)}] = \dim \mathbf{T}_2(O_{(0,1)}) = 1.$$

The proof of $\dim \mathbf{T}_2[O_{(0,0)}] = 0$: We start with the construction of a nonzero distribution in $\mathbf{T}_2(O_{(0,0)})$. Associated to the longest Weyl group element $w_{(0,0)}$, one has an intertwining operator $\mathcal{M}_{(0,0)}(s)$ for generic values of s from $I_{3,\infty}^{4,sm}(s)$ to $I_{3,\infty}^{4,sm}(-s)$, which is defined by following integral:

$$\mathcal{M}_{(0,0)}(s)(f_s)(g) = \int_{N_3^4} f_s(w_{(0,0)}ng)dn,$$

for any section $f_s \in I_{3,\infty}^{4,sm}(s)$. This integral converges absolutely for $Re(s)$ large and continues to a meromorphic function on the whole complex plane (s-plane). Further, there is a normalizing factor $c_{(0,0)}(s) = \frac{\zeta(s)\zeta(s-1)\zeta(s-2)\zeta(2s-1)}{\zeta(s+1)\zeta(s+2)\zeta(s+3)\zeta(2s+2)}$ (where $\zeta(s) = \Gamma(\frac{s}{2})$), so that the normalized intertwining operator $\mathcal{M}_{(0,0)}^*(s) := c_{(0,0)}^{-1}(s) \cdot \mathcal{M}_{(0,0)}(s)$ will be holomorphic in s for $Re(s)$ positive and takes the unique normalized spherical section in $I_{3,\infty}^{4,sm}(s)$ to the unique normalized spherical section in $I_{3,\infty}^{4,sm}(-s)$.

Combining $\mathcal{M}_{(0,0)}(s)$ with pr, we obtain a $\mathbb{C}_s$-distribution $\tilde{\mathcal{M}}_{(0,0)}(s)$ of $C_c^\infty(G)$, which is defined by

$$\tilde{\mathcal{M}}_{(0,0)}(s)(\varphi) := \mathcal{M}_{(0,0)}(s)(pr(\varphi))(e).$$

Note that this distribution $\tilde{\mathcal{M}}_{(0,0)}(s)$ is valid only for $s \neq 0, \frac{1}{2}, 1, 2$ since the integral $\mathcal{M}_{(0,0)}(s)$ has a simple pole at $s = 0, \frac{1}{2}, 1, 2$. On the other hand, $\tilde{\mathcal{M}}_{(0,0)}(2)$ is a well defined distribution of $C_c^\infty(\Omega_{(0,0)})$. Moreover, one can show that $\tilde{\mathcal{M}}_{(0,0)}(s)$ belongs to $\mathbf{T}_s(O_{(0,0)})$ for all values of s. In fact, for $p_1, p_2 \in P_3^4$ and $\varphi \in C_c^\infty(\Omega_{(0,0)})$,

$$
\begin{aligned}
(p_1, p_2) \circ \tilde{\mathcal{M}}_{(0,0)}(s)(\varphi) &= \int_{N_3^4} \int_{P_3^4} ((p_1, p_2)^{-1} \circ \varphi)(pw_{(0,0)}n)|a(p)|^{-(s+3)} dp dn \\
&= \int_{N_3^4} \int_{P_3^4} \varphi(p_1 p w_{(0,0)} n p_2^{-1})|a(p)|^{-(s+3)} dp dn \\
&= |a(p_1 p_2)|^{s-3} \tilde{\mathcal{M}}_{(0,0)}(s)(\varphi).
\end{aligned}
$$

In other words, the distribution $\tilde{\mathcal{M}}_{(0,0)}(s)$ satisfies the condition of quasi-invariance:

$$
(p_1, p_2) \circ \tilde{\mathcal{M}}_{(0,0)}(s) = |a(p_1 p_2)|^{s-3} \tilde{\mathcal{M}}_{(0,0)}(s).
$$

Note that this quasi-invariant property of $\tilde{\mathcal{M}}_{(0,0)}(s)$ is also valid for generic values of s, i.e. $Re(s) > 0$ and $s \neq \frac{1}{2}, 1, 2$, as a distribution of $C_c^\infty(G)$.

Next we shall prove that the distribution $\tilde{\mathcal{M}}_{(0,0)}(2)$ in $\mathbf{T}_2(O_{(0,0)})$ can not be extended to be a distribution in $\mathbf{T}_2$, that is, $\dim \mathbf{T}_2[O_{(0,0)}] = 0$.

First of all, the (Cauchy) principal value of $\tilde{\mathcal{M}}_{(0,0)}(s)$ at the singular point $s = 2$ is one of the extensions of $\tilde{\mathcal{M}}_{(0,0)}(2)$ from a distribution of $C_c^\infty(\Omega_{(0,0)})$ to a distribution of $C_c^\infty(G)$. However, in general, the principal value extension will 'loss' quasi-invariant properties. More precisely, we consider, for $\varphi \in C_c^\infty(G)$, the Laurent expansion and principal value of the distribution $\tilde{\mathcal{M}}_{(0,0)}(s)(\varphi)$ at $s = 2$:

$$
\tilde{\mathcal{M}}_{(0,0)}(s)(\varphi) = \frac{1}{s-2} Res_{s=2}(\tilde{\mathcal{M}}_{(0,0)}(s))(\varphi) + PV_{s=2}(\tilde{\mathcal{M}}_{(0,0)}(s))(\varphi) + \cdots
$$

and

$$
|a(p_1 p_2)|^{s-3} = |a(p_1 p_2)|^{-1} + |a(p_1 p_2)|^{-1} \ln |a(p_1 p_2)| \cdot (s - 2) + \cdots.
$$

Hence the residue $\Lambda_{(0,0)}(2) := Res_{s=2}(\tilde{\mathcal{M}}_{(0,0)}(s))$, as a distribution of $C_c^\infty(G)$, satisfies the quasi-invariant property:

$$
(p_1, p_2) \circ \Lambda_{(0,0)}(2) = |a(p_1 p_2)|^{-1} \Lambda_{(0,0)}(2).
$$

In other words, the restriction $\Lambda_{(0,0)}(2)|_{\Omega_{(0,1)}}$ belongs to $\mathbf{T}_2[O_{(0,1)}]$. Since the dimension of $\mathbf{T}_s[O_{(0,1)}]$ is one and the dimension of $\mathbf{T}_s[O_{(i,j)}]$ for $i \geq 1$ is zero, the distribution $\Lambda_{(0,0)}(2)$ must be proportional to the distribution $\tilde{\mathcal{M}}_{(0,1)}(2)$, which has its support $\overline{O_{(0,1)}}$, the closure of the orbit $O_{(0,1)}$. Thus the support of the distribution $\Lambda_{(0,0)}(2)$ is the closed subset $\overline{O_{(0,1)}}$. However, the principal value (the second term) $PV_{(0,0)}(2) := PV_{s=2}(\tilde{\mathcal{M}}_{(0,0)}(s))$, as a distribution of $C_c^\infty(G)$, satisfies following inhomogeneous property:

$$
(p_1, p_2) \circ PV_{(0,0)}(2) = |a(p_1 p_2)|^{-1} PV_{(0,0)}(2) + |a(p_1 p_2)|^{-1} \ln |a(p_1 p_2)| \cdot \Lambda_{(0,0)}(2).
$$

Notice that the principal value $PV_{(0,0)}(2)$ is an extension of $\tilde{\mathcal{M}}_{(0,0)}(2)$ as a distribution of $C_c^\infty(\Omega_{(0,0)})$ to a distribution of $C_c^\infty(G)$, but $PV_{(0,0)}(2)$ is not $P_3^4 \times P_3^4$-quasi-invariant.

Now we suppose that the distribution $\tilde{\mathcal{M}}_{(0,0)}(2)$ in $\mathbf{T}_2(O_{(0,0)})$ do have an extension in $\mathbf{T}_2$. This means that the subspace $\mathbf{T}_2[O_{(0,0)}]$ is not trivial. Without loss of generality, we may assume that there exists a nonzero quasi-invariant distribution Λ in $\mathbf{T}_2$ such that its restriction to $\Omega_{(0,0)}$ is equal to $\tilde{\mathcal{M}}_{(0,0)}(2)$, i.e., for $\varphi \in C_c^\infty(\Omega_{(0,0)})$,

$$\Lambda(\varphi) = \tilde{\mathcal{M}}_{(0,0)}(2)(\varphi).$$

Set $T := \Lambda - PV_{(0,0)}(2)$. Then T is a distribution of $C_c^\infty(G)$ with support in $G - \Omega_{(0,0)} = \overline{O_{(0,1)}}$ satisfying following properties: for $p_1, p_2 \in P_3^4$ and $n_1, n_2 \in N_3^4$,

(a) $(p_1, p_2) \circ T = |a(p_1 p_2)|^{-1}T - |a(p_1 p_2)|^{-1}\ln|a(p_1 p_2)| \cdot \Lambda_{(0,0)}(2)$, and
(b) $(n_1, n_2) \circ T = T$.

By taking differentiation, any differential operator X from the universal enveloping algebra of $\mathfrak{g}$ can act on a distribution T at the left (or the right) hand side, which will be denoted by $L_X * T$ (or $R_X * T$, resp.). Hence we obtain the following:

(a) for $X \in \mathfrak{n}_3^4$, $L_X * T = R_X * T = 0$,
(b) for $H \in \mathfrak{p}_3^4$,

$$
\begin{aligned}
&L_H * T \\
&= \frac{d}{dt}[|a(\exp(tH))|^{-1}]_{t=0}T - \frac{d}{dt}[|a(\exp(tH))|^{-1}\ln|a(\exp(tH))|]_{t=0}\Lambda_{(0,0)}(2) \\
&= R_H * T,
\end{aligned}
$$

(c) the support of T contains that of $\Lambda_{(0,0)}(2)$ and then is equal to $\overline{O_{(0,1)}}$.

After being restricted to the open subset $\Omega_{(0,1)}$, T and $\Lambda_{(0,0)}(2)$ become nonzero distributions of $C_c^\infty(\Omega_{(0,1)})$, and because T and $\Lambda_{(0,0)}(2)$ are supported off the open orbit $\Omega_{(0,0)}$, both of them are going to be nonzero distributions of $C_c^\infty(O_{(0,1)})$. Note that $\Lambda_{(0,0)}(2)$ is $B^4 \times B^4$-quasi-invariant, but T is not (B^4 is the standard Borel subgroup of G).

Let $\Omega := N^4 A^4 w_1 N_{w_1}^4$, where N^4 is the unipotent radical of the standard Borel subgroup B^4 and T^4 is the maximal split torus of B^4, and w_1 is a Weyl group element of following form:

$$
w_1 = \begin{pmatrix} 0 & & I_2 & \\ & 0 & & J_2 \\ -I_2 & & 0 & \\ & -J_2 & & 0 \end{pmatrix}, \text{ with } J_2 = \begin{pmatrix} 0 & 1 \\ 1 & 0 \end{pmatrix}.
$$

Then the Bruhat cell Ω is a (Zariski) open subset of $O_{(0,1)}$ and both T and $\Lambda_{(0,0)}(2)$ become (by restriction) nontrivial distributions of $C_c^\infty(\Omega)$ satisfying following properties: for $b_1, b_2 \in B^4$,

$$
\begin{aligned}
(b_1, b_2) \circ \Lambda_{(0,0)}(2) &= |a(b_1 b_2)|^{-1}\Lambda_{(0,0)}(2), \\
(b_1, b_2) \circ T &= |a(b_1 b_2)|^{-1}T - |a(b_1 b_2)|^{-1}\ln|a(b_1 b_2)| \cdot \Lambda_{(0,0)}(2).
\end{aligned}
$$

According to the decomposition $\Omega = N^4 A^4 w_1 N^4_{w_1}$, any function $\varphi \in C^\infty_c(\Omega)$ can be written as

$$\varphi(\omega) := \varphi(n, a, n').$$

By the standard computation, we obtain that for vector field X_α associated to a positive root α, $L_{X_\alpha} * T = \frac{\partial T}{\partial X_\alpha} = 0$, and for vector field Y_α associated to a positive root α in $\mathfrak{n}^4_{w_1}$, $R_{Y_\alpha} * T = \frac{\partial T}{\partial Y_\alpha} = 0$.

Now we pick two elements, for $k = 1, 2$,

$$H_k = diag(0, 0, \epsilon_k, \epsilon'_k; 0, 0, -\epsilon_k, -\epsilon'_k)$$

in the lie algebra $\mathfrak{t}$. Then we have

$$L_{H_1} * T = -\epsilon_1 T - \epsilon_1 \Lambda_{(0,0)}(2) \text{ and } R_{H_2} * T = -\epsilon_2 T - \epsilon_2 \Lambda_{(0,0)}(2).$$

Further, by definition, for $\varphi \in C^\infty_c(\Omega)$, one has $L_{H_1} * T(\varphi) = T(\frac{d}{dt}[(\exp(-tH_1), e) \circ \varphi]_{t=0})$ and

$$\begin{aligned}
\frac{d}{dt}[(\exp(-tH_1), e) \circ \varphi]_{t=0}(n, a, n') &= \frac{d}{dt}\varphi(ad(e^{tH_1})(n), e^{tH_1}a, n')_{t=0} \\
&= X * \varphi + (\epsilon_1 a_3 \frac{\partial}{\partial a_3} + \epsilon'_1 a_4 \frac{\partial}{\partial a_4})\varphi,
\end{aligned}$$

and $R_{H_2} * T(\varphi) = T(\frac{d}{dt}[(e, \exp(-tH_1)) \circ \varphi]_{t=0})$ and

$$\begin{aligned}
\frac{d}{dt}[(e, \exp(-tH_1)) \circ \varphi]_{t=0}(n, a, n') &= \frac{d}{dt}\varphi(n, ae^{-tw_1 H_2 w_1^{-1}}, ad(e^{tH_1})(n'))_{t=0} \\
&= Y * \varphi + (\epsilon'_2 a_3 \frac{\partial}{\partial a_3} + \epsilon_2 a_4 \frac{\partial}{\partial a_4})\varphi,
\end{aligned}$$

where X is certain differential operator from $\mathfrak{n}^4_3$ and Y is from $\mathfrak{n}^4_{w_1}$. Hence we have that

$$\begin{aligned}
L_{H_1} * T &= T((\epsilon_1 a_3 \frac{\partial}{\partial a_3} + \epsilon'_1 a_4 \frac{\partial}{\partial a_4})\varphi) \\
R_{H_2} * T &= T((\epsilon'_2 a_3 \frac{\partial}{\partial a_3} + \epsilon_2 a_4 \frac{\partial}{\partial a_4})\varphi).
\end{aligned}$$

Let $\epsilon_1 = \epsilon_2 = 1$. Then we get $L_{H_1} * T = R_{H_2} * T$. In other words, we obtain following identity:

$$(a_3 \frac{\partial}{\partial a_3} + \epsilon'_1 a_4 \frac{\partial}{\partial a_4})(T) = (\epsilon'_2 a_3 \frac{\partial}{\partial a_3} + a_4 \frac{\partial}{\partial a_4})(T)$$

for any values of ϵ'_1 and ϵ'_2. By plugging different values of ϵ'_1 and ϵ'_2, we conclude that

$$a_3 \frac{\partial T}{\partial a_3} = 0 \quad \text{and} \quad a_4 \frac{\partial T}{\partial a_4} = 0.$$

This implies that $L_{H_1} * T = R_{H_2} * T = 0$, that is, $T = \Lambda_{(0,0)}(2)$. But this is a contradiction since T and $\Lambda_{(0,0)}(2)$ have different $B^4 \times B^4$-invariant properties. Therefore we prove that $\mathbf{T}_2[O_{(0,0)}] = 0$ and $\dim Hom_G(I_{3,\infty}^{4,sm}(2), I_{3,\infty}^{4,sm}(-2)) = 1$.

PROPOSITION 2.3.1. *The dimension of the space* $Hom_G(I_{3,\infty}^{4,sm}(2), I_{3,\infty}^{4,sm}(-2))$ *is one.*

2.4. Dimension of $Hom_G(I_{3,\infty}^{4,sm}(1), I_{3,\infty}^{4,sm}(-1))$. . This case is going to be proved in the same way as case $s = 2$, but the argument goes more complicated. According to the estimates by means of Bruhat's theory, we have the following:

$$\dim \mathbf{T}_1(O_{(i,j)}) \leq \begin{cases} 1, & \text{if } (i,j) = (0,0), (1,1), \text{ or } (1,2), \\ 0, & \text{otherwise.} \end{cases}$$

Hence we obtain following identities:

$$\begin{aligned}
\dim \mathbf{T}_1 \quad &\leq \dim \mathbf{T}_1[O_{(0,0)}] + \dim \mathbf{T}_1[O_{(1,1)}] + \dim \mathbf{T}_1[O_{(1,2)}] \\
&\leq \dim \mathbf{T}_1(O_{(0,0)}) + \dim \mathbf{T}_1(O_{(1,1)}) + \dim \mathbf{T}_1(O_{(1,2)}) \\
&\leq 1 + 1 + 1,
\end{aligned}$$

where $\mathbf{T}_1[O_{(i,j)}]$ and $\mathbf{T}_1(O_{(i,j)})$ are defined as in §2.2.

In this case, there are three cells involved.. We first construct a typical distribution for each cell $O_{(0,0)}$, $O_{(1,1)}$, and $O_{(1,2)}$, respectively. This will prove that

$$\dim \mathbf{T}_1(O_{(0,0)}) = \dim \mathbf{T}_1(O_{(1,1)}) = \dim \mathbf{T}_1(O_{(1,2)}) = 1.$$

Then we shall prove that the distributions in the spaces $\mathbf{T}_1(O_{(0,0)})$ and $\mathbf{T}_1(O_{(1,1)})$ have no extensions in $\mathbf{T}_1$. This implies that $\dim \mathbf{T}_1[O_{(0,0)}] = 0$ and $\dim \mathbf{T}_1[O_{(1,1)}] = 0$ and then that $\dim \mathbf{T}_1 = \dim \mathbf{T}_1[O_{(1,2)}] = 1$.

The proof of $\dim \mathbf{T}_1[O_{(1,2)}] = 1$: We are going to prove that $\dim \mathbf{T}_1[O_{(1,2)}] = 1$ by constructing a convergent integral with required properties. Attached to the Weyl group element $w_{(1,2)}$, there is a canonical intertwining operator $\mathcal{M}_{w_{(1,2)}}(s)$ given by, for any section f_s in $I_{3,\infty}^4(s)$,

$$\mathcal{M}_{w_{(1,2)}}(s)(f_s)(g) := \int_{N_{w_{(1,2)}}^4} f_s(w_{(1,2)}ng)dn.$$

This integral converges absolutely for $Re(s)$ large and has a meromorphic continuation as a function of s to the whole complex plane. At the point where the integral is well defined, the intertwining operator $\mathcal{M}_{w_{(1,2)}}(s)$ maps from $I_{3,\infty}^4(s)$ to the (normalized) induced representation $Ind_{P_3^4}^{Sp(4)}([|\det|^{-1} \otimes Ind_B^{GL(3)}(|\frac{a}{c}|^s)] \otimes I_1^1(s))$. Let $\mathcal{M}_{\overline{w_\circ}}(s)$ be another intertwining operator defined as follows:

$$\mathcal{M}_{\overline{w_\circ}}(s)(f_s)(g) := \int_{F^4} f_s(\overline{w_\circ} \begin{pmatrix} (x,y,z) & & & \\ & 1 & & w \\ & & {}^t(x,y,z)^{-1} & \\ & & & 1 \end{pmatrix} g)dn,$$

where f_s is in $Ind_{P_3^4}^{Sp(4)}([|\det|^{-1} \otimes Ind_B^{GL(3)}(|\frac{a}{c}|^s)] \otimes I_1^1(s))$, and J_3 is the longest element

in the Weyl group of $GL(3)$ and $(x, y, z) = \begin{pmatrix} 1 & x & z \\ & 1 & y \\ & & 1 \end{pmatrix}$ is the standard maximal unipo-

tent subgroup of $GL(3)$, and $\overline{w_o} = \begin{pmatrix} J_3 & & \\ & 0 & & 1 \\ & & J_3 & \\ -1 & & 0 \end{pmatrix}$. It is easy to check that at the

value of s where the integral is well defined, $\mathcal{M}_{\overline{w_o}}(s)$ is an intertwining operator from
$Ind_{P_3^4}^{Sp(4)}([|\det|^{-1} \otimes Ind_B^{GL(3)}(|\frac{a}{c}|^s)] \otimes I_1^1(s))$ to $Ind_{P_3^4}^{Sp(4)}([|\det|^{-1} \otimes Ind_B^{GL(3)}(|\frac{a}{c}|^{-s})] \otimes$
$I_1^1(-s))$.

Now we set $\mathcal{M}_{(1,2)}(s) := \mathcal{M}_{\overline{w_o}}(s) \circ \mathcal{M}_{w_{(1,2)}}(s)$, which is an intertwining opera-
tor from $I_{3,\infty}^4(s)$ to $Ind_{P_3^4}^{Sp(4)}([|\det|^{-1} \otimes Ind_B^{GL(3)}(|\frac{a}{c}|^{-s})] \otimes I_1^1(-s))$ for those values
of s where the operator is well defined. Furthermore, by meromorphic continu-
ation and suitable normalization, the normalized intertwining operator $\mathcal{M}_{(1,2)}^* :=$
$c_{1,2}^{-1} \cdot \mathcal{M}_{(1,2)}(s)$, with $c_{(1,2)} = \frac{\zeta_v(2s)\zeta_v^3(s)}{\zeta_v(2s+2)\zeta_v(s+1)\zeta_v(s+2)\zeta_v(s+3)}$, is holomorphic for $Re(s)$ positive. In other words, the intertwining operator $\mathcal{M}_{(1,2)}(s)$ is holomorphic for $Re(s)$
positive. Moreover, $\mathcal{M}_{(1,2)}(s)$ maps from $I_{3,\infty}^4(1)$ to $I_{3,\infty}^4(-1)$. Therefore the distribution $\tilde{\mathcal{M}}_{(1,2)}(1)$ associated to the intertwining operator $\mathcal{M}_{(1,2)}(1)$, via $\tilde{\mathcal{M}}_{(1,2)}(1)(\varphi) :=$
$\mathcal{M}_{(1,2)}(1)(pr(\varphi))(e)$ for φ in $C_c^\infty(G)$, belongs to the subspace $\mathbf{T}_1[O_{(1,2)}]$ since $\tilde{\mathcal{M}}_{(1,0)}(1)$
has the required quasi-invariant properties and has support $\overline{O_{(1,2)}}$. This proves that
the dimension of the subspace $\mathbf{T}_1[O_{(1,2)}]$ is one.

The proof of $\dim \mathbf{T}_1[O_{(1,1)}] = 0$: Associated to the Weyl group element $w_{(1,1)}$,
there is an intertwining operator $\mathcal{M}_{(1,1)}(s)$ defined by following integral:

$$\mathcal{M}_{(1,1)}(s)(f_s)(g) := \int_{N_{w_{(1,1)}'}^4} f_s(w_{(1,1)}' ng)dn,$$

where f_s is a section in $I_{3,\infty}^{4,sm}(s)$ and $w_{(1,1)}'$ is the Weyl group element of form:

$$\begin{pmatrix} 0 & & I_2 & \\ & I_2 & & 0 \\ -I_2 & & 0 & \\ & 0 & & I_2 \end{pmatrix}.$$

It is easy to check that for generic values of s,

$$\mathcal{M}_{(1,1)}(s) : I_{3,\infty}^{4,sm}(s) \to Ind_{P_3^4}^{Sp(4),sm}(|\det|^{\frac{-s-2}{2}} \otimes Ind_{P_{2,1}}^{GL(3),sm}(|\frac{\det A}{d^2}|^{\frac{-4s+1}{6}}))$$

where $P_{2,1} = \begin{pmatrix} A & * \\ 0 & d \end{pmatrix}$. Note that the above integral has a simple pole at $s = 1$. The normalized intertwining operator $\mathcal{M}^*_{(1,1)}(s) := c^{-1}_{(1,1)'}(s)\mathcal{M}_{(1,1)}(s)$, with $c_{(1,1)'}(s) = \frac{\zeta(2s-1)\zeta(s)\zeta(s-1)}{\zeta(2s+2)\zeta(s+2)\zeta(s+3)}$, is holomorphic at $s = 1$ and maps from $I^{4,sm}_{3,\infty}(1)$ to $I^{4,sm}_{3,\infty}(-1)$.

Set $\tilde{\mathcal{M}}_{(1,1)}(s)(\varphi) := \mathcal{M}_{(1,1)}(s)(pr(\varphi))(e)$ for any $\varphi \in C^\infty_c(G)$, which defines the distribution associated to the intertwining operator $\mathcal{M}_{(1,1)}(s)$. Since the integral $\mathcal{M}_{(1,1)}(s)$, when $Re(s)$ positive, has simple poles at $s = \frac{1}{2}, 1$, as a distribution of $C^\infty_c(G)$, $\tilde{\mathcal{M}}_{(1,1)}(s)$ is valid only for $Re(s) > 0$ and $s \neq \frac{1}{2}, 1$. By the definition, at $s = 1$, $\tilde{\mathcal{M}}_{(1,1)}(s)$ can be actually viewed as a distribution of $C^\infty_c(\Omega_{(1,1)})$ with support $O_{(1,1)}$ and $P^4_3 \times P^4_3$-quasi-invariant property:

$$(p_1, p_2) \circ \tilde{\mathcal{M}}_{(1,1)}(1)(\varphi) = |a(p_1 p_2)|^{-2}\tilde{\mathcal{M}}_{(1,1)}(1)(\varphi),$$

for $\varphi \in C^\infty_C(\Omega_{(1,1)})$. In other words, $\tilde{\mathcal{M}}_{(1,1)}(1)$ belongs to $\mathbf{T}_1(O_{(1,1)})$. However, for $Re(s) > 0$ and $\neq \frac{1}{2}, 1$, $\tilde{\mathcal{M}}_{(1,1)}(s)$, as a distribution of $C^\infty_c(G)$, only satisfies following properties of quasi-invariance:

$$(p_1, e) \circ \tilde{\mathcal{M}}_{(1,1)}(s) = |a(p_1)|^{s-3}\tilde{\mathcal{M}}_{(1,1)}(s),$$
$$(e, p_2) \circ \tilde{\mathcal{M}}_{(1,1)}(s) = |\det A|^{s-3}|d|^{-s-1}\tilde{\mathcal{M}}_{(1,1)}(s).$$

Hence the principal value at $s = 1$ $PV_{(1,1)}(1) := PV_{s=1}(\tilde{\mathcal{M}}_{(1,1)}(s))$ of the distribution $\tilde{\mathcal{M}}_{(1,1)}(s)$ will enjoy the following properties:

$$(p_1, e) \circ PV_{(1,1)}(1) = |a(p_1)|^{-2}PV_{(1,1)}(1) + |a(p_1)|^{-2}\ln|a(p_1)| \cdot \Lambda_{(1,1)}(1),$$
$$(e, p_2) \circ PV_{(1,1)}(1) = |\det A|^{-2}|d|^{-2}PV_{(1,1)}(1) + |\det Ad|^{-2}\ln|\frac{\det A}{d}| \cdot \Lambda_{(1,1)}(1)$$

where $\Lambda_{(1,1)}(1) := Res_{s=1}(\tilde{\mathcal{M}}_{(1,1)}(s))$, the residue at $s = 1$ of the distribution $\tilde{\mathcal{M}}_{(1,1)}(s)$, which is a distribution of $C^\infty_c(G)$ with support off the open subset $\Omega_{(1,1)}$ and satisfying the quasi-invariant properties:

$$(p_1, p_2) \circ \Lambda_{(1,1)}(1) = |a(p_1 p_2)|^{-2}\Lambda_{(1,1)}(1).$$

In other words, we have that the restriction $\Lambda_{(1,1)}(1)|_{\Omega_{(1,2)}}$ belongs to $\mathbf{T}_1[O_{(1,2)}]$. Note that the restriction $\Lambda_{(1,1)}(1)|_{\Omega_{(1,2)}}$ does not vanish. Otherwise, the distribution $\Lambda_{(1,1)}(1)$ will be supported off the open subset $\Omega_{(1,2)}$. According to our estimate (1.42), there is no such distributions at all, except the zero distribution. It is sure that the distribution $\Lambda_{(1,1)}(1)$ itself can not be zero. Hence $\Lambda_{(1,1)}(1)$ is proportional to the distribution $\tilde{\mathcal{M}}_{(1,2)}(1)$ and the support of $\Lambda_{(1,1)}(1)$ is equal to the closed subset $\overline{O_{(1,2)}}$.

Now if the space $\mathbf{T}_1[O_{(1,1)}]$ is not zero, we may assume that there exists a nonzero distribution Λ in $\mathbf{T}_1$ such that the restriction to $\Omega_{(1,1)}$ of Λ will be equal to $\tilde{\mathcal{M}}_{(1,1)}(1)$, that is, for $\varphi \in C^\infty_C(\Omega_{(1,1)})$, one has $\Lambda(\varphi) = \tilde{\mathcal{M}}_{(1,1)}(1)(\varphi)$.

Set $T := \Lambda - PV_{(1,1)}(1)$. Then T is a distribution of $C_c^\infty(G)$ with support off the open subset $\Omega_{(1,1)}$, i.e., in $G - \Omega_{(1,1)} = \overline{O_{(1,2)}}$, and satisfies following functional equations: for $p_1 \in P_3^4$ and $p_2 \in P_{2,1}^4$,

(a) $(p_1, e) \circ T = |a(p_1)|^{-2}T - |a(p_1)|^{-2}\ln|a(p_1)|\Lambda_{(1,1)}(1)$,
(b) $(e, p_1) \circ T = |\det A|^{-2}|d|^{-2}T - |\det A|^{-2}|d|^{-2}\ln|\frac{\det A}{d}|\Lambda_{(1,1)}(1)$.

By taking differentiation along a vector field H attached to certain semisimple element in the split Cartan subalgebra, we can achieve a differential equation of following form:

$$H * T = aT + b \cdot \Lambda_{(1,1)}(1),$$

where a and b are nonzero constants. This implies that the support of T contains that of $\Lambda_{(1,1)}(1)$, which is $\overline{O_{(1,2)}}$. Hence both distributions T and $\Lambda_{(1,1)}(1)$ are nonzero after restricted to the open subset $\Omega_{(1,2)}$. Since both of them vanish on the open subset $\Omega_{(1,1)}$, they can be viewed as nontrivial distributions of $C_c^\infty(O_{(1,2)})$.

In the same way as in the case $s = 2$, we choose the a nice Bruhat cell $\Omega := N^4 A^4 w_2 N_{w_2}^4$, which is a (Zariski) open subset in the orbit $O_{(1,2)}$, where w_2 is a Weyl group element of following form:

$$\begin{pmatrix} 0 & & & & 1 & & & \\ & 0 & & & & 0 & & 1 \\ & & 1 & & & & 0 & \\ & & & 0 & & 1 & & 0 \\ -1 & & & & 0 & & & \\ & 0 & & -1 & & 0 & & \\ & & 0 & & & & 1 & \\ & & -1 & & 0 & & & 0 \end{pmatrix}.$$

Then the restrictions to the open subset Ω of T and $\Lambda_{(1,1)}(1)$ are nonzero distributions of $C_c^\infty(\Omega)$ satisfying following conditions:

(a) for $\varphi \in C_c^\infty(\Omega)$, $\varphi(\omega) = \varphi(n, a, n')$, and for vector field X_α associated to a positive root α, $L_{X_\alpha} * T = \frac{\partial T}{\partial X_\alpha} = 0$, and for vector field Y_α associated to a positive root α in $\mathfrak{n}_{w_2}^4$, $R_{Y_\alpha} * T = \frac{\partial T}{\partial Y_\alpha} = 0$,
(b) picking $H_k := diag(0, \epsilon_k, 0, \epsilon_k'; 0, -\epsilon_k, 0, -\epsilon_k')$ $(k = 1, 2)$ in the Lie algebra $\mathfrak{t}$ $(\epsilon_k = \pm 1)$, we have

$$(\exp(tH_1), e) \circ T = e^{-2\epsilon_1 t}T - e^{-2\epsilon_1 t}(\epsilon_1 t)\Lambda_{(1,1)}(1).$$
$$(e, \exp(tH_2)) \circ T = e^{-2\epsilon_2 t}T - e^{-2\epsilon_2 t}(\epsilon_2 t)\Lambda_{(1,1)}(1).$$

Let $\epsilon_1 = \epsilon_2 = 1$. Then by taking differentiation, we obtain that

$$L_{H_1} * T = R_{H_2} * T = -2T - \Lambda_{(1,1)}(1).$$

For $\varphi \in C_c^\infty(\Omega)$,

$$\frac{d}{dt}[(\exp(-tH_1), e) \circ \varphi]_{t=0}(n, a, n') = \frac{d}{dt}[\varphi(Ad(e^{tH_1})(n), e^{tH_1}a, n')]_{t=0}$$

$$= X * \varphi + (a_2 \frac{\partial \varphi}{\partial a_2} + \epsilon_1' a_4 \frac{\partial \varphi}{\partial a_4})(n, a, n')$$

and

$$\frac{d}{dt}[(e, \exp(-tH_2)) \circ \varphi]_{t=0}(n, a, n') = \frac{d}{dt}[\varphi(n, ae^{-tw_2 H_2 w_2^{-1}}, Ad(e^{tH_1})(n'))]_{t=0}$$

$$= Y * \varphi + (\epsilon_2' a_2 \frac{\partial \varphi}{\partial a_2} + a_4 \frac{\partial \varphi}{\partial a_4})(n, a, n'),$$

where X is certain differential operator from $\mathfrak{N}_3^4$ and Y is such from $\mathfrak{n}_{w_2}^4$. Thus we obtain formulas for the distribution T as follows:

$$(L_{H_1} * T)(\varphi) = T(a_2 \frac{\partial \varphi}{\partial a_2} + \epsilon_1' a_4 \frac{\partial \varphi}{\partial a_4});$$

$$(R_{H_2} * T)(\varphi) = T(\epsilon_2' a_2 \frac{\partial \varphi}{\partial a_2} + a_4 \frac{\partial \varphi}{\partial a_4}).$$

In other words, we have

$$(a_2 \frac{\partial T}{\partial a_2} + \epsilon_1' a_4 \frac{\partial T}{\partial a_4}) = (\epsilon_2' a_2 \frac{\partial T}{\partial a_2} + a_4 \frac{\partial T}{\partial a_4}).$$

By choosing different values for ϵ_1' and ϵ_2', we conclude that

$$\frac{\partial T}{\partial a_2} = 0 \text{ and } \frac{\partial T}{\partial a_4} = 0,$$

that is, $L_{H_1} * T = R_{H_2} * T = 0$. This implies that $2T = \Lambda_{(1,1)}(1)$. However, this is impossible since those two distributions $2T$ and $\Lambda_{(1,1)}(1)$ have different quasi-invariant properties. Therefore the space $\mathbf{T}_1[O_{(1,1)}]$ is trivial.

The proof of $\dim \mathbf{T}_1[O_{(0,0)}] = 0$: The argument here will go slightly different from the case $s = 2$ since the support of the residue of the canonical distribution attached to the Weyl group element $w_{(0,0)}$ is in the closure of the 'two step' lower orbit.

Let $\tilde{\mathcal{M}}_{(0,0)}(s)$ be the distribution canonically attached to $w_{(0,0)}$ as in case $s = 2$, see (1.24), which is in $\mathbf{T}_s$ for $s \neq 0, \frac{1}{2}, 1, 2$, and for $s = 1$, $\tilde{\mathcal{M}}_{(0,0)}(s)$ belongs to $\mathbf{T}_1(O_{(0,0)})$. The residue at $s = 1$, denoted by $\Lambda_{(0,0)}(1); +Res_{s=1}(\tilde{\mathcal{M}}_{(0,0)}(s))$, as a distribution of $C_c^\infty(G)$, belongs to $\mathbf{T}_1$, the space of quasi-invariant distributions on G as defined in (1.6). However, the principal value at $s = 1$, denoted by $PV_{(0,0)}(1) := PV_{s=1}(\tilde{\mathcal{M}}_{(0,0)}(s))$ of $\tilde{\mathcal{M}}_{(0,0)}(s)$, as a distribution of $C_c^\infty(G)$, enjoys the following inhomogeneous equation:

$$(p_1, p_2) \circ PV_{(0,0)}(1) = |a(p_1 p_2)|^{-2} PV_{(0,0)}(1) + |a(p_1 p_2)|^{-2} \ln|a(p_1 p_2)| \cdot \Lambda_{(0,0)}(1).$$

Before our argument goes on, we have to make sure where the residue $\Lambda_{(0,0)}(1)$ is supported. First we notice that $\Lambda_{(0,0)}(1)$ is supported off the open subset $\Omega_{(1,1)}$ since, by definition, the residue $\Lambda_{(0,0)}(1)$ must be supported off the open orbit $\Omega_{(0,0)} = O_{(0,0)}$, and we have proved that both spaces $\mathbf{T}_1[O_{(0,1)}]$ and $\mathbf{T}_1[O_{(1,1)}]$ are zero. Then the restriction to the open subset $\Omega_{(1,2)}$ of the residue $\Lambda_{(0,0)}(1)$ can not be zero since we already know that all spaces $\mathbf{T}_1[O_{(i,j)}]$ with $i > 2$ are zero. Thus the residue $\Lambda_{(0,0)}(1)$ belongs to $\mathbf{T}_1[O_{(1,2)}]$. Because the space $\mathbf{T}_1[O_{(1,2)}]$ is of one dimension, the nonzero distribution $\Lambda_{(0,0)}(1)$ is proportional to the nonzero distribution $\tilde{\mathcal{M}}_{(1,2)}(1)$, which is given by a convergent integral. Thus the support of $\Lambda_{(0,0)}(1)$ is the closure $\overline{O_{(1,2)}}$ of the orbit $O_{(1,2)}$. Note that $\Lambda_{(0,0)}(1)$ and $\Lambda_{(1,1)}(1)$ are also proportional to each other as distributions in $\mathbf{T}_1$.

Let us assume that $\mathbf{T}_1[O_{(0,0)}]$ is not trivial. Then there exists a nonzero distribution Λ in $\mathbf{T}_1$, the restriction to the open orbit $\Omega_{(0,0)}$ of which is nonzero, that is, $\Lambda|_{\Omega_{(0,0)}}$ is nonzero in $\mathbf{T}_1[O_{(0,0)}]$. Without loss of generality, we may assume that, for any $\varphi \in C_c^\infty(\Omega_{(0,0)})$,

$$\Lambda(\varphi) = \tilde{\mathcal{M}}_{(0,0)}(1)(\varphi).$$

Set $T := \Lambda - PV_{(0,0)}(1)$. Then T is a nonzero distribution of $C_c^\infty(G)$ with support in the closed subset $G - O_{(0,0)} = \overline{O_{(0,1)}}$ and satisfies the following functional equation:

$$(p_1, p_2) \circ T = |a(p_1 p_2)|^{-2} T - |a(p_1 p_2)|^{-2} \ln|a(p_1 p_2)| \Lambda_{(0,0)}(1).$$

Since the residue $\Lambda_{(0,0)}(1)$ is supported off the open subset $\Omega_{(1,1)}$, the restriction to the open subset $\Omega_{(1,1)}$ of the distribution T will enjoy the $P_3^4 \times P_3^4$-quasi-invariant property:

$$(p_1, p_2) \circ T|_{\Omega_{(1,1)}} = |a(p_1 p_2)|^{-2} T|_{\Omega_{(1,1)}}.$$

Hence $T|_{\Omega_{(0,1)}}$ belongs to $\mathbf{T}_1(O_{(0,1)})$, but according to our estimate in the previous subsection via Bruhat's theory, the space $\mathbf{T}_1(O_{(0,1)})$ is trivial. This means that the distribution T is supported off the open subset $\Omega_{(0,1)}$ and the restriction to the open $\Omega_{(1,1)}$ of T belongs to $\mathbf{T}_1(O_{(1,1)})$, which is of one dimension and generated by the restriction to $\Omega_{(1,1)}$ of the principal value $PV_{(1,1)}(1)$. In other words, there is a nonzero constant c so that

$$T|_{\Omega_{(1,1)}} = c \cdot PV_{(1,1)}(1)|_{\Omega_{(1,1)}}.$$

As distributions of $C_c^\infty(G)$, we set

$$S := T - c \cdot PV_{(1,1)}(1).$$

Then S is supported off the open subset $\Omega_{(1,1)}$ and satisfies following functional equations: for $p_1 \in P_3^4$ and $p_2 \in P_{2,1}^4$,

$$(p_1, e) \circ S = |a(p_1)|^{-2} S - |a(p_1)|^{-2} \ln|a(p_1)| (\Lambda_{(0,0)}(1) + c\Lambda_{(1,1)}(1)).$$

$$(e, p_2) \circ S = |\det A|^{-2} |d|^{-2} S - |\det A|^{-2} |d|^{-2} \ln|\det A| (\Lambda_{(0,0)}(1) + c\Lambda_{(1,1)}(1))$$
$$- |\det A|^{-2} |d|^{-2} \ln|d| (\Lambda_{(0,0)}(1) - c\Lambda_{(1,1)}(1)).$$

From (1.65), we notice that the support of the distribution S is equal to that of either $(\Lambda_{(0,0)}(1) + c\Lambda_{(1,1)}(1))$ or $(\Lambda_{(0,0)}(1) - c\Lambda_{(1,1)}(1))$, which is the closed subset $\overline{O_{(1,2)}}$.

If $\Lambda_{(0,0)}(1) + c\Lambda_{(1,1)}(1) \neq 0$ as a distribution of $C_c^\infty(\Omega_{(1,2)})$, then the same argument as in the proof of the fact that $\dim \mathbf{T}_1[O_{(1,1)}] = 0$ works here and implies that $2S = \Lambda_{(0,0)}(1) + c\Lambda_{(1,1)}(1)$. This contradicts the quasi- invariant property of S.

If $\Lambda_{(0,0)}(1) + c\Lambda_{(1,1)}(1) = 0$ as a distribution of $C_c^\infty(\Omega_{(1,2)})$, then $\Lambda_{(0,0)}(1) - c\Lambda_{(1,1)}(1) \neq 0$. In this case, the distribution S satisfies following functional equation: for $p_1 \in P_3^4$ and $p_2 \in P_{2,1}^4$,

$$
\begin{aligned}
(p_1, e) \circ S &= |a(p_1)|^{-2}S, \\
(e, p_2) \circ S &= |\det A|^{-2}|d|^{-2}S - |\det A|^{-2}|d|^{-2}\ln|d|(\Lambda_{(0,0)}(1) - c\Lambda_{(1,1)}(1)).
\end{aligned}
$$

Choosing the same Bruhat cell $\Omega = N^4 A^4 w_2 N_{w_2}^4$ which is (Zariski) open in $O_{(1,2)}$ and the same elements H_1 and H_2, we obtain via similar computation that $L_{H_1} * S = R_{H_2} * S = -2S$ ($\epsilon_1 = \epsilon_2 = 1$) and

$$
L_{H_1} * S(\varphi) = S(a_2\frac{\partial\varphi}{\partial a_2} + \epsilon_1' a_4\frac{\partial\varphi}{\partial a_4}) = S(\epsilon_2' a_2\frac{\partial\varphi}{\partial a_2} + a_4\frac{\partial\varphi}{\partial a_4}) = R_{H_2} * S(\varphi).
$$

By picking different values of ϵ_1' and ϵ_2', we obtain that $L_{H_1} * S = R_{H_2} * S = 0$. This means that the restriction to Ω of S is zero, which contradicts the fact that the support of S is the closed subset $\overline{O_{(1,2)}}$. Therefore we finally prove that the space $\mathbf{T}_1[O_{(0,0)}]$ is zero. This is equivalent to that $\mathbf{T}_1$ is of one dimension.

2.5. Dimension of $Hom_G(I_{3,\infty}^{4,sm}(\frac{1}{2}), I_{3,\infty}^{4,sm}(-\frac{1}{2}))$**.** . From the estimate of §2.2, we have

$$
\begin{aligned}
\dim \mathbf{T}_{\frac{1}{2}} &\leq \dim \mathbf{T}_{\frac{1}{2}}[O_{(0,0)}] + \dim \mathbf{T}_{\frac{1}{2}}[O_{(2,2)}] \\
&\leq \dim \mathbf{T}_{\frac{1}{2}}(O_{(0,0)}) + \dim \mathbf{T}_{\frac{1}{2}}(O_{(2,2)}) \\
&\leq 1 + 1.
\end{aligned}
$$

Now let $\mathcal{M}_{(2,2)}(s) := \mathcal{U}_{w_2,\infty}^4(s)$ as defined in §2.2 of Chapter II. Then one knows from there that the intertwining operator $\mathcal{M}_{(2,2)}(s)$ is holomorphic for $Re(s) > 0$ and its induced quasi-invariant distribution $\tilde{\mathcal{M}}_{(2,2)}(\frac{1}{2})$ belongs to the space $\mathbf{T}_{\frac{1}{2}}[O_{(2,2)}]$. This implies that the dimension of the space $\mathbf{T}_{\frac{1}{2}}[O_{(2,2)}]$ is one.

As known in the proof of the case of $s = 2$, the distribution $\tilde{\mathcal{M}}_{(0,0)}(s)$ associated t the longest Weyl group element $w_{(0,0)}$ has a simple pole at $s = \frac{1}{2}$. By the same argument as that used in the cases of $s = 1, 2$, the dimension of the space $\mathbf{T}_{\frac{1}{2}}[O_{(0,0)}]$ can be proved to be zero (although the dimension of the space $\mathbf{T}_{\frac{1}{2}}(O_{(0,0)})$ is one). In other words, we obtain that the dimension of the space $\mathbf{T}_{\frac{1}{2}}$ is one.

Combining results from §2.2, 2.3 , 2.4, and 2.5, we obtain following theorem.

THEOREM 2.5.1. *The dimension of the space* $Hom_G(I_{3,\infty}^{4,sm}(s), I_{3,\infty}^{4,sm}(-s))$ *is one for all real number s greater than zero.*

3. Residues of Eisenstein Series $E_3^4(g, s; f_s)$

Now we come back to the global situation. We are concerned with the residue representations of the Eisenstein series $E_3^4(g, s; f_s)$ at $s = 1, 2$. According to Theorem 4.0.1 in Chapter III or Proposition 1.1.2 in §1.1, $E_3^4(g, s; f_s)$ has a simple pole at $s = 2$ and has a pole of order two at $s = 1$. Taking the Laurent expansion of $E_3^4(g, s; f_s)$ at $s = 1$ and $s = 2$, respectively, one has

$$
E_3^4(g, s; f_s) = \frac{\Lambda_{-2}^{4,3}(1, f_s)}{(s-1)^2} + \frac{\Lambda_{-1}^{4,3}(1, f_s)}{s-1} + \cdots.
$$

$$
E_3^4(g, s; f_s) = \frac{\Lambda_{-1}^{4,3}(2, f_s)}{s-2} + \Lambda_0^{4,3}(2, f_s) + \cdots.
$$

$$
E_4^4(g, s; \phi_s) = \frac{\Lambda_{-1}^{4,4}(\frac{1}{2}, \phi_s)}{s - \frac{1}{2}} + \Lambda_0^{4,4}(\frac{1}{2}, \phi_s) + \cdots.
$$

$$
E_1^4(g, s; \varphi_s) = \Lambda_0^{4,1}(1, \varphi_s) + \Lambda_1^{4,1}(1, \varphi_s)(s - 1) + \cdots.
$$

We are going to understand the first terms $\Lambda_{-2}^{4,3}(1, f_s)$ and $\Lambda_{-1}^{4,3}(2, f_s)$ from the representation theoretic point of view.

3.1. Square Integrability. To determine the square integrability of an automorphic form there is a criterion in terms of its automorphic exponents, [**Lan**], [**Jac**], and [**KRS**]. We shall use this effective criterion to determine the square integrability of the residue representations of the Eisenstein series $E_3^4(g, s; f_s)$ at $s = 1, 2$, that is, the first terms $\Lambda_{-2}^{4,3}(1, f_s)$ and $\Lambda_{-1}^{4,3}(2, f_s)$. Note that the Eisenstein series $E_3^4(g, s; f_s)$ for any section $f_s \in I_3^4(s)$ is concentrated on the Borel subgroup B_4 of G_4 in the terminology of [**Lan**] and [**Jac**]. Hence we have to compute the automorphic exponents of the residue representations $\Lambda_{-2}^{4,3}(1, f_s)$ and $\Lambda_{-1}^{4,3}(2, f_s)$ along the Borel subgroup B_4.

According to the definition of the automorphic exponents of an automorphic form formula (9) in p.186 of [**Jac**] and formula (6.16) in p. 520 of [**KRS**], the automorphic exponents of the residue representations $\Lambda_{-2}^{4,3}(1, f_s)$ and $\Lambda_{-1}^{4,3}(2, f_s)$ will be a finite set $\{\mu_1(s), \mu_2(s), \mu_3(s), \mu_4(s)\}$. By the criterion, the representation $\Lambda_{-2}^{4,3}(1, f_s)$ (or $\Lambda_{-1}^{4,3}(2, f_s)$) is square integrable if and only if $\mu_i(j) \in -\mathcal{O}(\mathfrak{a}_{B_4})$ for $i = 1, 2, 3, 4$ ($j = 1$ or 2). The definition of $\mathcal{O}(\mathfrak{a}_{B_4})$ is as follows: let $\{\alpha_1, \alpha_2, \alpha_3, \alpha_4\}$ be the set of all simple roots of G_4 determined by the fixed Borel subgroup B_4. then, following p. 187 [**Jac**], we have

$$
\mathcal{O}(\mathfrak{a}_{B_4}) = \{\sum_{i=1}^4 x_i \alpha_i \in X^*(T_4) \otimes_{\mathbb{Z}} \mathbb{R} \ : \ x_i > 0\}.
$$

By computing the constant term of the Eisenstein series $E_3^4(g, s; f_s)$ along the Borel subgroup B_4, we obtain the following results on the automorphic exponents of $\Lambda_{-2}^{4,3}(1)$ and $\Lambda_{-1}^{4,3}(2)$:

(a) The automorphic exponents of $\Lambda_{-2}^{4,3}(1, f_s)$ along B_4 are

$$
\begin{array}{cccc}
(-1,0,-2,1) & (-1,0,-2,-1) & (-1,-2,0,1) & (-1,-2,0,-1) \\
(-1,-2,-1,0) & (-2,0,1,-1) & (-2,0,-1,-1) & (-2,0,-1,1) \ . \\
(-2,-1,0,1) & (-2,-1,0,-1) & (-2,-1,-1,0) &
\end{array}
$$

It is easy to check that all of these exponents are in $-\mathcal{O}(\mathfrak{a}_{B_4})$.

(b) The automorphic exponents of $\Lambda_{-1}^{4,3}(2, f_s)$ along B_4 are

$$
\begin{array}{cccc}
(1,-3,-1,2) & (1,-3,-1,-2) & (-1,-3,-2,-1) & (-3,1,-2,-1) \\
(-3,-1,-2,-1) & (-3,-2,1,-1) & (-3,-2,-1,-1) & (-3,-2,-1,1).
\end{array}
$$

Notice that the first exponent can be written as $(1,-3,-1,2) = \alpha_1 - 2\alpha_2 - 3\alpha_3 - \frac{1}{2}\alpha_4$, which does not belong to $-\mathcal{O}(\mathfrak{a}_{B_4})$.

PROPOSITION 3.1.1. *The residue representation $\Lambda_{-2}^{4,3}(1, f_s)$ of the Eisenstein series $E_3^4(g, s; f_s)$ at $s = 1$ is square integrable, but the residue representation $\Lambda_{-1}^{4,3}(2, f_s)$ of $E_3^4(g, s; f_s)$ at $s = 2$ is not square integrable.*

3.2. Residue at $s = 1$. Before studying the residue representation $\Lambda_{-2}^{4,3}(1, f_s)$ we are going to recall some relevant results of Kudla and Rallis in [**KuRa**]. Let $(V, (,))$ be a 6-dimensional non-degenerate quadratic vector space over the totally real number field F with Witt index $r = 3$ and $H = O(V) = O(3,3)$ the group of all isometries of $(V, (,))$. Let $(W, <, >)$ be a non-degenerate symplectic vector space over the field F and $G = Sp(4) = Sp(W)$. Then (G, H) forms a dual reductive pair in $Sp(V \otimes W)$ in the sense of Howe. Let $G(\mathbb{A})$ (resp. $H(\mathbb{A})$) denote the adele group of G (resp. H) and, foe fixed non-trivial additive character $\psi : \mathbb{A}/F \longrightarrow \mathbb{C}^\times$, let $\omega = \omega_\psi$ denote the Weil representation of the group $G(\mathbb{A}) \times \mathbb{H}(\mathbb{A})$ on the space $\mathcal{S}(\mathcal{V}(\mathbb{A})^{\natural})$ of Schwartz-Bruhat functions on $V(\mathbb{A})^{\natural}$. For $\varphi \in \mathcal{S}(\mathcal{V}(\mathbb{A})^{\natural})$, $g \in G(\mathbb{A})$ and $h \in H(\mathbb{A})$, define the theta function

$$
\theta(g, h; \varphi) = \sum_{x \in V(F)^4} \omega(g)\varphi(h^{-1}x) = \sum_{w \in W(F)^3} \omega(h)\hat{\varphi}(wg),
$$

where $\hat{\varphi} \in \mathcal{S}(\mathcal{W}(\mathbb{A})^{\natural})$, which is another model of the Weil representation ω obtained by taking partial Fourier transformation from the model $\mathcal{S}(\mathcal{V}(\mathbb{A})^{\natural})$ of ω.

As usual, the theta integral is defined by

$$
I(g, \varphi) = \int_{H(F) \backslash H(\mathbb{A})} \theta(g, h; \varphi) dhv
$$

where dh is an invariant measure on $H(F) \backslash H(\mathbb{A})$, normalized to have total volume one. Since the quadratic space V is splitting and $m - r = 3 < n + 1 = 5$, the theta integral may diverge for an arbitrary $\varphi \in \mathcal{S}(\mathcal{V}(\mathbb{A})^{\natural})$ (or $\hat{\varphi} \in \mathcal{S}(\mathcal{W}(\mathbb{A})^{\natural})$) according to Weil's convergence condition [**W**]. Kudla and Rallis, in [**KuRa**], regularized the theta integral by applying an differential operator $\omega(z)$ to φ (or $\hat{\varphi}$).

We will describe the regularized theta integral in the case we concerned,i.e., the case of $m = 6$ and $r = 3$. Let Q_3 be the Siegel parabolic subgroup of H and $h = nm(a)k$ corresponding to the Iwasawa decomposition $H(\mathbb{A}) = \mathbb{Q}_{\not{k}}(\mathbb{A})\mathbb{K}_{\mathbb{H}}$, here

$$m(a) = \begin{pmatrix} a & 0 \\ 0 & {}^t a^{-1} \end{pmatrix}.$$

Let $|a(h)| = |det a|$ and $\Psi(h, s) = |a(h)|^{s+1}$. Then an auxiliary Eisenstein series is defined as

$$E(h, s) = \sum_{\gamma \in Q_3(F) \backslash H(F)} \Psi(\gamma h, s).$$

By Proposition 5.4.1. in [**KuRa**], this Eisenstein series has a simple pole at $s = 1$ with constant residue $Res_{s=1} E(h, s) = \kappa$. The regularized theta integral is defined as follows:

$$I(g, s; \omega(z)\varphi) = \int_{H(F) \backslash H(\mathbb{A})} \theta(g, h; \omega(z)\varphi) E(h, s) dh.$$

Since the regularized theta function $\theta(g, h; \omega(z)\varphi)$ is rapidly decreasing on $H(F) \backslash H(\mathbb{A})$, the regularized theta integral $I(g, s; \omega(z)\varphi)$ is absolutely convergent whenever the Eisenstein series $E(h, s)$ is holomorphic and hence defines a meromorphic function of s.

The regularized Siegel-Weil formula is an identity between the regularized theta integral and an Eisenstein series on $G(\mathbb{A})$. More precisely, let $\varphi \in \mathcal{S}(\mathcal{V}(\mathbb{A})^{\not{2}})$ and $\hat{\varphi}$ its Fourier transformation in $\mathcal{S}(\mathcal{W}(\mathbb{A})^{\not{k}})$, define

$$F(g, s; \varphi) = \int_{GL(3, \mathbb{A})} \omega(g)\hat{\varphi}({}^t a w_\circ) |det a|^{s+3} da$$

where $w_\circ = (0_{3\times 4}, I_3, 0_{3\times 1}) \in M_{3\times 8}(F) \cong W(F)^3$. It was proved in [**KuRa**] that for $Re(s) > 0$, the integral converges absolutely and defines a $G(\mathbb{A})$-intertwining operator from $\mathcal{S}(\mathcal{V}(\mathbb{A})^{\not{2}})$ into $I_3^4(s)$, and for all φ,

$$F(g, s; \omega(z)\varphi) = P(s)F(g, s; \varphi)$$

with $P(s) = (s - 3)(s - 2)(s - 1)(s + 1)(s + 2)(s + 3)$. Then the regularized theta integral is equal to an Eisenstein series, that is,

$$I(g, s; \omega(z)\varphi) = P(s)\mathcal{E}(\int, \}; \varphi),$$

where the Eisenstein series is defined by

$$\mathcal{E}(\int, \}; \varphi) = \sum_{\gamma \in \mathcal{P}_3(F) \backslash \mathcal{G}(F)} \mathcal{F}(\gamma\}, \int; \varphi_{\mathcal{K}_{\mathcal{H}}}).$$

The Eisenstein series $\mathcal{E}(\int, \}; \varphi)$ is actually our Eisenstein series $E_3^4(g, a; f_s)$ attached to a special section

$$f_s = F(g, s; \varphi_{K_H}) \in I_3^4(s).$$

Applying Kudla-Rallis' theory of first term identity to this special case, we have the following identity

$$\Lambda_{-1}^{4,4}(\frac{1}{2};\Phi) = c \cdot \Lambda_{-2}^{4,3}(1;\varphi),$$

where c is a nonzero constant and Φ is a standard section in $_4^4(\frac{1}{2})$ related to the Schwartz-Bruhat function φ in $\mathcal{S}(V^4(\mathbb{A}))$. Note that the right hand side of the identity is only valid for certain special sections in $I_3^4(1)$ determined by Kudla-Rallis' regularized theta integral [**KuRa**]. Our purpose here is to prove that the above identity is valid for general sections in $I_3^4(s)$. Let $\mathcal{T}_s^*$ be an intertwining operator defined by the following integral, for $Re(s) > 0$

$$\mathcal{T}_s^*(f_s)(g) := \frac{\zeta(s+3)}{\zeta(s)} \int_{\mathbb{A}^3} f_s(n(x,y,z)g)dxdydz,$$

where

$$n(x,y,z) = \begin{pmatrix} 1 & & & & & & & \\ & 1 & & & & & & \\ & & 1 & & & & & \\ x & y & z & 1 & & & & \\ & & & & 1 & & & -x \\ & & & & & 1 & & -y \\ & & & & & & 1 & -z \\ & & & & & & & 1 \end{pmatrix}.$$

LEMMA 3.2.1. *The intertwining operator $\mathcal{T}_s^*$ is holomorphic at $Re(s) > 0$ and at $s = 1$, $\mathcal{T}_1^*$ is an intertwining map from $I_3^4(1)$ to $I_4^4(\frac{1}{2})$, which takes the normalized spherical section in $I_3^4(1)$ to that in $I_4^4(\frac{1}{2})$.*

THEOREM 3.2.1. *For any holomorphic section f_s belonging to $I_3^4(s)$, there is a nonzero constant c such that*

$$\Lambda_{-2}^{4,3}(1;f_s) = c \cdot \Lambda_{-1}^{4,4}(\frac{1}{2};\mathcal{T}_s^*(f_s)).$$

PROOF. Since the residue representation $\Lambda_{-2}^{4,3}(1;f_s)$ is square integrable as an automorphic representation by Proposition 3.1.1, $\Lambda_{-2}^{4,3}(1;f_s)$ gives an intertwining map from $I_3^4(1)$ to $\mathcal{A}_2(G_4(\mathbb{A}))$, the space of all square integrable automorphic forms over $G_4(\mathbb{A})$. Kudla-Rallis' first term identity implies that $\Lambda_{-1}^{4,4}(\frac{1}{2};I_4^4(\frac{1}{2})) \subset \Lambda_{-2}^{4,3}(1;I_3^4(1))$. Note that $\Lambda_{-1}^{4,4}(\frac{1}{2};I_4^4(\frac{1}{2}))$ is an irreducible automorphic representation (Theorem 4.9 in [**KuRa**]).

Now assume that $\Lambda_{-1}^{4,4}(\frac{1}{2};I_4^4(\frac{1}{2})) \neq \Lambda_{-2}^{4,3}(1;I_3^4(1))$. By the completely reducibility of $\mathcal{A}_2(G_4(\mathbb{A}))$, we have the following direct sum decomposition:

$$\Lambda_{-2}^{4,3}(1;I_3^4(1)) = \Lambda_{-1}^{4,4}(\frac{1}{2};I_4^4(\frac{1}{2})) \oplus X \oplus \cdots,$$

where X is an irreducible piece in $\Lambda_{-2}^{4,3}(1; I_3^4(1))$. By our assumption, X is not zero and not equal to $\Lambda_{-1}^{4,4}(\frac{1}{2}; I_4^4(\frac{1}{2}))$. Hence both $\Lambda_{-1}^{4,4}(\frac{1}{2}; I_4^4(\frac{1}{2}))$ and X are irreducible quotient representations of $I_3^4(1)$. Because $I_3^4(1)$ has a unique (up to isomorphism) irreducible quotient representation (Corollary 1.2.1), we conclude that

$$\Lambda_{-1}^{4,4}(\frac{1}{2}; I_4^4(\frac{1}{2})) \cong X,$$

as automorphic representations of $G_4(\mathbb{A})$. By the strong uniqueness theorem (Theorem 3.1 in [**KuRa**]), $\Lambda_{-1}^{4,4}(\frac{1}{2}; I_4^4(\frac{1}{2}))$ must be equal to X as subspaces in $\mathcal{A}_2(G_4(\mathbb{A}))$. This is a contradiction to our assumption. Therefore we prove the following equality $\Lambda_{-1}^{4,4}(\frac{1}{2}; I_4^4(\frac{1}{2})) = \Lambda_{-2}^{4,3}(1; I_3^4(1))$. It is routine to check that the following map

$$\Lambda_{-2}^{4,3}(1; f_s) \to \Lambda_{-1}^{4,4}(\frac{1}{2}; \mathcal{T}_s^*(f_s))$$

is $G_4(\mathbb{A})$-intertwining. By the irreducibility of $\Lambda_{-1}^{4,4}(\frac{1}{2}; I_4^4(\frac{1}{2}))$, the above intertwining map is actually the multiplication by scalar. This proves the identity. $\square$

Note that the nonzero constant c can be explicitly expressed in terms of the residue and value of the global zeta function at some points.

3.3. Residue at $s = 2$. In this subsection, we shall prove a new type of 'first term identity' which is not related to the Eisenstein series of Siegel type. Let us first introduce an intertwining operator as follows

$$\mathcal{M}_s^*(f_s)(g) := \frac{\zeta(2s+2)\zeta(s+2)\zeta(s+3)}{\zeta(2s+1)\zeta(s-1)\zeta(s)} \int_{N_w} f_s(wng)dn$$

where

$$w_{(1,1)}^4 = \begin{pmatrix} 1 & & 0 & & \\ & 0 & & I_2 & \\ & & 1 & & 0 \\ 0 & & & 1 & \\ & -I_2 & & 0 & \\ & & & & 1 \end{pmatrix}$$

and

$$N_{(1,1)}^4 = \begin{pmatrix} 1 & & 0 & & \\ & I_2 & X & W & Y \\ & & 1 & {}^tY & 0 \\ & & & 1 & \\ & & & I_2 & \\ & & & -{}^tX & 1 \end{pmatrix}.$$

LEMMA 3.3.1. *The intertwining operator $\mathcal{M}_s^*(f_s)$ is holomorphic in s, when $Re(s)$ is positive, for any holomorphic section f_s in $I_3^4(s)$. At $s = 2$, $\mathcal{M}_2^*$ is an intertwining* *map from $I_3^4(2)$ onto $I_1^4(1)$ and takes the normalized spherical section in $I_3^4(2)$ to that in $I_1^4(1)$. (Note that $I_1^4(1)$ is irreducible).*

THEOREM 3.3.1 (First Term Identity). *For any holomorphic section f_s in $I_3^4(s)$, we have*

$$\Lambda_{-1}^{4,3}(2; f_s) = \frac{\zeta(2)Res_{s=1}\zeta(s)}{\zeta(4)\zeta(6)}\Lambda_0^{4,1}(1; \mathcal{M}_s^*(f_s)).$$

PROOF. We shall first check that the identity holds for the normalized spherical section f_s^0 in $I_3^4(s)$. Considering the constant term along the maximal parabolic subgroup P_1^4 of the Eisenstein series $E_3^4(g, s; f_s^0)$ and Eisenstein series $E_1^4(g, s; \varphi_s^0)$, respectively, we have

$$\begin{aligned}
E_{3,P_1^4}^4(m(t,g), s; f_s^0) &= |t|^{s+3}E_2^3(g, s+\frac{1}{2}; f_{s,r}^0) \\
&\quad + |t|^3 E_3^3(g, s; f_{s,1}^0)\frac{\zeta(s)}{\zeta(s+3)} \\
&\quad + |t|^{-s+3}E_2^3(g, s-\frac{1}{2}; f_{s,2}^0)\frac{\zeta(2s)\zeta(s)}{\zeta(2s+2)\zeta(s+3)}
\end{aligned}$$

and

$$\begin{aligned}
E_{3,P_1^4}^4(m(t,g), s; \varphi_s^0) &= |t|^{s+4}(1) \\
&\quad + |t|E_1^3(g, s; \varphi_{s,1}^0)\frac{\zeta(s+3)}{\zeta(s+4)} \\
&\quad + |t|^{-s+4}\varphi_{s.2}^0(1)\frac{\zeta(s-3)}{\zeta(s+4)}.
\end{aligned}$$

It reduces to prove the following three identities:

$$\begin{aligned}
Res_{s=2}E_2^3(g, s+\frac{1}{2}; f_{s,r}^0) &= \frac{\zeta(2)Res_{s=1}\zeta(s)}{\zeta(4)\zeta(6)} \\
Res_{s=2}E_3^3(g, s; f_{s,1}^0) &= \frac{\zeta(3)Res_{s=1}\zeta(s)}{\zeta(4)\zeta(6)} \\
Res_{s=2}E_2^3(g, s-\frac{1}{2}; f_{s,2}^0) &= \frac{Res_{s=1}\zeta(s)}{\zeta(4)}E_1^3(g, s; \varphi_{s,1}^0).
\end{aligned}$$

All these three identities can be checked directly by computing their constant terms along the maximal parabolic subgroup P_1^3. We omit the computation here, see [**Jia**].

Since the identity holds for spherical sections in $I_3^4(s)$, the images $\Lambda_{-1}^{4,3}(2; I_3^4(2))$ and $\Lambda_0^{4,1}(1; I_1^4(1))$ has nonzero intersection in $\mathcal{A}(G_4(\mathbb{A}))$, the space of all automorphic forms of $G_4(\mathbb{A})$. Further, we conclude that $\Lambda_{-1}^{4,3}(2; I_3^4(2))$ must contain $\Lambda_0^{4,1}(1; I_1^4(1))$

as a subrepresentation because $\Lambda_0^{4,1}(1; I_1^4(1))$ is irreducible. We claim that

$$\Lambda_{-1}^{4,3}(2; I_3^4(2)) = \Lambda_0^{4,1}(1; I_1^4(1)).$$

In fact, we notice first that, at each local place v, the unique irreducible quotient representation of $I_{3,v}^4(2)$ is generated by the normalized spherical section $f_{2,v}^0$. Then any section f_s in $I_3^4(s)$ which is holomorphic at $s = 2$ can be expressed as a finite sum as follows

$$f_s = \sum_i a_i(g_i, f_2)\rho_s(g_i)f_s^0 + (s-2)f_s'$$

where $a_i(g_i, f_2)$ are constants and f_s' is a section in $I_3^4(s)$ which is holomorphic at $s = 2$. Therefore we obtain that

$$Res_{s=2}E_3^4(g, s; f_s) = \sum_i a_i(g_i, f_2)Res_{s=2}E_3^4(gg_i, s; f_s^0).$$

In other words, we have proved our claim. Following the same argument as in the case $s = 1$ in §3.2, we finally prove the identity. $\qquad\square$

4. Analytic Properties of the Global Integral

As one of the applications of our results on the Eisenstein series $E_3^4(g, s; f_s)$, we can determine the analytic properties of our global zeta integral $\mathcal{Z}(s, \phi_1, \phi_2, f_s)$ for $Re(s) \geq 0$. According to Theorem 4.0.1 in Chapter III and Proposition 1.1.2, the global integral $\mathcal{Z}(s, \phi_1, \phi_2, f_s)$ is holomorphic for all s except for $s \in \{\pm 1, \pm 2, \pm 3\}$ where the integral may achieve a pole of degree at most two. Actually, we can prove the following more precise result.

THEOREM 4.0.2. *The global integral $\mathcal{Z}(s, \phi_1, \phi_2, f_s)$ is holomorphic for all s except for $s = 1$ where it may achieve a pole of degree two.*

PROOF. We have to the holomorphy at $s = 2$ and 3 of the global integral $\mathcal{Z}(s, \phi_1, \phi_2, f_s)$. At these two points the Eisenstein series achieves a simple pole, respectively. So the global integral $\mathcal{Z}(s, \phi_1, \phi_2, f_s)$ may has a pole at $s = 2, 3$ of order at most one. At $s = 3$, which is the end point of the critical strip, the residue of the Eisenstein series $E_3^4(g, s; f_s)$ is the constant representation. Hence the residue of the global integral

$$= Res_{s=3}\int_{C(\mathbb{A})H(F)\backslash H(\mathbb{A})} \phi_1(g_1)\phi_2(g_2)E_3^4(g, s; f_s)dg_1dg_2$$

$$= c \cdot \int_{C(\mathbb{A})H(F)\backslash H(\mathbb{A})} \phi_1(g_1)\phi_2(g_2)dg_1dg_2.$$

The later integral vanishes since both ϕ_1 and ϕ_2 are cusp forms. This proves the holomorphy of the global integral $\mathcal{Z}(s, \phi_1, \phi_2, f_s)$ at $s = 3$.

Next we consider the holomorphy of the global integral $\mathcal{Z}(s, \phi_1, \phi_2, f_s)$ at $s = 2$. By means of the first term identity in Theorem 3.3.1, we have

$$= Res_{s=2} \int_{C(\mathbb{A})H(F)\backslash H(\mathbb{A})} \phi_1(g_1)\phi_2(g_2)E_3^4(g, s; f_s)dg_1dg_2$$

$$= \int_{C(\mathbb{A})H(F)\backslash H(\mathbb{A})} \phi_1(g_1)\phi_2(g_2)Res_{s=2}E_3^4(g, s; f_s)dg_1dg_2$$

$$= \int_{C(\mathbb{A})H(F)\backslash H(\mathbb{A})} \phi_1(g_1)\phi_2(g_2)c \cdot E_1^4(g, 1; \mathcal{M}_2^*(f_2))dg_1dg_2.$$

The last integral vanishes because the H-orbits on the flag variety $P_1^4 \setminus Sp(4)$ are negligible in the sense of [**PSRa**]. $\qquad\square$

Note that at $s = 1$ the global integral $\mathcal{Z}(s, \phi_1, \phi_2, f_s)$ may achieve the second order pole. One of the important applications of Kudla-Rallis' theory of first term identity is to prove the following Theorem.

THEOREM 4.0.3. *Let π_1 and π_2 be generic cusp forms of $GSp(2, \mathbb{A})$ with trivial central character.*

(a) *If the global integral $\mathcal{Z}(s, \phi_1, \phi_2, f_s)$ achieves the second order pole at $s = 1$, then $\pi_2 = \pi_1^\vee$, the contragredient representation of π_1.*

(b) *Let $\pi = \pi_1$ and $\phi_1 \in \pi$ and $\phi_2 \in \pi^\vee$. Then $\mathcal{Z}(s, \phi_1, \phi_2, f_s)$ achieves the second order pole at $s = 1$ if and only if the theta lifting of π to $GO(2, 2)$ is nonzero.*

PROOF. The Laurent expansion at $s = 1$ of the global integral $\mathcal{Z}(s, \phi_1, \phi_2, f_s)$ is

$$\frac{Z_{-2}(1; \phi_1, \phi_2, f_s)}{(s-1)^2} + \frac{Z_{-1}(1; \phi_1, \phi_2, f_s)}{(s-1)} + \cdots.$$

By the identity in Theorem 3.2.1, we have

$$Z_{-2}(1; \phi_1, \phi_2, f_s) = \int_{C(\mathbb{A})H(F)\backslash H(\mathbb{A})} \phi_1(g_1)\phi_2(g_2)\Lambda_{-2}^{4,3}(1; f_s)dg_1dg_2$$

$$= c \cdot \int_{C(\mathbb{A})H(F)\backslash H(\mathbb{A})} \phi_1(g_1)\phi_2(g_2)\Lambda_{-1}^{4,4}(\frac{1}{2}; \Phi_s)dg_1dg_2$$

$$= c \cdot Res_{s=\frac{1}{2}} \int_{C(\mathbb{A})H(F)\backslash H(\mathbb{A})} \phi_1(g_1)\phi_2(g_2)E_4^4(g, s; \Phi_s)dg_1dg_2.$$

Since the central characters of representations π_1 and π_2 are trivial, the global integral $\int_{C(\mathbb{A})H(F)\backslash H(\mathbb{A})} \phi_1(g_1)\phi_2(g_2)E_4^4(g, s; \Phi_s)dg_1dg_2$ is exactly the global integral established in [**PSRa**] and studied extensively in [**KuRa**]. Hence part (1) follows from [**PSRa**] and part (2) follows from section 7 in [**KuRa**]. $\qquad\square$

It is important to mention that, after we establish the local theory of the Rankin-Selberg convolutions for our local zeta integral $\mathcal{Z}_v(s, W_1, W_2, f_s)$, the condition on the existence of the second order pole at $s = 1$ of the global integral $\mathcal{Z}(s, \phi_1, \phi_2, f_s)$ will be replaced by that of the degree 16 standard L-function $L^S(s, \pi_1 \times \pi_2, \rho_1 \otimes \rho_2)$. In other words, the existence of the second order pole at $s = 1$ of $L^S(s, \pi_1 \times \pi_2, \rho_1 \otimes \rho_2)$ should be a criteria of the vanishing of the theta lifting of the cusp forms of $GSp(2, \mathbb{A})$ to the cusp forms of $GO(2, 2; \mathbb{A})$.

CHAPTER 5

Local Theory of Rankin-Selberg Convolution

The analytic properties of the global integral $\mathcal{Z}(s, \phi_1, \phi_2, f_s)$ is in general determined by those of the normalized Eisenstein series and the representation-theoretic properties of π_1 and π_2. Following the fundamental identity (Theorem 5.0.1), the analytic properties of the L-function $L^S(s, \pi_1 \otimes \pi_2, \rho_1 \otimes \rho_1)$ will be determined if we have sufficient acknowledge about those ramified local zeta integrals $\mathcal{Z}_v(s, W_{1v}, W_{2v}, f_{sv})$, both archimedean and non-archimedean. It is our goal here to study those ramified local zeta integrals

$$\mathcal{Z}_v(s, W_1, W_2, f_s) = \int_{CN^{2,\triangle}(Z_2 \times I_4)\backslash H} f_s(\gamma_\circ(g_1, g_2)) W_1(g_1) W_2(g_2) dg_1 dg_2. \tag{151}$$

More precisely, we will prove that those local integrals absolutely converge for $Re(s)$ large, and have, in nonarchimedean case, a meromorphic continuation to the whole complex plane, and also for an appropriate choice of the integrand data, those local integrals can be made to be a constant independent of s (in archimedean case, we assume that the section f_s is smooth). To this end, we have to estimate Whittaker functions on the splitting torus, similar to [**JPSS**], [**JaSh**], and [**Sou**], and estimate an integral of mixed type in the sense that the special case of it will be either an intertwining integral or certain partial 'Fourier Coefficient'. Since we only consider local situations, we will drop v from the subscripts of all notations. Therefore, the underline field F is the real archimedean field or any nonarchimedean one with characteristics zero.

1. Some Estimates

The estimates to be established in this section will be used to study those ramified local integrals later. The estimates of Whittaker functions are made in the standard way, while the estimates of the integral of mixed type will be more complicated.

1.1. Estimates of Whittaker Functions: Let π be any irreducible admissible representation of $GSp(2, F)$ and $\mathcal{W}(\pi, \psi)$ the Whittaker model of π with respect to the generic additive unitary character ψ of N^2, the standard maximal unipotent subgroup of $GSp(2)$ in form:

$$N^2 = \{u(x,y,w,z) = \begin{pmatrix} 1 & x & z & w \\ & 1 & w' & y \\ & & 1 & \\ & -x & & 1 \end{pmatrix}\}.$$

As usual, the generic additive unitary characters ψ can be assumed in following form: $\psi(u(x,y,w,z)) = \psi_\circ(x+y)$, where $\psi_\circ$ is any additive character of F.

Recall from [**JaSh1**] that a *finite* function on a locally compact abelian group is a continuous function whose translates span a vector space of finite dimension. When the locally compact abelian group is $(F^\times)^2$, any finite function on $(F^\times)^2$ is a finite linear combination of functions of the following types:

$$\chi(a,b) = \chi_1(a)\chi_2(b)|a|^{u_1}|b|^{u_2}(\log|a|^{n_1})(\log|a|^{n_1}), \qquad (152)$$

where χ_1 and χ_2 are characters module 1, u_1 and u_2 are real, and n_1 and n_2 are nonnegative integers.

We consider first the case that the underlying field F is a nonarchimedean local field. We shall use the methods from [**JPSS**], [**JaSh**], and [**Sou**] to estimate the Whittaker functions W in $\mathcal{W}(\pi,\psi)$ on the splitting torus. Since the representations are assumed to have trivial central characters, one has an identity

$$W(h(abc, bc, a^{-1}c, c)) = W(h(ab, b, a^{-1}, 1))$$

for $W \in \mathcal{W}(\pi,\psi)$ and $h(abc, bc, a^{-1}c, c) \in T_2$, the maximal split torus of $GSp(2)$. So we can assume that any element in the maximal torus T_2 is of form $h(ab, b, a^{-1}, 1)$ when deal with Whittaker functions. We denote $T_2' = \{h(ab, b, a^{-1}, 1) \in T_2\}$. As in [**Sou**], we have

LEMMA 1.1.1. *Let F be a nonarchimedean local field. Then we have*

(1) *The function $(a,b) \mapsto W(h(ab, b, a^{-1}, 1))$ vanishes when $|a|$ or $|b|$ is sufficiently large, and*

(2) *the function $(a,c,d) \mapsto W(h(cd, c, c^{-1}d^{-1}a, c^{-1}a))$ vanishes when $|d|$ or $|c^2a^{-1}|$ is sufficiently large.*

PROOF. The proof will be similar to that of Lemma 2.1 in [**Sou**]. Let $u(x,y,w,z)$ be an element in N^2. By the admissibility of $\mathcal{W}(\pi,\psi)$, we can pick up an element $u = u(x,y,w,z)$ so close to (but not equal to) I_4 that $W(gu) = W(g)$ for $g \in GSp(2)$ and $W \in \mathcal{W}(\pi_1,\psi)$. Now for any $t \in A_2'$, that is, $t = h(ab, a, b^{-1}, 1)$, we have

$$W(t) = W(tu) = W(tut^{-1}t) = \psi(bx + ay)W(t).$$

For a given $u = u(x,y,w,z)$, if $|a|$ and $|b|$ are sufficiently large, $\psi(bx + ay)$ will not be one and then $W(t)$ must vanish. This proves part (1). part (2) can be proved in exactly same way and will be omitted here. $\qquad\square$

Actually, Whittaker functions can be estimated more precisely by a formula analogue to that in [**JPSS**]. The argument to obtain such a formula is also from [**JPSS**]. In the case under consideration, the reductive group is $GSp(2)$. We have two maximal parabolic subgroup P_1^2 and P_2^2, up to conjugation. These are $P_1^2 = M_1^2 N_1^2$ with the Levi part $M_1^2 = (GL(1) \times GSp(1))^\circ$ and $P_2^2 = M_2^2 N_2^2$(Siegel parabolic). Let H_i be the center of M_i, $i = 1, 2$, respectively. Then by the same reason, we can assume that $H_1 = \{h(a, 1, a^{-1}, 1)\}$ and $H_2 = \{h(b, b, 1, 1)\}$. It is clear that $A_2' = H_1 H_2$. Let $\mathcal{W}(\pi, \psi)(N_i) = \{\pi_i(n_i)W - W : n_i \in N_i, W \in \mathcal{W}(\pi, \psi)\}$. It is easy to check as in [**Sou**] that for $W \in \mathcal{W}(\pi, \psi)(N_i)$ and $\underline{a}_i \in H_i$, $W(\underline{a}_i) = 0$ when $|\underline{a}_i|$ is sufficiently small or sufficiently large. The Jacquet module $\mathcal{W}(\pi, \psi)_{N_i} = \mathcal{W}(\pi, \psi)/\mathcal{W}(\pi, \psi)(N_i)$ is a finitely generated, admissible M_i-module for $i = 1, 2$. We denote by τ_i the representation of H_i on $\mathcal{W}(\pi, \psi)_{N_i}$. Then the algebra spanned by $\{\tau_i(\underline{a}_i) : \underline{a}_i \in H_i\}$ is of finite dimension. Now let $\mathcal{V}$ be the A_2'-module obtained by restricting $\mathcal{W}(\pi, \psi)$ to A_2' as in [**JPSS**], and $\mathcal{V}_i$ the subspace of vectors $v \in \mathcal{V}$ which vanish for $|\underline{a}_i|$ small enough and σ_i the representation of H_i on $\mathcal{V}/\mathcal{V}_i$. It is not difficult to see as in [**JPSS**] that $\mathcal{W}(\pi, \psi)(N_i)|_{H_i} \subset \mathcal{V}_i$ for $i = 1, 2$. Therefore the representation σ_i is a quotient of the representation τ_i and the algebra spanned by $\{\sigma_i(\underline{a}_i) : \underline{a}_i \in H_i\}$ is also finitely dimensional. Following the argument and Lemma (2,2,1) in [**JPSS**], we obtain an analogue of estimates for Whittaker functions.

PROPOSITION 1.1.1. *There exists a finite set X of finite functions on $(F^\times)^2$, so that for any Whittaker function W in $\mathcal{W}(\pi, \psi)$, there is for every $\chi \in X$, a function $\phi_\chi \in \mathcal{S}(F^2)$ satisfying*

$$W(\underline{t}) = \sum_{\chi \in X} \phi_\chi(a, b)\chi(a, b), \tag{153}$$

where $\underline{t} = h(ab, b, a^{-1}, 1)$ in A_2'.

Next we are going to estimate the Whittaker functions when the underlying field F is the real archimedean field. Since the results and their proofs are exactly same as those in [**Sou1**], which are the analogies of the corresponding results in [**JaSh**]. For convenience, we shall introduce the relevant notions and state the results. The proofs of those results will be omitted. Let $\mathfrak{g}$ be the Lie algebra of the reductive group $G = GSp(2, \mathbb{R})$. According to Casselman [**Cas1**], any irreducible admissible $(\mathfrak{g}, K_2)$-module V^0 can be realized as the $(\mathfrak{g}, K_2)$-module of a continuous irreducible representation π of G on a Frechet space V_π, which is smooth and of moderate growth in the sense of [**Cas1**]. We assume that π is generic. In other words, there is a continuous functional λ on V_π with respect to the standard maximal unipotent subgroup N^2 and the standard generic unitary character ψ of N^2. If we set $W_v(g) := \lambda(\pi(g)v)$ for $v \in V_\pi$, the space of all those $W_v(g)$ is called the Whittaker model of π and denoted by $\mathcal{W}(\pi, \psi)$, which is unique after [**Shl**].

One can define a gauge ξ on G as follows: Let $G = N^2 T_2 K_2$ be the Iwasawa decomposition of G, χ a sum of positive quasicharacters of T_2' as defined in subsection

1.1, and $\phi(a, b)$ a positive smooth and rapidly decreasing function in $\mathcal{S}(F^2)$, that is, given integers M, N, there is a positive constant C, so that

$$\phi(a, b) \leq C(1 + a^2)^M (1 + b^2)^N.$$

Then the gauge ξ satisfies

$$\xi(uak) = \chi(a)\phi(a, b), \tag{154}$$

see [**JPSS**].

LEMMA 1.1.2. *Let $f \in C_c^\infty(G)$. There is a gauge ξ on G and a continuous seminorm p on V_π, such that*

$$|W_{\pi(f)v}(g)| \leq p(v)\xi(g)$$

for all $g \in G, v \in V_\pi$.

PROPOSITION 1.1.2. *There is a finite set X of finite functions on $F^{\times 2}$, so that for any $W \in \mathcal{W}(\pi, \psi)$,*

(1) *there exist, for every $\alpha \in X$, a function $\phi_\alpha \in \mathcal{S}(F^2)$ satisfying*

$$W(h(ab, a, b^{-1}, 1)) = \sum_{\alpha \in X} \phi_\alpha(a, b)\alpha(a, b);$$

(2) *there exist, for every $\alpha \in X$, a function $\phi_\alpha \in \mathcal{S}(F^2 \times K_2)$ such that*

$$W(h(ab, a, b^{-1}, 1)k) = \sum_{\alpha \in X} \phi_\alpha(a, b; k)\alpha(a, b).$$

1.2. Estimate of the Partial 'Fourier' Transform. According to Chapter II, it is important to estimate the following integral:

$$I(u, v, \alpha, \beta; f, \psi) = \int_{F^3} f(\chi(w, y, x, u, v))\psi(\alpha x + \beta y)dwdxdy. \tag{155}$$

In this subsection, we shall consider two special cases: (1) $\alpha = \beta = 0$ over a local field and (2) over a nonarchimedean local field, the following integral

$$I(\alpha, \beta; f_s, \psi) = \int_{F^3} f_s(\chi(w, y, x))\psi(\alpha x + \beta y)dwdydx, \tag{156}$$

where α and β are in $F^\times$, and $\chi(w, y, x)$ is a unipotent subgroup of $GSp(4, F)$ of following form:

$$\chi(w, y, x) := \chi_{-\varepsilon_3 - \varepsilon_4}(w)\chi_{-2\varepsilon_3}(y)\chi_{-\varepsilon_2 - \varepsilon_4}(x). \tag{157}$$

Case (2) can be viewed as a partial Fourier transform of schwartz function $f_s(\chi(w, y, x))$ with respect to the additive character $\psi(\alpha x + \beta y)$, while case (1) is actually an intertwining operator as in the following Lemma.

LEMMA 1.2.1. *Let $f_s \in I_3^4(s)$. Then the following integral*

$$M_{w^*}(f_s)(g) = \frac{\zeta(s+2)\zeta(s+3)}{\zeta^2(s+1)} \int_{F^3} f_s(w^* \begin{pmatrix} 1 & & & & & & \\ & 1 & x & & & & w \\ & & 1 & & & & \\ & & & 1 & & w & y \\ & & & & 1 & & \\ & & & & & 1 & \\ & & & -x & & & 1 \end{pmatrix} g) dw dy dx \tag{158}$$

defines an intertwining operator from $I_{3v}^4(s)$ to

$$I_{B_4}^4(s) := ind_{B_4}^{GSp(4,F)}(|t_1|^{s+3}|t_2|^2|t_3|^{s+2}|t_4|^{-s}|t_4 t_4'|^{-\frac{1}{2}s - \frac{7}{2}})$$

and takes the normalized spherical section f_s° in $I_3^4(s)$ to the normalized spherical section Φ^0 in $I_{B_4}^4(s)$, where

$$w^* = \begin{pmatrix} 1 & & & 0 & & & \\ & 0 & 1 & & 0 & & \\ & 0 & 0 & & & 0 & 1 \\ & 1 & & 0 & & 0 & 0 \\ & & & & 1 & & \\ & & & & & 0 & 1 \\ & & -1 & & & 0 & 0 \\ & & & & & 1 & 0 \end{pmatrix}.$$

Now we consider case (2), here we assume the underlying field F is nonarchimedean.

LEMMA 1.2.2. *Let ψ be any additive unitary character of the field F with conduct δ_ψ, dx a Haar measure on F, and e any integer. For any $\alpha \in F^\times$, one has following formulas:*

(1)

$$\int_{|x| \leq q^{-e}} \psi(\alpha x) dx = \begin{cases} 0 & \text{if } q^{-e}|\alpha| > |\delta_\psi|; \\ q^{-e} & \text{if } q^{-e}|\alpha| \leq |\delta_\psi|, \end{cases}$$

(2)

$$\int_{\mathcal{O}^\times} \psi(\alpha x) dx = \begin{cases} 1 - q^{-1} & \text{if } ord_v(\alpha) \geq ord_v(\delta_\psi), \\ -q^{-1} & \text{if } ord_v(\alpha) = ord_v(\delta_\psi) - 1, \\ 0 & \text{if } ord_v(\alpha) < ord_v(\delta_\psi) - 1. \end{cases}$$

(3) *If $ord_v(\alpha) \geq e + ord_v(\delta_\psi) - 1$, the integral $\int_{|x| \geq q^e} |x|^{-s-2}\psi(\alpha x) dx$ equals a linear combination of 1 and $|\alpha|^{s+1}$ with coefficients in $\mathbb{C}(q^{-s})$, the field of rational functions in q^{-s} with complex coefficients; otherwise the integral vanishes.*

PROOF. Formulas (1) and (2) can be verified easily. We omit the proof here. We are going to check formula (3). Since

$$\int_{|x| \geq q^e} |x|^{-s-2}\psi(\alpha x) dx = \sum_{i \geq e} q^{-(s+1)i} \int_{\mathcal{O}^\times} \psi(\pi^{ord_v(\alpha)-i}x) dx.$$

It is easy to see that if $ord_v(\alpha) - e < ord_v(\delta_\psi) - 1$, then the last integral vanishes for any $i \geq e$. So we may assume that $e \leq ord_v(\alpha) - ord_v(\delta_\psi) + 1$. In this case, the integral can be computed as follows:

$$\int_{|x| \geq q^e} |x|^{-s-2} \psi(\alpha x) dx$$

$$\sum_{i=e}^{ord_v(\alpha) - ord_v(\delta_\psi)+1} q^{-(s+1)i} \int_{\mathcal{O}^\times} \psi(\pi^{ord_v(\alpha)-i}x) dx$$

$$\sum_{i=e}^{ord_v(\alpha) - ord_v(\delta_\psi)+1} q^{-(s+1)i} [\int_{\mathcal{O}} \psi(\pi^{ord_v(\alpha)-i}x) dx - q^{-1} \int_{\mathcal{O}} \psi(\pi^{ord_v(\alpha)-i+1}x) dx]$$

$$\sum_{i=e}^{ord_v(\alpha) - ord_v(\delta_\psi)} q^{-(s+1)i}(1 - q^{-1}) - q^{-(s+1)(ord_v(\alpha)-ord_v(\delta_\psi)+1)-1}.$$

It is not difficult to conclude that the last expression equals a linear combination of 1 and $|\alpha|^{s+1}$ with coefficients in $\mathbb{C}(q^{-s})$, the field of rational functions in q^{-s} with complex coefficients. $\qquad\qquad\qquad\qquad\qquad\qquad\qquad\qquad\qquad\qquad\qquad\qquad$ $\square$

LEMMA 1.2.3. *For any section* $f_s \in I_3^4(s)$ *and* $\chi(w) = \chi_{-\varepsilon_3-\varepsilon_4}(w)$, *we have*

$$\int_F f(\chi(w)) dw$$

$$= [\sum_{i_w} q^{-e}|w_{i_w}|f(\chi(w_{i_w})) + \sum_{i_w} |w_{i_w}|^{s+2} f(k(w_{i_w})) \frac{(1-q^{-1})q^{-(s+2)e}}{1 - q^{-(s+2)}}],$$

where the summation is over a finite set of indexes $\{i_w\}$. *In other words, the integral is equal to a rational function of* q^{-s}.

PROOF. Since the degenerate principal series representation $I_3^4(s)$ is admissible, there is a positive integer e so that $f(\chi(w)) = f(1)$ for $|w| \leq q^{-e}$. Let us denote by $\{w_{i_w}\}$ the set of representatives of the cosets of $\pi^e\mathcal{O}$ in $\mathcal{O}$. Then the integral can be reduced as follows:

$$\int_F f(\chi(w)) dw$$

$$= \int_{|w| \leq 1} f(\chi(w)) dw + \int_{|w| > 1} f(\chi(w)) dw$$

$$= \sum_{i_w} q^{-e}|w_{i_w}|f(\chi(w_{i_w})) + \int_{|w| > 1} |w|^{-(s+2)} f(k(w^{-1})) d^\times w.$$

By changing the variable $w \mapsto w^{-1}$, the last integral is equal to

$$
\begin{aligned}
\int_{|w|>1} |w|^{-(s+2)} f(k(w^{-1})) d^\times w &= \int_{0<|w|<1} |w|^{(s+2)} f(k(w)) d^\times w \\
&= \sum_{i_w} |w_{i_w}|^{s+2} f(k(w_{i_w})) \int_{0<|w|<q^{-e}} |w|^{s+2} d^\times w \\
&= \sum_{i_w} |w_{i_w}|^{s+2} f(k(w_{i_w})) q^{-(s+2)e} \int_{\mathcal{O}} |w|^{s+1} dw \\
&= \sum_{i_w} |w_{i_w}|^{s+2} f(k(w_{i_w})) \frac{(1-q^{-1}) q^{-(s+2)e}}{1-q^{-(s+2)}}
\end{aligned}
$$

$\square$

Recall that ψ is the additive character of F that defines the standard generic additive character ψ of the standard maximal unipotent subgroup N^2 of $GSp(2)$. We set $\psi_i(\alpha) := \psi(x_i \alpha)$ and $\psi_j(\beta) := \psi(y_j \beta)$, where x_i and y_j are elements in F. The main result of this section is

THEOREM 1.2.1. *For any $f_s \in I_3^4(s)$ and any additive unitary character ψ of the p-adic field F, the following integral*

$$
I(\alpha, \beta; f_s, \psi) = \int_{F^3} f(\chi(w, y, x)) \psi(\alpha x + \beta y) dw dy dx
$$

is, if $ord_v(\alpha) \geq e$ and $ord_v(\beta) \geq e$ for a positive integer e, a linear combination of $\psi_i(\alpha)\psi_j(\beta)$, $\psi_j(\beta)|\alpha|^{s+1}$, $\psi_i(\alpha)|\beta|^{s+1}$, and $|\alpha|^{s+1}|\beta|^{s+1}$, where i, j varies in a finite set of indexes, with coefficients in $\mathbb{C}(q^{-s})$, the field of rational functions in q^{-s} with complex coefficients; otherwise the integral $I(\alpha, \beta; f_s, \psi)$ will be zero.

PROOF. Basically, we deduce each variable once by means of the admissibility of the representation ρ_s, the degenerate principal series representation $I_3^4(s)$. First of all, let $w_{\varepsilon_2+\varepsilon_4}$ be a Weyl group element of following form:

$$
\begin{pmatrix}
1 & & & 0 & & \\
& 0 & & & 0 & 1 \\
& & 1 & & 0 & \\
& & & 0 & 1 & 0 \\
0 & & & 1 & & \\
& 0 & -1 & 0 & & 1 \\
& 0 & & & 1 & \\
-1 & & 0 & & & 0
\end{pmatrix}.
$$

Then there exists a positive integer e_x so that

$$
f(g\chi_{-\varepsilon_2-\varepsilon_4}(x)) = f(g) \text{ and } \rho_s(w_{\varepsilon_2+\varepsilon_4})(f)(g\chi_{-\varepsilon_2-\varepsilon_4}(x)) = \rho_s(w_{\varepsilon_2+\varepsilon_4})(f)(g),
$$

for any $|x| \leq q^{-e_x}$ by the admissibility of ρ_s. The integral $I(\alpha, \beta, ; f, \psi)$ can be deduced as follows.

$$
\begin{aligned}
& I(\alpha, \beta, ; f, \psi) \\
= {} & \int_{F^2} \psi(\beta y) \int_{|x| \leq q^{e_x}} f(\chi(w, y) \chi_{-\varepsilon_2 - \varepsilon_4}(x)) \psi(\alpha x) dx dw dy \\
& + \int_{F^2} \psi(\beta y) \int_{|x| > q^{e_x}} f(\chi(w, y) \chi_{-\varepsilon_2 - \varepsilon_4}(x)) \psi(\alpha x) dx dw dy \qquad (159) \\
= {} & I(\leq q^{e_x}) + I(> q^{e_x}).
\end{aligned}
$$

Then we consider the integral

$$
I(w, y; f_s) := \int_{|x| > q^{e_x}} f(\chi(w, y) \chi_{-\varepsilon_2 - \varepsilon_4}(x)) \psi(\alpha x) dx. \qquad (160)
$$

By Bruhat decomposition, one has

$$
\begin{aligned}
& \chi_{-\varepsilon_2 - \varepsilon_4}(x) \qquad\qquad\qquad\qquad\qquad\qquad\qquad\qquad (161) \\
= {} & h(1, x^{-1}, 1, x^{-1}, 1, x, 1, x) \chi_{\varepsilon_2 + \varepsilon_4}(x) \chi_{-\varepsilon_2 - \varepsilon_4}(-x^{-1}) w_{\varepsilon_2 + \varepsilon_4}.
\end{aligned}
$$

Plugging in the integral $I(w, y; f_s)$, we have

$$
\begin{aligned}
I(w, y; f_s) & = \int_{|x| > q^{e_x}} |x|^{-s-3} f_s(\chi(x^{-1} w, y) \chi_{-\varepsilon_2 - \varepsilon_4}(-x^{-1}) w_{\varepsilon_2 + \varepsilon_4}) \psi(\alpha x) dx \\
& = \int_{|x| > q^{e_x}} |x|^{-s-3} f_s(\chi(x^{-1} w, y)) w_{\varepsilon_2 + \varepsilon_4}) \psi(\alpha x) dx. \qquad (162)
\end{aligned}
$$

So, the integral $I(> q^{e_x})$ becomes

$$
\begin{aligned}
& \int_{F^2} \psi(\beta y) \int_{|x| > q^{e_x}} |x|^{-s-3} f(\chi(x^{-1} w, y) w_{\varepsilon_2 + \varepsilon_4}) \psi(\alpha x) dx dw dy \\
= {} & \int_{F^2} \psi(\beta y) \int_{|x| > q^{e_x}} |x|^{-s-2} f(\chi(w, y) w_{\varepsilon_2 + \varepsilon_4}) \psi(\alpha x) dx dw dy \\
= {} & \int_{|x| > q^{e_x}} |x|^{-s-2} \psi(\alpha x) dx \int_{F^2} \psi(\beta y) f(\chi(w, y) w_{\varepsilon_2 + \varepsilon_4}) dw dy. \qquad (163)
\end{aligned}
$$

Denote by $\{x_i\}$ the set of the representatives of $\mathcal{P}^{-e_x} / \mathcal{P}^{e_x}$, where $\mathcal{P}$ denotes the maximal ideal in the ring $\mathcal{O}$ of integers in the p-adic field F. Then we obtain that

$$
\begin{aligned}
I(\leq q^{e_x}) & = \sum_i \int_{F^2} \psi(\beta y) \int_{|x| \leq q^{-e_x}} f(\chi(w, y, x_i + x)) \psi(\alpha(x_i + x)) dx dw dy \\
& = \sum_i \psi(\alpha x_i) \int_{|x| \leq q^{-e_x}} \psi(\alpha x) dx \int_{F^2} (\rho_s(\chi(x_i)) f)(\chi(w, y)) \psi(\beta y) dw dy.
\end{aligned}
$$

In other words, we obtain that

$$
I(\alpha, \beta, ; f, \psi) \tag{164}
$$
$$
= \int_{|x|>q^{e_x}} |x|^{-s-2}\psi(\alpha x)dx \int_{F^2} \psi(\beta y)f(\chi(w,y)w_{\varepsilon_2+\varepsilon_4})dwdy
$$
$$
+ \sum_i \psi(\alpha x_i) \int_{|x|\leq q^{-e_x}} \psi(\alpha x)dx \int_{F^2} \rho_s(\chi_{-\varepsilon_2-\varepsilon_4}(x_i))(f)(\chi(w,y))\psi(\beta y)dwdy.
$$

Next we have to consider the integrals of following type:

$$
I(\beta; f_s, \psi) := \int_{F^2} f_s(\chi(w,y))\psi(\beta y)dwdy, \tag{165}
$$

for any section f_s in $I_3^4(s)$. Applying the same procedure to the integrating variable y, we obtain

$$
I(\beta, ; f, \psi)
$$
$$
= \int_{|y|>q^{e_y}} |y|^{-s-2}\psi(\beta y)dy \int_F f(\chi(w)w_{2\varepsilon_3})dw
$$
$$
+ \sum_j \psi(\beta y_j) \int_{|y|\leq q^{-e_y}} \psi(\beta y)dy \int_F \rho_s(\chi_{-2\varepsilon_3}(y_j)(f)(\chi(w))dw. \tag{166}
$$

Note here that the positive integer e_y depends on more sections f's than e_x does and $\{y_j\}$ is the set of all representatives of $\mathcal{P}^{-e_y}/\mathcal{P}^{e_y}$. Finally we obtain that

$$
I(\alpha, \beta, ; f, \psi)
$$
$$
= \int_{|x|>q^{e_x}} |x|^{-s-2}\psi(\alpha x)dx \int_{|y|>q^{e_y}} |y|^{-s-2}\psi(\beta y)dy \int_f f'(\chi(w))dw
$$
$$
+ \sum_j \int_{|x|>q^{e_x}} |x|^{-s-2}\psi(\alpha x)dx\psi(\beta y_j) \int_{|y|\leq q^{-e_y}} \psi(\beta y)dy \int_F f_j(\chi(w))dw
$$
$$
+ \sum_i \psi(\alpha x_i) \int_{|x|\leq q^{-e_x}} \psi(\alpha x)dx \int_{|y|>q^{e_y}} |y|^{-s-2}\psi(\beta y)dy \int_f f_i(\chi(w))dw \tag{167}
$$
$$
+ \sum_{i,j} \psi(\alpha x_i) \int_{|x|\leq q^{-e_x}} \psi(\alpha x)dx\psi(\beta y_j) \int_{|y|\leq q^{-e_y}} \psi(\beta y)dy \int_F f_{i,j}(\chi(w))dw
$$

where sections f', f_j, f_i, and $f_{i,j}$ are as follows:

$$
\begin{aligned}
f' &:= \rho_s(w_{2\varepsilon_3}w_{\varepsilon_2+\varepsilon_4})(f_s) \\
f_j &:= \rho_s(\chi_{-2\varepsilon_3}(y_j)w_{\varepsilon_2+\varepsilon_4})(f_s) \\
f_i &:= \rho_s(w_{2\varepsilon_3}\chi_{-\varepsilon_2-\varepsilon_4}(x_i))(f_s) \\
f_{i,j} &:= \rho_s(\chi_{-2\varepsilon_3}(y_j)\chi_{-\varepsilon_2-\varepsilon_4}(x_i))(f_s).
\end{aligned}
$$

According to Lemma 1.2.2 and 1.2.3 above, there is a positive integer e such that when $ord_v(\alpha) < e$ or $ord_v(\beta) < e$, the integral $I(\alpha, \beta, ; f, \psi)$ will vanish. On the other hand, if $ord_v(\alpha) \geq e$ and $ord_v(\beta) \geq e$, the integrals $\int_{|x|\leq q^{-e_x}} \psi(\alpha x)dx$, $\int_{|y|\leq q^{-e_y}} \psi(\beta y)dy$, and

$\int_F f(\chi(w))dw$ are rational functions in q^{-s} and the integrals $\int_{|x|\geq q^{e_x}} |x|^{-s-2}\psi(\alpha x)dx$ and $\int_{|y|\geq q_y^e} |y|^{-s-2}\psi(\alpha y)dy$ are equal to linear combinations of 1 and $|\alpha|^{s+1}$ with coefficients in $\mathbb{C}(q^{-s})$, the field of rational functions in q^{-s} with complex coefficients. Therefore, if $ord_v(\alpha) \geq e$ and $ord_v(\beta) \geq e$ and if we set $\psi_i(\alpha) := \psi(x_i\alpha)$ and $\psi_j(\beta) := \psi(y_j\beta)$, the integral $I(\alpha, \beta, ; f, \psi)$ is a linear combination of $\psi_i(\alpha)\psi_j(\beta)$, $\psi_j(\beta)|\alpha|^{s+1}$, $\psi_i(\alpha)|\beta|^{s+1}$, and $|\alpha|^{s+1}|\beta|^{s+1}$ with coefficients in $\mathbb{C}(q^{-s})$, the field of rational functions in q^{-s} with complex coefficients. $\qquad\square$

2. Nonvanishing of Local Zeta Integrals

The local zeta integral we are going to study in this section is

$$\mathcal{Z}(s, W_1, W_2, f) = \int_{\mathcal{D}(F)} f(\gamma_\circ(g_1, g_2), s)W_1(g_1)W_2(g_2)dg_1dg_2, \tag{168}$$

where $\mathcal{D} := CN^{2,\triangle}(Z_2 \times I_4) \setminus H$ and f_s is any smooth section in $I_3^4(s)$ and W_1, W_2 are smooth Whittaker functions of $GSp(2, F)$ in the Whittaker models $\mathcal{W}(\pi_1, \psi)$, $\mathcal{W}(\pi_2, \overline{\psi})$, respectively. We shall prove that one can make an appropriate choice of (f_s, W_1, W_2) so that the integral $\mathcal{I}(s, W_1, W_2, f)$ is a constant independent of s.

2.1. Nonarchimedean Cases: We begin with proving the nonvanishing property for a certain type of integral similar to the integrals studied extensively by Jacquet and Shalika in [**JaSh**] and [**JaSh1**].

We first recall some notations. Let P_1^2 be the maximal parabolic subgroup of $GSp(2)$ with Levi decomposition $P_1^2 = M_1^2N_1^2$. The Levi factor M_1^2 and the unipotent radical N_1^2 are in forms:

$$P_1^2 = \left\{ \begin{pmatrix} a & & & \\ & \alpha & \beta & \\ & & a' & \\ & \gamma & & \delta \end{pmatrix} \right\} \text{ and } N_1^2 = \left\{ n(x,w,z) = \begin{pmatrix} 1 & x & z & w \\ & 1 & w & 0 \\ & & 1 & \\ & & -x & 1 \end{pmatrix} \right\}.$$

We choose such an embedding of $GL(2)$ into the Levi part M_1^2 of the parabolic subgroup P_1^2 of $GSp(2)$ as

$$g = \begin{pmatrix} \alpha & \beta \\ \gamma & \delta \end{pmatrix} \longmapsto m(detg, g) = \begin{pmatrix} detg & & & \\ & \alpha & & -\beta \\ & & 1 & \\ & -\gamma & & \delta \end{pmatrix}.$$

Let B be the standard Borel subgroup of $GL(2)$ and $B = NT$ in the usual sense. We are going to study the following integral

$$I(s, W_1, W_2, \Phi) = \int_{U_2 \backslash GL(2)} W_1(m(detg, g))W_2(m(detg, g))\Phi((0,1)g)|detg|^s dg, \tag{169}$$

the properties of which will be used to prove the nonvanishing property of the integral $\mathcal{W}(s, W_1, W_2, f_s)$ for an appropriately chosen data (W_1, W_2, f_s).

LEMMA 2.1.1. (a) *For any Whittaker functions $W_1 \in \mathcal{W}(\pi_1, \psi)$ and $W_2 \in \mathcal{W}(\pi_2, \overline{\psi})$, and any Schwartz-Bruhat function $\Phi \in \mathcal{S}(F^2)$, $I(s, W_1, W_2, \Phi)$ converges absolutely for $Re(s) >> 0$.*

(b) *For a suitable choice of Whittaker functions $\tilde{W}_1 \in \mathcal{W}(\pi_1, \psi)$ and $W_2 \in \mathcal{W}(\pi_2, \overline{\psi})$, and for a suitable choice of a Schwartz-Bruhat function $\Phi \in \mathcal{S}(F^2)$, the integral $I(s, \tilde{W}_1, W_2, \Phi)$ is a nonzero constant independent of s.*

PROOF. The proof of the absolute convergence of the integral will follow as a special case from that of absolute convergence of the local zeta integral, which will be given in the next section. Here we only give the proof for (b). In terms of Iwasawa decomposition of $GL(2)$, $U_2 \backslash GL(2) = [U_2 \backslash P] F^\times K'$, where $K' = GL(2, \mathcal{O})$, $F^\times$ is the center of $GL(2)$, and $U_2 \backslash P = \begin{pmatrix} \alpha & 0 \\ 0 & 1 \end{pmatrix}$. Then the integral $I(s, W_1, W_2, \Phi)$ equals

$$\int_{U_2 \backslash GL(2)} W_1(m(detg, g)) W_2(m(detg, g)) \Phi((0,1)g) |detg|^s dg$$

$$= \int_{K'} \int_{F^\times} \int_{U_2 \backslash P} |detp|^{s_1} W_1(pak') W_2(pak') dp \Phi((0,1)ak') |a|^{s_2} d^\times a dk'$$

for any Whittaker functions W_1 and W_2, and any Schwartz-Bruhat function Φ. We denote the inner integral by $J(g, s, W_1, W_2)$, that is,

$$J(g, s, W_1, W_2) = \int_{U_2 \backslash P} |detp|^s W_1(pg) W_2(pg) dp. \tag{170}$$

Then this integral enjoys following two properties:

(1) $J(pg, s, W_1, W_2) = J(g, s, W_1, W_2)$ for $p \in P \cap K'$.
(2) We can choose Whittaker functions $\tilde{W}_1$ and W_2 such that the integral

$$J(1, s, \tilde{W}_1, W_2) = 1.$$

Property (1) is straightforward, while property (2) is not so direct to see. We are going to verify property (2).

Note that there are Whittaker functions $\tilde{W}_1 \in \mathcal{W}(\pi_1, \psi)$ and $\tilde{W}_2 \in \mathcal{W}(\pi_2, \overline{\psi})$ so that $\tilde{W}_i(1) = 1$ for $i = 1, 2$. Then we can pick up a Schwartz-Bruhat function $\Phi' \in \mathcal{S}(F)$ with its compact support supp(Φ') in a so small neighborhood of 1 that the following integral can be made one, that is,

$$\int_{F^\times} |a|^{s_1} \tilde{W}_1(h_2(a, a, 1, 1)) \tilde{W}_2(h_2(a, a, 1, 1)) \Phi'(a) d^\times a = 1 \tag{171}$$

where $h_2(a, b, c, d)$ denote a diagonal matrix element in $GSp(2)$.

We then consider the Fourier transform $\hat{\Phi}'$ of Φ' with respect to the additive character $\psi_\circ$ of F, i.e.,

$$\hat{\Phi}'(y) = \int_F \Phi'(x) \psi_\circ(yx) dx, \tag{172}$$

which is also a Schwartz-Bruhat function on F. We define a convolution of the Whittaker function $\tilde{W}_2$ with the Schwartz-Bruhat function $\hat{\Phi}'$ by

$$W_2(g) := \cdot \int_F \tilde{W}_2(gu(0, y, 0, 0))\hat{\Phi}'(y)dy \tag{173}$$

for $g \in GSp(2)$. It is easy to check that $W_2(g)$ is again a Whittaker function in $\mathcal{W}(\pi_2, \overline{\psi})$. With those choice of Whittaker functions $\tilde{W}_1$ and W_2, the integral $J(g, s, \tilde{W}_1, W_2)$ at $g = 1$ will equal one since

$$J(1, s_1, \tilde{W}_1, W_2)$$
$$\int_{F^\times} |a|^{s_1} \tilde{W}_1(h_2(a, a, 1, 1))W_2(h_2(a, a, 1, 1))d^\times a$$
$$= \int_{F^\times} |a|^{s_1} \tilde{W}_1(h_2(a, a, 1, 1)) \int_F \tilde{W}_2(h_2(a, a, 1, 1)u(0, y, 0, 0))\hat{\Phi}'(y)dyd^\times a$$
$$= \int_{F^\times} |a|^{s_1} \tilde{W}_1(h_2(a, a, 1, 1))\tilde{W}_2(h_2(a, a, 1, 1)) \int_F \hat{\Phi}'(y)\overline{\psi}(ay)dyd^\times a$$
$$= \int_{F^\times} |a|^{s_1} \tilde{W}_1(h_2(a, a, 1, 1))\tilde{W}_2(h_2(a, a, 1, 1))\Phi'(a)d^\times a$$
$$= 1.$$

This prove that the integral $J(g, s, W_1, W_2)$ enjoys property (2). Note that with the chosen Whittaker functions $\tilde{W}_1$ and W_2, the integral $J(g, s, \tilde{W}_1, W_2)$ defines a locally constant complex-valued function on $GSp(2, F)$, which we denote by $J(g)$. Then there is a small open compact subgroup $K'_\circ$ contained in K' so that the restriction of $J(g)$ to $K'_\circ$ is a nonzero constant independent of s and $(0, 1)K'_\circ \subset \mathcal{O}^{\times 2}$. We choose the Schwartz-Bruhat function $\Phi = ch_{(0,1)K'_\circ}$ the characteristic function of $(0, 1)K'_\circ$. Then the desired property of the integral $I(s, \tilde{W}_1, W_2, \Phi)$ for the chosen data $(\tilde{W}_1, W_2, \Phi)$ can be deduced as follows:

$$\begin{aligned} I(s, \tilde{W}_1, W_2, \Phi) &= \int_{K'} \int_{F_v^\times} J(ak')\Phi((0, 1)ak')|a|^{s_2}d^\times adk' \\ &= m_\Phi \int_{K'} J(k')\Phi((0, 1)k')dk' \\ &= m_\Phi \int_{[P \cap K']K'_\circ} J(pk')dpdk' \\ &= m_1 \int_{K'_\circ} J(k')dk'. \end{aligned}$$

It is sure that the last integral is a nonzero constant independent of s. This proves the lemma. $\qquad \square$

Now we turn to deal with the local zeta integral $\mathcal{I}(s, W_1, W_2, f_s)$. We need more notations. The center Z_2 of N_1^2 is the one-parameter subgroup $\chi_{2\varepsilon_1}(z)$. By means of the isomorphism $Z_2 \backslash N_1^2 \cong F^2$ as algebraic varieties, it is well known that there is a surjection from the space $\mathcal{S}(N_1^2)$ of Schwartz-Bruhat functions on N_1^2 onto the space $\mathcal{S}(F^2)$ of Schwartz-Bruhat functions on F^2. The surjection is defined by the integral:

$$\rho_z \ : \ \varphi \longmapsto \Phi(x,w) = \int_F \varphi(n(x,w,z))dz. \tag{174}$$

We also define the Fourier transform $\hat{\Phi}$ of Φ with respect to the additive unitary character $\psi_\circ(-uw - vx)$ of F^2 by

$$\hat{\Phi}(x,w,) = \int_{F^2} \Phi(u,v)\psi(-uw - vx)dudv. \tag{175}$$

Note that we can choose a Schwartz-Bruhat functions $\varphi \in \mathcal{S}(N_1^2)$ and $\Phi \in \mathcal{S}(F^2)$ so that $\rho_z(\varphi) = \hat{\Phi}(x,w)$.

THEOREM 2.1.1 (Non-archimedean Cases). *We can choose an appropriate section $f_s \in I_3^4(s)$, and Whittaker functions $W_1 \in \mathcal{W}(\pi_1, \psi)$ and $W_2 \in \mathcal{W}(\pi_2, \overline{\psi})$, so that the (unnormalized) local zeta integral $\mathcal{Z}(s, W_1, W_2, f_s)$ equals a nonzero constant independent of s.*

PROOF. Let $\tilde{W}_1$, W_2 be the Whittaker functions and Φ the Schwartz Bruhat function on F^2 as chosen in the proof of the previous lemma. We define now another Whittaker function W_1 in $\mathcal{W}(\pi_1, \psi)$ as follows: for $g \in GSp(2)$ and $\varphi \in \mathcal{S}(N_1^2)$,

$$W_1(g) := \int_{N_1^2} \tilde{W}_1(gu)\varphi(u)du. \tag{176}$$

Since the representations π_1 and π_2 are admissible, there is a small open and compact subgroup $K_\circ$ contained in the maximal open and compact subgroup K_2 of $GSp(2)$ so that W_1 and W_2 are right $K_\circ$-invariant.

Next, we are going to make the choice of the section f_s. Since $P_3^4\gamma_\circ H$ is the unique open H-orbit in $P_3^4 \backslash GSp(4)$, it is easy to check that the following isomorphism of H-modules holds

$$\{f \in I_3^4(s) \ : \ supp(f) \subset P_3\gamma_\circ H\} \cong ind_{P_1^{2,\triangle}(Z_2\times I_4)}^H (|a|^{\frac{s+3}{2}}|a'|^{-\frac{s+3}{2}})$$

$$f(\gamma_\circ(g_1,g_2)) = f^*(g_1, g_2).$$

Here $P_1^{2,\triangle}(Z_2 \times I_4)$ is the stabilizer of $\gamma_\circ$ in H and f^* has compact support modulo $P_1^{2,\triangle}(Z_2 \times I_4)$.

Applying the Iwasawa decomposition of $H = (GSp(2)\times GSp(2))^\circ$ to our situation, we can decompose the group H into $H = P_1^{2,\triangle}(Z_2 \times I_4)([Z_2\backslash P_1] \times I_4)^\circ(K_2 \times K_2)^\circ$. We claim that

$$P_1^{2,\triangle}(Z_2 \times I_4) \cap [([Z_2\backslash P_1] \times I_4)^\circ(K_2 \times K_2)^\circ] = P_1^{2,\triangle}(Z_2 \times I_4) \cap (K_2 \times K_2)^\circ. \tag{177}$$

This will be proved at the end of this proof.

We denote $Q = P_1^{2,\triangle}(Z_2 \times I_4) \cap (K_2 \times K_2)^\circ$ and $Q_\circ = P_1^{2,\triangle}(Z_2 \times I_4) \cap (K_\circ \times K_\circ)^\circ$. Then the quotient $Q/Q_\circ$ is a finite set, which is denoted by $\{q_1, q_2, \cdots, q_t\}$. Let

$\Omega_o = \cup_{i=1}^{t} q_i (K_o \times K_o)^\circ$. Then Ω_o is an open compact subset in $(Z_2 \backslash P_1 \times I_4)^\circ (K_2 \times K_2)^\circ$. It is evident that there exists a section $f^* \in ind_{P_1^{2,\triangle}(Z_2 \times I_4)}^{H}(|a|^{\frac{s+3}{2}}|a'|^{-\frac{s+3}{2}})$ so that the restriction of f^* to $(Z_2 \backslash P_1 \times I_4)^\circ (K_2 \times K_2)^\circ$ is equal to the characteristic function of Ω_o and its support $supp(f^*)$ is $P_1^{2,\triangle}(Z_2 \times I_4)(K_o \times K_o)^\circ$. In this way, we obtain the section f in $I_3^4(s)$ via the correspondence indicated above.

With the data (W_1, W_2, f) as chosen above, the local zeta integral $\mathcal{Z}(s, W_1, W_2, f)$ equals

$$\int_{CN^{2,\triangle}(Z_2 \times I_4) \backslash H} f(\gamma_o(g_1, g_2), s) W_1(g_1) W_2(g_2) dg_1 dg_2$$

$$= \int_{[C_2 N^2 \backslash P_1](K_o \times K_o)^\circ} |a|^{\frac{3}{2}s} W_1(pk_1) W_2(pk_2) dp dk_1 dk_2$$

$$= \mu((K_o \times K_o)^\circ) \int_{U_2 \backslash GL(2)} |detg|^{s'} W_1(m(detg, g)) W_2(m(detg, g)) dg$$

where $\mu((K_o \times K_o)^\circ)$ denote the volume of $(K_o \times K_o)^\circ$ with respect to the Haar measure on the group $GSp(2)$. Replacing W_1 by the corresponding integral, we obtain that

$$\int_{U_2 \backslash GL(2)} |detg|^{s'} W_1(m(detg, g)) W_2(m(detg, g)) dg$$

$$= \int_{U_2 \backslash GL(2)} |detg|^{s'} \int_{N_1^2} \tilde{W}_1(m(detg, g)u) \varphi(u) du W_2(m(detg, g)) dg$$

$$= \int_{U_2 \backslash GL(2)} |detg|^{s'} \tilde{W}_1(m(detg, g)) W_2(m(detg, g))$$

$$\cdot \int_{N_1^2} \varphi(u(x, w, z)) \psi(\gamma w + \delta x) dz dx dw dg$$

$$= \int_{U_2 \backslash GL(2)} |detg|^{s'} \tilde{W}_1(m(detg, g)) W_2(m(detg, g)) \int_{F^2} \hat{\Phi}(x, w) \psi(\gamma w + \delta x) dx dw dg$$

$$= \int_{U_2 \backslash GL(2)} |detg|^{s'} \tilde{W}_1(m(detg, g)) W_2(m(detg, g)) \Phi((0, 1)g) dg,$$

where we let $g = \begin{pmatrix} \alpha & \beta \\ \gamma & \delta \end{pmatrix}$. This implies that $\Phi((\gamma, \delta)) = \Phi((0,1)g)$. According to the Lemma 2.1.1 (b), with the chosen data (W_1, W_2, f_s), the last integral equals a nonzero constant independent of s and so does the local zeta integral $\mathcal{Z}(s, W_1, W_2, f_s)$. Finally we are going to prove the *claim*: '$P_1^{2,\triangle}(Z_2 \times I_4) \cap [(Z_2 \backslash P_1 \times I_4)^\circ (K_2 \times K_2)^\circ] = P_1^{2,\triangle}(Z_2 \times I_4) \cap (K_2 \times K_2)^\circ$'. In fact, for any $x \in P_1^{2,\triangle}(Z_2 \times I_4) \cap [(Z_2 \backslash P_1 \times I_4)^\circ (K_2 \times K_2)^\circ]$, x can be written as

$$x = (pz, p) = (p_1 k_1, k_2).$$

Then we have $p = k_2$ and $pz = p_1 k_1$. This implies that $k_1 \in P_1 \cap K_2$ and the Levi part m_1 of p_1 belongs to $M_1^2 \cap K_2$. Hence we obtain that $x = (pz, p) = (m_1 u, p)$ with $u \in Z_2 \backslash U_1$ and $p \in P_1 \cap K_2$. This implies that $u \in U_1 \cap K_2$ and $z \in Z_2 \cap K_2$. Therefore $x \in P_1^{2,\triangle}(Z_2 \times I_4) \cap (K_2 \times K_2)^\circ$. Our claim is true. $\qquad \square$

COROLLARY 2.1.1. *With an appropriate choice of the data (W_1, W_2, f_s), we can make the local zeta integral $\mathcal{Z}(s, W_1, W_2, f_s) = 1$.*

2.2. Archimedean Cases: We assume in this subsection that the local field F is the real archimedean field. We denote by K_2 the standard maximal compact subgroup of $GSp(2, F)$. In other words, $K_2 = GU(2)$. Let π_1 and π_2 be irreducible admissible representations of $GSp(2, F)$ and $\pi_{1,\circ}$ and $\pi_{2,\circ}$ the K_2-finite vectors of π_1 and π_2, respectively. We denote by $\mathcal{W}_\circ(\pi_1, \psi)$ and $\mathcal{W}_\circ(\pi_2, \overline{\psi})$ the K_2-finite Whittaker functions in $\mathcal{W}(\pi_1, \psi)$ and $\mathcal{W}(\pi_1, \overline{\psi})$, respectively. However, we assume that the degenerate principal series representation $I_3^4(s)$ consists of all *smooth* sections (not necessarily K_4-finite). Our local zeta integral is

$$\mathcal{Z}(s, W_1, W_2, f) = \int_{[CN^{2,\triangle}(Z_2 \times I_4)\backslash H]} f(\gamma(g_1, g_2), s)W_1(g_1)W_2(g_2)dg_1 dg_2. \tag{178}$$

It is the goal of this subsection to prove an archimedean version of the nonvanishing property of local zeta integrals. The statement will be

THEOREM 2.2.1 (Archimedean Cases). *For a given value s_0 of s, there are K_2-finite Whittaker functions $W_1 \in \mathcal{W}_\circ(\pi_1, \psi)$ and $W_2 \in \mathcal{W}_\circ(\pi_2, \overline{\psi})$, and then there is a smooth function $f_s \in I_3^4(s)$, so that the zeta integral $\mathcal{Z}(s_0, W_1, W_2, f_s)$ dose not vanish.*

The idea to prove this theorem is similar to that used in our proof of non-archimedean cases. In other words, we have to prove the theorem for an integral of the following type

$$I(s, W_1, W_2, \Phi) = \int_{U_2 \backslash GL(2)} |detg|^s W_1(m(detg, g))W_2(m(detg, g))\Phi((0, 1)g)dg. \tag{179}$$

Again such types of archimedean integrals were studied extensively by Jacquet and Shalika in [**JaSh**] and [**JaSh1**].

LEMMA 2.2.1.   (a) *For K_2-finite Whittaker functions W_1 and W_2 and Schwartz function $\Phi(x, y) \in C_c^\infty(F^2)$, the integral $I(s, W_1, W_2, \Phi)$ converges absolutely for $Re(s) >> 0$.*
   (b) *For a given value s, we can choose K_2-finite Whittaker functions W_1 and W_2, and a function Φ in $C_c^\infty(F^2)$ so that $I(s, W_1, W_2, \Phi)$ does not vanish.*

PROOF. Statement (a) will follow from the estimates of Whittaker functions. The proof for more general cases will be given in the next section. We are going to prove (b) here.

Let K' be the standard maximal compact subgroup of $GL(2, F)$, i.e., $K' = O(2)$. Applying the Iwasawa decomposition, the quotient $U_2 \backslash GL(2)$ can be written as $[U_2 \backslash P]F^\times K'$ where $F^\times$ is the center of $GL(2)$ and $U_2 \backslash P = \begin{pmatrix} \alpha & 0 \\ 0 & 1 \end{pmatrix}$. Thus the integral $I(s, W_1, W_2, \Phi)$ can be deduced as follows:

$$
\begin{aligned}
& I(s, W_1, W_2, \Phi) \\
= & \int_{U_2 \backslash GL(2)} |detg|^s W_1(m(detg, g)) W_2(m(detg, g)) \Phi((0,1)g) dg \\
= & \int_{K'} \int_{F^\times} \int_{U_2 \backslash P} |detp|^{s_1} W_1(pak') W_2(pak')) dp \Phi((0,1)ak') |a|^{s_2} d^\times a dk'. \quad (180)
\end{aligned}
$$

As in the non-archimedean cases, for a fixed s_1, we denote by $J(g, s_1, W_1, W_2)$ the inner integral of (176), that is,

$$
J(g, s_1, W_1, W_2) = \int_{U_2 \backslash P} |detp|^{s_1} W_1(pg) W_2(pg)) dp. \quad (181)
$$

Note that the function $J(g, s_1, W_1, W_2)$ is K'-finite.

We are going to pick up some Whittaker functions W_1 and W_2 so that the integral $J(1, s_1, W_1, W_2) \neq 0$. We choose K_2-finite Whittaker functions $W_1(g)$ in $\mathcal{W}(\pi_1, \psi)$ and $\tilde{W}_2(g)$ in $\mathcal{W}(\pi_2, \tilde{\psi})$ so that $W_1(e) = \tilde{W}_2(e) = 1$. We choose a function $\Phi'(t) \in C_c^\infty(F^\times)$ whose support is such a small neighborhood of 1 that the following integral

$$
\int_{F^\times} |t|^{s_1} W_1(h_2(t, t, 1, 1)) \tilde{W}_2(h_2(t, t, 1, 1)) \Phi'(t) d^\times t \quad (182)
$$

does not vanish. For the sake of convenience, we can make it equal to 1. We define the Fourier transform of Φ' with respect to the additive character $\psi_\circ$ as

$$
\hat{\Phi}'(y) = \int_F \Phi'(x) \psi_\circ(yx) dx, \quad (183)
$$

which is again in $C_c^\infty(F^\times)$. Then a convolution of the Whittaker function $\tilde{W}_2$ and the Schwartz function $\hat{\Phi}'$ is defined as

$$
W_2(g) := \int_F \tilde{W}_2(gu(0, y, 0, 0)) \hat{\Phi}'(y) dy, \quad (184)
$$

which is a Whittaker function in $\mathcal{W}(\pi_2, \overline{\psi})$. For such chosen Whittaker functions $W_1(g)$ and $W_2(g)$, the integral $J(g, s_1, W_1, W_2)$ at $g = 1$ can be computed as

$$J(1, s_1, W_1, W_2)$$
$$= \int_{U_2 \backslash P} |det p|^{s_1} W_1(p) W_2(p) dp$$
$$= \int_{F^\times} |t|^{s_1'} W_1(h_2(t, t, 1, 1)) W_2(h_2(t, t, 1, 1)) dt^\times$$
$$= \int_{F^\times} |t|^{s_1'} W_1(h_2(t, t, 1, 1)) \int_F \tilde{W}_2(h_2(t, t, 1, 1) u(0, y, 0, 0)) \hat{\Phi}'(y) dy dt^\times$$
$$= \int_{F^\times} |t|^{s_1'} W_1(h_2(t, t, 1, 1)) \tilde{W}_2(h_2(t, t, 1, 1)) \int_F \hat{\Phi}'(y) \overline{\psi}(ty) dy dt^\times$$
$$= \int_{F^\times} |t|^{s_1'} W_1(h_2(t, t, 1, 1)) \tilde{W}_2(h_2(t, t, 1, 1)) \Phi'(t) dt^\times$$
$$= 1. \tag{185}$$

With the chosen Whittaker functions W_1 and W_2 and the fixed s_1, we set $J(g) := J(g, s_1, W_1, W_2)$, which is a K'-finite function on $GL(2)$ and $J(1) = 1$.

With the preparation above, we can prove the statement (b) as follows:

$$(176) = \int_{K'} \int_{F^\times} \int_{U_2 \backslash P} |det p|^{s_1} W_1(pak') W_2(pak')) dp \Phi((0, 1) ak') |a|^{s_2} d^\times a dk'$$
$$= \int_{K'} \int_{F^\times} J(ak') \Phi((0, 1) ak') |a|^{s_2} d^\times a dk'$$
$$= \int_{F^\times} |a|^{s_2} \int_{K'} J(k'a) \Phi((0, 1) k'a) dk' d^\times a. \tag{186}$$

We define $\Phi((0, 1) k'a) := \phi(a) \overline{J}(k')$ with $\phi(a) \in C_c^\infty(F^\times)$ and $\phi(1) = 1$, $\overline{J}$ the complex conjugation of J. Then $\Phi((0, 1) k'a)$ belongs to $\in C_c^\infty(F^2)$ since $J(g)$ is K'-finite. Plugging the function $\Phi((0, 1) k'a) := \phi(a) \overline{J}(k')$ into the integral (183), we obtain that

$$(183) = \int_{F^\times} |a|^{s_2} \int_{K'} J(k'a) \overline{J}(k') dk' \phi(a) d^\times a. \tag{187}$$

Note that if we set $\varphi(a) = \int_{K'} J(k'a) \overline{J}(k') dk'$, then the function $\varphi(a)$ is continuous in a and $\varphi(1) = 1$. We can assume that the support of the function $\phi(a)$ is in such a small neighborhood of 1 that the restriction of $\varphi(a)$ to the support of ϕ is always positive and the integral (183) does not vanish. This finishes the proof. $\square$

PROOF. **(The Proof of Theorem 2.2.1)** For a given complex number s, we choose W_1 and W_2 as in Lemma 2.2.1. As in the p-adic case, we have the decomposition of the group H,

$$H = (P_1 \times P_1)^\circ (K_2 \times K_2)^\circ = P_1^{2, \triangle} (Z_2 \times I_4)([Z_2 \backslash P_1] \times I_4)^\circ (K_2 \times K_2)^\circ, \tag{188}$$

and the following isomorphism of H-modules

$$\{f \in ind_{P_3}^G(\delta^s) \; : \; supp(f) \subset P_3\gamma_\circ H\} \cong ind_{P_1^{2,\triangle}(Z_2\times I_4)}^H(|a|^{\frac{s+3}{2}}|a'|^{-\frac{s+3}{2}})$$

$$f(\gamma_\circ(g_1,g_2)) = f^*(g_1,g_2).$$

Let ${}^0M_1 = P_1 \cap K_2$. Let k_2 and ${}^0\mathsf{m}_1$ be the Lie algebras of ${}^0M_1$ and K_2, respectively. Then we know that $\mathsf{k}_2 = {}^0\mathsf{m}_1 \oplus \mathsf{k}_\circ$ (orthogonal direct sum with respect to the Killing form on k_2) and $\mathsf{k}_\circ$ is locally homeomorphic to the quotient ${}^0M_1 \setminus K_2$. Choose an open neighborhood d_o of 0 in $\mathsf{k}_\circ$ so that the exponential image D_o of d_o is an open neighborhood of the distinguished element e in the quotient ${}^0M_1 \setminus K_2$, (the size of which will be determined later), and $P_1^{2,\triangle}(Z_2 \times I_4)([Z_2 \setminus P_1] \times I_4)^\circ(D_o \times D_o)^\circ$ is an open subset in H. Therefore we can choose a section f_s^* in the following form:

$$f_s^*((p,I_4)(d_1.d_2)) = \phi_0(p)\varphi_0(d_1,d_2), \tag{189}$$

where $\phi_0(p)$ is smooth and compactly supported over $Z_2 \setminus P_1$ and $\varphi_0(d_1,d_2)$ is smooth and compactly supported over $(D_o \times D_o)^\circ$ (whose size of the supports will be determined later). Since $P_1 = M_1U_1$, we may require that $\phi_0(m_1u_1) = \phi_0'(m_1)\phi_0''(u_1)$ where $\phi_0'(m_1)$ is in $C_c^\infty(M_1)$ and $\phi_0''(u_1)$ is in $C_c^\infty(U_1)$ (the size of the supports will be determined later). Thus the restriction of f_s^* to the subset $([Z_2\setminus P_1] \times I_4)^\circ(D_o \times D_o)^\circ$ is independent of s. For such a section f_s^* (or f_s), the local zeta integral is equal to

$$
\begin{aligned}
&\mathcal{Z}(s,W_1,W_2,f_s) \\
&= \int_{[CN^{2,\triangle}(Z_2\times I_4)\setminus H]} f(\gamma_\circ(g_1,g_2),s)W_1(g_1)W_2(g_2)dg_1dg_2 \\
&= \int_{(D_o\times D_o)^\circ} \int_{([Z_2\setminus P_1]\times I_4)^\circ} f_s^*((p_1,I_4)(d_1,d_2)) \\
&\qquad \cdot \int_{CN^2\setminus P_1} |det p|^{s'} W_1(pp_1d_1)W_2(pd_2)dpdp_1d(d_1,d_2) \\
&= \int_{(D_o\times D_o)^\circ} \varphi_0(d_1,d_2) \int_{([Z_2\setminus P_1]\times I_4)^\circ} \phi_0'(m_1)\phi_0''(u_1) \\
&\qquad \cdot \int_{CN^2\setminus P_1} |det p|^{s'} W_1(pu_1m_1d_1)W_2(pd_2)dpdp_1d(d_1,d_2). \tag{190}
\end{aligned}
$$

and the inner integral equals

$$
\begin{aligned}
&\int_{([Z_2\setminus P_1]\times I_4)^\circ} \phi_0'(m_1)\phi_0''(u_1) \int_{CN^2\setminus P_1} |det p|^{s'} W_1(pu_1m_1d_1)W_2(pd_2)dpdp_1 \\
&= \int_{(M_1\times I_4)^\circ} |det m_1|^\alpha \phi_0'(m_1) \int_{U_2\setminus GL(2)} |det g|^{s'} W_1(m(det g,g)m_1d_1)W_2(m(det g,g)d_2) \\
&\qquad \cdot \int_{Z_2\setminus U_1} \phi_0''(u_1)\psi(Ad(m(det g,g)u_1)du_1dgdm_1d.
\end{aligned}
$$

We set $\Phi((0,1)g) = \int_{Z_2 \backslash U_1} \phi_0''(u_1)\psi(Ad(m(detg,g)u_1))du_1$. Then the last integral becomes

$$\int_{(M_1 \times I_4)^\circ} |detm_1|^\alpha \phi_0'(m_1)$$

$$\cdot \int_{U_2 \backslash GL(2)} |detg|^{s'} W_1(m(detg,g)m_1 d_1) W_2(m(detg,g)d_2)\Phi((0,1)g)dgdm_1.$$

Let denote the inner integral in the above integral by $\mathcal{A}(s', (m_1 d_1, d_2); W_1, W_2, \Phi)$, that is,

$$\mathcal{A}(s', (m_1 d_1, d_2); W_1, W_2, \Phi)$$

$$= \int_{U_2 \backslash GL(2)} |detg|^{s'} W_1(m(detg,g)m_1 d_1) W_2(m(detg,g)d_2)\Phi((0,1)g)dg.$$

Then it is easy to figure out that $\mathcal{A}(s', (1,1); W_1, W_2, \Phi) = I(s', W_1, W_2, \Phi)$. According to Lemma 2.2.1, if we pick up such a function $\phi_0''(u_1)$ in $C_c^\infty(U_1)$ that its Fourier transform $\Phi((0,1)g)$ is the required function in the lemma and also pick up Whittaker functions as in the same lemma, the integral $\mathcal{A}(s', (1,1); W_1, W_2, \Phi)$ does not vanish. By means of this nonvanishing property, we may choose the function $\varphi_0(d_1, d_2)$ in $C^\infty((D_o \times D_o)^\circ$ and the function $\phi_0'(m_1)$ in $C_c^\infty(M_1)$ with their compact support in such small neighborhoods of the identities in the group M_1 and the group $(D_o \times D_o)^\circ$, respectively, that the integral

$$\int_{(D_o \times D_o)^\times} \varphi_0(d_1, d_2) \int_{(M_1 \times I_4)^\circ} |detm_1|^\alpha \phi_0'(m_1)\mathcal{A}(s', (m_1 d_1, d_2); W_1, W_2, \Phi)dm_1 d(d_1, d_2)$$

does not vanish. The theorem is finally proved. $\qquad\square$

In our set-up, we made an assumption that at a real archimedean place, the sections f_s be $K_{4,\infty}$-finite in the degenerate principal series representation $I_{3,\infty}^4(s)$ when we established our Rankin-Selberg global integral in the first chapter. In other words, we have to refine Theorem 2.2.1 so that the same result holds for any $K_{4,\infty}$-finite section f_s in $I_{3,\infty}^4(s)$ before we use the theorem to study the analytic properties of the degree 16 L-function $L^S(s', \pi \otimes \pi, \rho_1 \otimes \rho_1)$. We will return to this later.

3. Absolutely Convergence of Local Zeta Integrals

We shall prove in this section that the local zeta integral $\mathcal{I}(s, W_1, W_2, f)$ converges absolutely for $Re(s)$ large– for both archimedean and nonarchimedean cases. We first prove some lemmas that hold for both cases.

LEMMA 3.0.2. Let $f_s, f_s^\circ \in I_3^4(s)$ and f_s° the normalized K_4-spherical section. Then one has following estimate:

$$|f_s(g)| \leq C_{K,f} \cdot f_s^\circ(g)$$

where $C_{K,f}$ is a constant independent of s.

LEMMA 3.0.3. *Let F be a nonarchimedean local field. Let Φ^0 be a right K_4-spherical section in $I_{B_4}^4(s)$ as in Lemma 1.2.1. Then we have following estimates:*

$$\Phi^0(\chi(u,v)) = \max\{1, |u|\}^{-s-1} \max\{1, |v|\}^{-2s-2}$$

where $\chi(u,v) = \chi_{-\varepsilon_1+\varepsilon_2}(u)\chi_{-\varepsilon_3+\varepsilon_4}(v)$

LEMMA 3.0.4. *Let F be a real archimedean field. Let Φ^0 be a right K_4-spherical section in $I_{B_4}^4(s)$ as in Lemma 1.2.1. Then we have following estimates:*

$$\Phi^0(\chi(u,v)) = (1+u^2)^{-\frac{1}{2}(s+1)}(1+v^2)^{-(s+1)}.$$

3.1. Nonarchimedean Cases: We are going to prove that the local zeta integral $\mathcal{Z}(s, W_1, W_2, f_s)$ converges absolutely for the real part of s large, that is, to prove the following theorem.

THEOREM 3.1.1 (Non-archimedean Cases). *Let ψ be any generic character of N^2. For any $f_s \in I_3^4(s)$, and any $W_1 \in \mathcal{W}(\pi_1, \psi)$ and $W_2 \in \mathcal{W}(\pi_2, \overline{\psi})$, the local zeta integral*

$$\mathcal{Z}(s, W_1, W_2, f) = \int_{CN^{2,\triangle}(Z_2 \times I_4)\backslash H} f(\dot{\gamma}_{\circ}(g_1, g_2), s)W_1(g_1)W_2(g_2)dg_1 dg_2 \tag{191}$$

absolutely converges for $Re(s) \gg 0$.

PROOF. In terms of the Iwasawa decomposition $H = (P_1 \times P_1)^\circ (K_2 \times K_2)^\circ$, we can write the quotient $CN^{2,\triangle}(Z_2 \times I_4)\backslash H$ as

$$CN^{2,\triangle}(Z_2 \times I_4)\backslash H = (Z_2\backslash N^2)[C\backslash (T_2 \times T_2)^\circ](K_2 \times K_2)^\circ. \tag{192}$$

Then the local zeta integral reduces as follows:

$$\mathcal{Z}(s, W_1, W_2, f_s)$$
$$= \int_{CN^{2,\triangle}(Z_2 \times I_4)\backslash H} f_s(\gamma_\circ(g_1, g_2), s)W_1(g_1)W_2(g_2)dg_1 dg_2$$
$$= \int_{(K_2 \times K_2)^\circ} \int_{C\backslash (T_2 \times T_2)^\circ} \delta_H^{-1}(t_1, t_2)W_1(t_1 k_1)W_2(t_2 k_2)$$
$$\int_{Z_2\backslash N^2} f_s(\gamma_\circ(ut_1, t_2)(k_1, k_2))\psi(u)du\underline{d}a\underline{d}k. \tag{193}$$

By the admissibility of the representations involved, we can pick up a small open compact subgroup K_2' in K_2 so that the Whittaker functions W_1 and W_2 are right K_2'-invariant and f_s is right $(K_2' \times K_2')^\circ$-invariant. We denoted by $\{(k_i', k_i'') : i = 1, 2, \cdots, t\}$ the finite quotient $(K_2 \times K_2)^\circ/(K_2' \times K_2')^\circ$. Then integral (189) can be reduced further and becomes a sum of integrals of nice shape.

$$(189) \quad = \quad \sum_{i=1}^{t} \mu_{K_2'} \int_{C \backslash (T_2 \times T_2)^{\circ}} \delta_H^{-1}(t_1, t_2) W_1(t_1 k_i') W_2(t_2 k_i'')$$

$$\times \int_{Z_2 \backslash N^2} f_s(\gamma_{\circ}(u t_1, t_2)(k_i', k_i'')) \psi(u) du d\underline{a}$$

$$= \quad \sum_{i=1}^{t} \mu_{K_2'} \int_{C \backslash (T_2 \times T_2)^{\circ}} \delta_H^{-1}(t_1, t_2) W_1^{(i)}(t_1) W_2^{(i)}(t_2)$$

$$\times \int_{Z_2 \backslash N^2} f_s^{(i)}(\gamma_{\circ}(u t_1, t_2)) \psi(u) du d\underline{a} \qquad (194)$$

where $\mu_{K_2'}$ is the volume of the open compact subgroup $(K_2' \times K_2')^{\circ}$ with respect to the Haar measure on the group H and we denote $W_1^{(i)} = \pi_1(k_i') W_1$ and $W_2^{(i)} = \pi_2(k_i'') W_2$, and $f^{(i)} = \rho_s((k_i', k_j'')) f$. It thus reduces to prove the absolute convergence of following type:

$$\mathcal{I}(s, W_1, W_2, f_s) := \int_{C \backslash (T_2 \times T_2)^{\circ}} W_1(t_1) W_2(t_2) \int_{Z_2 \backslash N^2} f_s(\gamma_{\circ}(u t_1, t_2)) \psi(u) du dt_1 dt_2, \qquad (195)$$

Taking absolute value and applying Lemma 1.1.1, we obtain that

$$|\mathcal{I}(s.W_1, W_2, f_s)| \quad \leq \quad \int_{C \backslash (T_2 \times T_2)^{\circ}} |W_1(t_1) W_2(t_2)| \int_{Z_2 \backslash N^2} |f_s(\gamma_{\circ}(u t_1, t_2))| du dt_1 dt_2$$

$$\leq \quad C_K \int_{C \backslash (T_2 \times T_2)^{\circ}} |W_1(t_1) W_2(t_2)| \int_{Z_2 \backslash N^2} f_s^{\circ}(\gamma_{\circ}(u t_1, t_2)) du dt_1 dt_2$$

$$= \quad I(s, |W_1|, |W_2|, f_s^{\circ}). \qquad (196)$$

Basically, we reduce the problem of absolute convergence to the case where the section f_s° is K_4-spherical. We set

$$I(t_1, t_2; f_s^{\circ}) := \int_{Z_2 \backslash N^2} f_s^{\circ}(\gamma_{\circ}(u t_1, t_2)) du. \qquad (197)$$

Applying the computation in the proof of Lemma 3.2.1 and Proposition 3.2.1 in Chapter II to the case of $\psi_{\circ} = 1$, Lemma 1.2.1, and Lemma 3.0.3, we have

$$I(t_1, t_2; f_s^{\circ}) \quad = \quad \delta^{\frac{1}{2}} (|a^{\frac{3}{2}} b c^{-1}|)^{s+1} \int_{F^3} f_s^{\circ}(\chi(w, y, x, u, v)) dw dy dx$$

$$= \quad \delta^{\frac{1}{2}} |a^{\frac{3}{2}} b c^{-1}|^{s+1} \frac{\zeta^2(s+1)}{\zeta(s+2) \zeta(s+3)} \Phi^0(\chi(u, v)) \qquad (198)$$

$$= \quad \delta^{\frac{1}{2}} |a^{\frac{3}{2}} b c^{-1}|^{s+1} \frac{\zeta^2(s+1)}{\zeta(s+2) \zeta(s+3)} \max\{1, |u|\}^{-s-1} \max\{1, |v|\}^{-2s-2},$$

where $u = abc^{-1}d^{-1}$ and $v = ac^{-1}$.

Recall that in the coordinates chosen on $C \setminus (T_2 \times T_2)^\circ$, $t_1 = h(ab, a, b^{-1}, 1)$ and $t_2 = h(cd, c, c^{-1}d^{-1}a, c^{-1}a)$. By Proposition 1.1.1, we have

$$W_1(t_1) = \sum_{\chi_1} \phi_{\chi_1}(a, b)\chi_1(a, b) \text{ and } W_2(t_2) = \sum_{\chi_2} \phi_{\chi_2}(c^2a^{-1}, d)\chi_2(c^2a^{-1}, d) \tag{199}$$

where the summation is taken over a finite set of finite functions χ's and ϕ_χ's are Schwartz-Bruhat functions over F^2. It reduces to prove the absolute convergence for integrals of following type:

$$\int \frac{\delta^{\frac{1}{2}}|a^{\frac{3}{2}}bc^{-1}|^{s+1}|\phi_{\chi_1}(a, b)\chi_1(a, b)\phi_{\chi_2}(c^2a^{-1}, d)\chi_2(c^2a^{-1}, d)|}{\max\{1, |u|\}^{s+1}\max\{1, |v|\}^{2s+2}}d(a, b, c, d) \tag{200}$$

where the integration is taken over the domain: $0 < |a|, |b|, |c|, |d| < \infty$.

Note that, in general, the absolute value of a finite function of two variables can be written as $\chi(a, b) = |a|^l|b|^k|\log|a|^m\log|b|^n|$, and when t goes to ∞, $|t^l\log|t|^m|$ is bounded by $|t|^{l+\epsilon}$ for any $\epsilon > 0$, and when t goes to 0, $|t^l\log|t|^m|$ is bounded by $|t|^{l-\epsilon}$ for any $\epsilon > 0$. Without loss of generality, we may assume that

$$\chi_1(a, b) = |a|^{u_1}|b|^{u_2} \text{ and } \chi_2(c^2a^{-1}, d) = |c^2a^{-1}|^{u_1}|d|^{u_2} \tag{201}$$

for our purpose to prove the absolute convergence of the integral in (196). Hence we have to show the convergence of following integral:

$$\int \frac{|a|^{\frac{3}{2}s+l_a}|b|^{s+l_b}|c|^{-s+l_c}|d|^{l_d}\phi_1(a, b)\phi_2(c^2a^{-1}, d)}{\max\{1, |u|\}^{s+1}\max\{1, |v|\}^{2s+2}}d(a, b, c, d) \tag{202}$$

where the integration is taken over the domain: $0 < |a|, |b|, |c|, |d| < \infty$ and the Schwartz-Bruhat functions are positive and s is a large real number. Changing the variables (a, b, c, d) to (a, b, u, v), the integral becomes

$$\int \frac{|a|^{\frac{1}{2}s+k_a}|b|^{s+k_b}|u|^{k_u}|v|^{s+k_v}\phi_1(a, b)\phi_2(av^{-2}, bvu^{-1})}{\max\{1, |u|\}^{s+1}\max\{1, |v|\}^{2s+2}}d(a, b, u, v) \tag{203}$$

where the integration is taken over the domain: $0 < |a|, |b|, |u|, |v| < \infty$. We write the integral as a sum of following two integrals:

$$I_1 := \int_{|a|, |b| > 0}^{\infty} |a|^{\frac{1}{2}s+k_a}|b|^{s+k_b}\phi_1(a, b)\int_{|v| > 0}^{\infty} \frac{|v|^{s+k_v}}{\max\{1, |v|\}^{2s+2}}$$
$$\cdot \int_{|u| \geq 1}^{\infty} |u|^{k_u-s}\phi_2(av^{-2}, bvu^{-1})dudvdadb \tag{204}$$

and

$$I_2 := \int_{|a|, |b| > 0}^{\infty} |a|^{\frac{1}{2}s+k_a}|b|^{s+k_b}\phi_1(a, b)\int_{|v| > 0}^{\infty} \frac{|v|^{s+k_v}}{\max\{1, |v|\}^{2s+2}}$$
$$\cdot \int_{|u| > 0}^{1} |u|^{k_u}\phi_2(av^{-2}, bvu^{-1})dudvdadb. \tag{205}$$

When $1 \leq |u| < \infty$, there is a constant C_1 so that

$$|u|^{k_u-s}\phi_2(av^{-2}, bvu^{-1}) \leq C_1|u|^{k_u-s}. \tag{206}$$

Thus we have

$$\int_{|u|\geq 1}^{\infty} |u|^{k_u-s}\phi_2(av^{-2}, bvu^{-1})du \leq C_1 \int_{|u|\geq 1}^{\infty} |u|^{k_u-s}du, \tag{207}$$

which converges for s large. Because of this estimate for the integration of the variable u, integral I_1 is bounded by

$$\int_{|a|,|b|>0}^{\infty} |a|^{\frac{1}{2}s+k_a}|b|^{s+k_b}\phi_1(a,b)dadb \int_{|v|>0}^{\infty} \frac{|v|^{s+k_v}}{\max\{1,|v|\}^{2s+2}}dv \int_{|u|\geq 1}^{\infty} C_1|u|^{k_u-s}du. \tag{208}$$

The integral $\int_{|a|,|b|>0}^{\infty} |a|^{\frac{1}{2}s+k_a}|b|^{s+k_b}\phi_1(a,b)dadb$ converges since ϕ_1 is compactly supported and s can be chosen suitably large. While the integral

$$\int_{|v|>0}^{\infty} \frac{|v|^{s+k_v}}{\max\{1,|v|\}^{2s+2}}dv = \int_{|v|\geq 1}^{\infty} |v|^{-s+k'_v}dv + \int_{|v|>0}^{1} |v|^{s+k_v}dv,$$

which is convergent for s positive and large. Hence integral I_1 is convergent.

On the other hand, when $0 < |u| \leq 1$, there is a positive integer e so that $\phi_2(av^{-2}, bvu^{-1})$ will vanish if $|bvu^{-1}| > q^e$. Thus there is a constant C_2 so that

$$\int_{|u|>0}^{1} |u|^{k_u}\phi_2(av^{-2}, bvu^{-1})du \leq C_2 \int_{q^{-e}|bv|\leq|u|\leq 1} |u|^{k_u}du \tag{209}$$

$$= C_2\frac{1 - q^{-(e+1)(k_u+1)}|bv|^{k_u+1}}{1 - q^{-(k_u+1)}}.$$

So integral I_2 is bounded by a two-term linear combination with complex coefficients of integrals of following type:

$$\int_{|a|,|b|>0}^{\infty} |a|^{\frac{1}{2}s+k'_a}|b|^{s+k'_b}\phi_1(a,b)dadb \int_{|v|>0}^{\infty} \frac{|v|^{s+k,v}}{\max\{1,|v|\}^{2s+2}}dv, \tag{210}$$

which is convergent following the same argument as used in the case $|u| \geq 1$. $\qquad\square$

3.2. Archimedean Cases: In this case, the proof of the absolute convergence of the local zeta integral will be very close to that of nonarchimedean case. The theorem we are going to prove is

THEOREM 3.2.1. *Let F be a real archimedean field. Let ψ be any generic character of N^2. For any smooth $f_s \in I_3^4(s)$, and any K_2-finite Whittaker functions $W_1 \in \mathcal{W}(\pi_1, \psi)$ and $W_2 \in \mathcal{W}(\pi_2, \overline{\psi})$, the local zeta integral*

$$\mathcal{Z}(s, W_1, W_2, f) = \int_{CN^2,\triangle(Z_2\times I_4)\backslash H} f(\gamma_o(g_1, g_2), s)W_1(g_1)W_2(g_2)dg_1dg_2 \tag{211}$$

absolutely converges for $Re(s) >> 0$.

PROOF. By Iwasawa decomposition, the local zeta integral $\mathcal{Z}(s, W_1, W_2, f_s)$ can be written as

$$\int_{(K_2 \times K_2)^\circ} \int_{C \backslash (T_2 \times T_2)^\circ} \delta_H^{-1}(\underline{a}) W_1(t_1 k_1) W_2(t_2 k_2) \int_{Z_2 \backslash N^2} f_s(\gamma_\circ(ut_1 k_1, t_2 k_2)) \psi(u) du d\underline{a} d\underline{k}.$$

Since W_1 and W_2 are assumed to be K_2-finite, there are a finite number of K_2-finite functions $\theta_i(k_1)$ and $\theta_j(k_2)$, and a finite number of Whittaker functions $W_1^{(i)}(g_1)$ and $W_2^{(j)}(g_2)$, so that

$$\pi_1(k_1) W_1(g_1) = \sum_i \theta_i(k_1) W_1^{(i)}(g_1) \text{ and } \pi_2(k_2) W_2(g_2) = \sum_j \theta_j(k_2) W_2^{(j)}(g_2).$$

We thus have

$$\mathcal{Z}(s, W_1, W_2, f_s)$$
$$= \sum_{i,j} \int_{C \backslash (T_2 \times T_2)^\circ} \delta_H^{-1}(t_1, t_2) W_1^{(i)}(t_1) W_2^{(j)}(t_2)$$
$$\cdot \int_{Z_2 \backslash N^2} \int_{(K_2 \times K_2)^\circ} \theta_i(k_1) \theta_j(k_2) f_s(\gamma_\circ(ut_1, t_2)(k_1, k_2)) dk_1 dk_2 \psi(u) du dt_1 dt_2$$
$$= \sum_{i,j} \int_{C \backslash (T_2 \times T_2)^\circ} \delta_H^{-1}(t_1, t_2) W_1^{(i)}(t_1) W_2^{(j)}(t_2)$$
$$\int_{Z_2 \backslash N^2} f_s^{(i,j)}(\gamma_\circ(ut_1, t_2)) \psi(u) du dt_1 dt_2, \tag{212}$$

where we set $f_s^{(i,j)}(g) := \int_{(K_2 \times K_2)^\circ} \theta_i(k_1) \theta_j(k_2) f_s(g(k_1, k_2)) dk_1 dk_2$. Note that $f_s^{(i,j)}$ is $(K_2 \times K_2)^\circ$-finite. It suffices to prove the absolute convergence for the integral of following type.

$$\mathcal{I}(s, W_1, W_2, f_s)$$
$$= \int_{C \backslash (T_2 \times T_2)^\circ} \delta_H^{-1}(t_1, t_2) W_1(t_1) W_2(t_2) \int_{Z_2 \backslash N^2} f_s(\gamma_\circ(ut_1, t_2)) \psi(u) du dt_1 dt_2.$$

By the same argument as that used in the proof of nonarchimedean case, it is enough to prove the convergence of the special case where $f_s = f_s^\circ$, the spherical section in $I_3^4(s)$, that is, the integral

$$\int_{C \backslash (T_2 \times T_2)^\circ} |W_1(t_1) W_2(t_2)| \int_{Z_2 \backslash N^2} f_s^\circ(\gamma_\circ(ut_1, t_2)) du dt_1 dt_2. \tag{213}$$

The proof here will go similarly to that of nonarchimedean case. Applying the computation in the proof of Lemma 3.2.1 and Proposition 3.2.1 in Chapter II, and Lemma

1.2.1 and Lemma 3.0.4, we have

$$
\int_{Z_2\backslash N^2} f_s^\circ(\gamma_\circ(ut_1, t_2))du
$$
$$
= \delta^{\frac{1}{2}}(|a^{\frac{3}{2}}bc^{-1}|)^{s+1} \int_{F^3} f_s^\circ(\chi(w, y, x, u, v))dwdydx
$$
$$
= \delta^{\frac{1}{2}}|a^{\frac{3}{2}}bc^{-1}|^{s+1} \frac{\zeta^2(s+1)}{\zeta(s+2)\zeta(s+3)}\Phi^0(\chi(u, v)) \tag{214}
$$
$$
= \delta^{\frac{1}{2}}|a^{\frac{3}{2}}bc^{-1}|^{s+1} \frac{\zeta^2(s+1)}{\zeta(s+2)\zeta(s+3)}(1 + u^2)^{-\frac{1}{2}(s+1)}(1 + v^2)^{-(s+1)},
$$

where $u = abc^{-1}d^{-1}$ and $v = ac^{-1}$. For Whittaker functions, we use proposition 2 and obtain

$$
W_1(t_1) = \sum_{\chi_1} \phi_{\chi_1}(a, b)\chi_1(a, b) \text{ and } W_2(t_2) = \sum_{\chi_2} \phi_{\chi_2}(c^2a^{-1}, d)\chi_2(c^2a^{-1}, d) \tag{215}
$$

where $t_1 = h(ab, a, b^{-1}, 1)$ and $t_2 = h(cd, c, c^{-1}d^{-1}a, c^{-1}a)$, the summation is taken over a finite set of finite functions χ's and ϕ_χ's are Schwartz functions over F^2. For our purpose of proving the convergence, we may assume as in the nonarchimedean case that the finite functions are in form:

$$
\chi_1(a, b) = |a|^{u_1}|b|^{u_2} \text{ and } \chi_2(c^2a^{-1}, d) = |c^2a^{-1}|^{u_3}|d|^{u_4}, \tag{216}
$$

and the Schwartz functions are positive and s is real. Thus it reduces to prove the convergence for following integral (when s is large)

$$
\int \frac{|a|^{\frac{3}{2}s+l_a}|b|^{s+l_b}|c|^{-s+l_c}|d|^{l_d}\phi_1(a, b)\phi_2(c^2a^{-1}, d)}{(1 + u^2)^{\frac{1}{2}(s+1)}(1 + v^2)^{s+1}}d(a, b, c, d) \tag{217}
$$

where the integration is taken over the domain: $0 < |a|, |b|, |c|, |d| < \infty$. Similarly, by changing the variables (a, b, c, d) to (a, b, u, v), the integral becomes

$$
\int \frac{|a|^{\frac{1}{2}s+k_a}|b|^{s+k_b}|u|^{k_u}|v|^{s+k_v}\phi_1(a, b)\phi_2(av^{-2}, bvu^{-1})}{(1 + u^2)^{\frac{1}{2}(s+1)}(1 + v^2)^{s+1}}d(a, b, u, v) \tag{218}
$$

where the integration is taken over the domain: $0 < |a|, |b|, |u|, |v| < \infty$.

We write the integral as a sum of following two integrals:

$$
I_1 := \int_{|a|,|b|>0}^\infty |a|^{\frac{1}{2}s+k_a}|b|^{s+k_b}\phi_1(a, b)\int_{|v|>0}^\infty \frac{|v|^{s+k_v}}{(1 + v^2)^{s+1}}
$$
$$
\cdot \int_{|u|\geq 1}^\infty \frac{|u|^{k_u}}{(1 + u^2)^{\frac{1}{2}(s+1)}}\phi_2(av^{-2}, bvu^{-1})dudvdadb \tag{219}
$$

and

$$I_2 := \int_{|a|,|b|>0}^{\infty} |a|^{\frac{1}{2}s+k_a} |b|^{s+k_b} \phi_1(a,b) \int_{|v|>0}^{\infty} \frac{|v|^{s+k_v}}{(1+v^2)^{s+1}}$$
$$\cdot \int_{|u|>0}^{1} \frac{|u|^{k_u}}{(1+u^2)^{\frac{1}{2}(s+1)}} \phi_2(av^{-2}, bvu^{-1}) du dv da db. \quad (220)$$

It is clear that the integral I_1 is bounded by

$$\int_{|a|,|b|>0}^{\infty} |a|^{\frac{1}{2}s+k_a} |b|^{s+k_b} \phi_1(a,b) da db \int_{|v|>0}^{\infty} \frac{|v|^{s+k_v}}{(1+v^2)^{s+1}} dv \int_{|u|\geq 1}^{\infty} C_1 \frac{|u|^{k_u}}{(1+u^2)^{\frac{1}{2}(s+1)}} du, \quad (221)$$

which is convergent by obvious reasons. When $0 < |u| \leq 1$, for any large integer T, there is a constant C_T so that

$$\phi_2(av^{-2}, bvu^{-1}) \leq C_T |bv|^{-T} |u|^T. \quad (222)$$

We can pick up T so large that

$$\int_{|u|>0}^{1} \frac{|u|^{k_u}}{(1+u^2)^{\frac{1}{2}(s+1)}} \phi_2(av^{-2}, bvu^{-1}) du \leq C_T |bv|^{-T} \int_{|u|>0}^{1} |u|^{T+k_u} du.$$

Therefore the integral I_2 is bounded by

$$C_T \int_{|a|,|b|>0}^{\infty} |a|^{\frac{1}{2}s+k'_a} |b|^{s+k'_b} \phi_1(a,b) da db \int_{|v|>0}^{\infty} \frac{|v|^{s+k'_v}}{(1+v^2)^{s+1}} dv \int_{|u|>0}^{1} |u|^{T+k_u} du,$$

which is again convergent. This proves the theorem. $\qquad \Box$

4. Meromorphic Continuation of Local Zeta Integrals

We will apply the estimates obtained in the first section of this Chapter to prove, in nonarchimedean case, that the local zeta integral $\mathcal{Z}(s, W_1, W_2, f_s)$ has a meromorphic continuation to the whole complex plane as a function of s. The archimedean version of such a result has not been proved as yet.

4.1. Nonarchimedean Case: Our proof of the meromorphic continuation of the nonarchimedean local zeta integral is based on the direct computation of the integral. In principal, such computation will be reduced to the integrals of one-variable integration as follows.

$$I(\alpha s + \beta; n; x, p; \lambda; \psi; l, k) \quad (223)$$
$$= \int_{q^{-k} \leq |t| \leq q^{-l}} |t|^{\alpha s + \beta} \lambda(t) \log |t|^n ch_{\{x+\mathcal{P}^p\}}(t) \psi(t) d^{\times} t$$

where λ is a character of the group $\mathcal{O}^{\times}$ of the units in the integral ring $\mathcal{O}$ with its conduct f_λ $(ker(\lambda) = 1 + \mathcal{P}^{f_\lambda})$, ψ is a additive unitary character of F with its conduct δ_ψ $(ker(\psi) = \mathcal{P}^{ord(\delta_\psi)})$, and $ch_{\{x+\mathcal{P}^p\}}(t)$ is the characteristic function of the subset $\{x + \mathcal{P}^p\}$.

LEMMA 4.1.1. *Let α, β be two real numbers and k, l, n, p integers such that $k \geq l$ and n is nonnegative. Let $ch_{\{x+\mathcal{P}^p\}}(t)$ be the characteristic function of $\{x + \mathcal{P}^p\}$ with $x \notin \mathcal{P}^p$, ψ any additive unitary character of F, and $\lambda(t)$ the ramified character of $\mathcal{O}^\times$. Then we have, for $p \geq ord(\delta_\psi)$, that*

$$I(\alpha s + \beta; n; x, p; \lambda; \psi; l, k)$$
$$= \begin{cases} 0 & \text{if } l > ord(x) \text{ or } ord(x) > k; \\ 0 & \text{if } l \leq ord(x) \leq k \text{ and } p - ord(x) < f_\lambda; \\ \psi(x)\lambda(x)q^{-p}|x|^{\alpha s + \beta}\log|x|^n & \text{if } l \leq ord(x) \leq k \text{ and } p - ord(x) \geq f_\lambda. \end{cases}$$

PROOF. Since the positive integer p is assumed to be larger than $ord(\delta_\psi)$, the integral $I(\alpha s + \beta, n, p; \lambda, \psi; l, k)$ is equal to $\psi(x)I(\alpha s + \beta, n, p; \lambda; l, k)$. It is obvious that $I(\alpha s + \beta, n, p; \lambda; l, k)$ will be zero if $l > ord(x)$ or $ord(x) > k$. When $l \leq ord(x) \leq k$, we have

$$I(\alpha s + \beta; n; x, p; \lambda; \psi; l, k) = \int_{\pi^{ord(x)}\mathcal{O}^\times} |t|^{\alpha s + \beta}\lambda(t)\log|t|^n ch_{\{x+\mathcal{P}^p\}}(t)d^\times t$$
$$= |x|^{\alpha s + \beta}\log|x|^n \int_{\pi^{ord(x)}\mathcal{O}^\times} \lambda(t)ch_{\{x+\mathcal{P}^p\}}(t)d^\times t$$
$$= |x|^{\alpha s + \beta}\log|x|^n \int_{x+\mathcal{P}^p} \lambda(t)d^\times t.$$

It is easy to see that $\int_{x+\mathcal{P}^p}\lambda(t)d^\times t$ is equal to zero if $p - ord(x) < f_\lambda$. When $p - ord(x) \geq f_\lambda$, one has

$$\int_{x+\mathcal{P}^p}\lambda(t)d^\times t = \lambda(x)q^{-p}.$$

This proves the lemma.   □

LEMMA 4.1.2. *Let α, β be two real numbers and k, l, n, p integers such that $k \geq l$ and n is nonnegative. Let $ch_{\mathcal{P}^p}(t)$ be the characteristic function of $\mathcal{P}^p$, ψ any additive unitary character of F, and $\lambda(t)$ the ramified character of $\mathcal{O}^\times$. Then we have, for $p \geq ord(\delta_\psi)$, that*

$$I(\alpha s + \beta; n; p; \lambda; \psi; l, k)$$
$$= \begin{cases} 0 & \text{if } \lambda \neq 1; \\ (1 - q^{-1})\log q^{-n}\sum_{i=\max\{l,p\}}^k iq^{-i(\alpha s + \beta)} & \text{if } \lambda = 1. \end{cases}$$

PROOF. Similarly, we may forget the additive character ψ upon the positive integer p being chosen large enough. Then the integral can be computed as follows:

$$I(\alpha s + \beta, n, p; \lambda, \psi; l, k)$$
$$= \int_{q^{-k} \leq |t| \leq q^{-l}} |t|^{\alpha s + \beta}\lambda(t)\log|t|^n ch_{\mathcal{P}^p}(t)d^\times t$$
$$= \sum_{i=\max\{l,p\}}^k iq^{-i(\alpha s + \beta)}\log q^{-n}\int_{\mathcal{O}^\times} \lambda(t)dt.$$

Hence, if λ is not trivial, the last integral is zero; otherwise, it is equal to $1 - q^{-1}$. The lemma is proved. $\square$

Note that one has the following obvious formula:

$$\sum_{i=l}^{k} iq^{-i(\alpha s+\beta)} \tag{224}$$

$$= \frac{lq^{-l(\alpha s+\beta)} - (k+1)q^{-(k+1)(\alpha s+\beta)} - (l-1)q^{-(l+1)(\alpha s+\beta)} + kq^{-(k+2)(\alpha s+\beta)}}{(1 - q^{-(\alpha s+\beta)})^2}.$$

The theorem we are going to prove is as follows:

THEOREM 4.1.1. *Let ψ be any generic character of N^2. There is a positive real number s_0, which is independent of the finite place v, so that for any $f_s \in I_3^4(s)$, and any $W_1 \in \mathcal{W}(\pi_1, \psi)$ and $W_2 \in \mathcal{W}(\pi_2, \overline{\psi})$, the local zeta integral*

$$\mathcal{Z}(s, W_1, W_2, f) = \int_{CN^{2,\triangle}(Z_2 \times I_4) \backslash H} f(\gamma_\circ(g_1, g_2), s) W_1(g_1) W_2(g_2) dg_1 dg_2 \tag{225}$$

is of form $P(q^{-s})$, $P(T)$ is a rational function in T with complex coefficients. In other words, the integral $\mathcal{Z}(s, W_1, W_2, f)$ has a meromorphic continuation to the whole complex s-plane.

PROOF. As in the proof of the absolute convergence of the local zeta integral, it suffices to prove the rationality of the following integral:

$$\mathcal{I}(s.W_1, W_2, f_s) := \int_{C \backslash (T_2 \times T_2)^\circ} W_1(t_1) W_2(t_2) I(t_1, t_2; f_s, \psi) dt_1 dt_2. \tag{226}$$

Applying the computation in the proof of Lemma 3.2.1 and Proposition 3.2.1 in Chapter II, we have

$$I(t_1, t_2; f_s, \psi) = \delta_H^{\frac{1}{2}} |a^{\frac{3}{2}} bc^{-1}|^{s+1} I(abc^{-1}d^{-1}, ac^{-1}, a^{-1}cd, a^{-1}c^2; f', \psi)$$

where $f' = \rho_s(w_\circ)(f)$. Following the same computation as in the proof of Lemma 3.2.3 in Chapter II, the local zeta integral $\mathcal{I}(s, W_1, W_2, f_s)$ can be rewritten as a linear combination of integrals of the following four types with complex coefficients:

$$\int_{\substack{|ac^{-1}| \leq q^{-e} \\ |abc^{-1}d^{-1}| \leq q^{-e}}} \delta_H^{-\frac{1}{2}}(t_1, t_2)(|a|^{\frac{3}{2}}|b||c|^{-1})^{s+1} W_1(t_1) W_2(t_2) I(a^{-1}cd, a^{-1}c^2; f, \psi) dt_1 dt_2, \tag{227}$$

$$\int_{\substack{|ac^{-1}| > q^e \\ |abc^{-1}d^{-1}| \leq q^{-e}}} \delta_H^{-\frac{1}{2}}(t_1, t_2)(|a|^{-\frac{1}{2}}|b||c|)^{s+1} W_1(t_1) W_2(t_2) I(-d, a; f, \psi) dt_1 dt_2, \tag{228}$$

$$\int_{\substack{|ac^{-1}| \leq q^{-e} \\ |abc^{-1}d^{-1}| > q^e}} \delta_H^{-\frac{1}{2}}(t_1, t_2)(|a|^{\frac{1}{2}}|d|)^{s+1} W_1(t_1) W_2(t_2) I(-b, a^{-1}c^2; f, \psi) dt_1 dt_2, \tag{229}$$

$$\int_{\substack{|ac^{-1}|>q^e \\ |abc^{-1}d^{-1}|>q^e}} \delta_H^{-\frac{1}{2}}(t_1,t_2)(|a|^{-\frac{3}{2}}|c|^2|d|)^{s+1}W_1(t_1)W_2(t_2)I(abc^{-1},a,f,\psi)dt_1dt_2.$$

(230)

Let $\epsilon_i = 0$, or 1 for $i = 1,2$ and $(t_1,t_2) = (h(ab,a,b^{-1},1), h(cd,c,c^{-1}d^{-1}a,c^{-1}a))$. Again by the estimates in Theorem 3.1.12, we have to prove the rationality for the following four types of integrals:

(1) The integrals

$$\int \psi_1^{1-\epsilon_1}(\alpha)\psi_2^{1-\epsilon_2}(\beta)[|a^{\frac{3}{2}}bc^{-1}||\alpha|^{\epsilon_1}|\beta|^{\epsilon_2}]^{s+1}W_1(t_1)W_2(t_2)dt_1dt_2$$

(231)

where the integration is taken over the domain: $(a,b,c,d) \in F^{\times 4}$ satisfying the following conditions: $|ab| < \varepsilon_1|cd|$, $|\alpha| = |a^{-1}cd|$, $|\beta| = |a^{-1}c^2| < \varepsilon_2$, $|a| < \varepsilon_1|c|$, and $|a|, |b|, |d| < \varepsilon_3$;

(2) The integrals

$$\int \psi_1^{1-\epsilon_1}(d)\psi_2^{1-\epsilon_2}(a)[|a^{-\frac{1}{2}}bc||d|^{\epsilon_1}|a|^{\epsilon_2}]^{s+1}W_1(t_1)W_2(t_2)dt_1dt_2$$

(232)

where the integration is taken over the domain: $(a,b,c,d) \in F^{\times 4}$ satisfying the following conditions: $|ab| < \varepsilon_1|cd|$, $|c| < \varepsilon_1|a|$, $|c^2| < \varepsilon_3|a|$, and $|a|, |b|, |d| < \varepsilon_3$;

(3) The integrals

$$\int \psi_1^{1-\epsilon_1}(b)\psi_2^{1-\epsilon_2}(a^{-1}c^2)[|a^{\frac{1}{2}}d||b|^{\epsilon_1}|a^{-1}c^2|^{\epsilon_2}]^{s+1}W_1(t_1)W_2(t_2)dt_1dt_2$$

(233)

where the integration is taken over the domain: $(a,b,c,d) \in F^{\times 4}$ satisfying the following conditions: $|cd| < \varepsilon_1|ab|$, $|a| < \varepsilon_1|c|$, $|c^2| < \varepsilon_3|a|$, and $|a|, |b|, |d| < \varepsilon_3$;

(4) The integrals

$$\int \psi_1^{1-\epsilon_1}(abc^{-1})\psi_2^{1-\epsilon_2}(a)[|a^{-\frac{3}{2}}c^2d||abc^{-1}|^{\epsilon_1}|a|^{\epsilon_2}]^{s+1}W_1(t_1)W_2(t_2)dt_1dt_2$$

(234)

where the integration is taken over the domain: $(a,b,c,d) \in F^{\times 4}$ satisfying the following conditions: $|cd| < \varepsilon_1|ab| < \varepsilon_1\varepsilon_2|c|$, $|c| < \varepsilon_1|a|$, $|c^2| < \varepsilon_2|a|$, and $|a|, |b|, |d| < \varepsilon_3$.

Note that the integration domains of the above four integrals are determined by the conditions of vanishing of Whittaker functions given in Lemma 1.1.1 and that of the integrals of type $I(\alpha,\beta;f_s,\psi)$, and the integration domains of four integrals from (227) to (230). In each of those four cases, the data (ϵ_1,ϵ_2) has four different choices: $(0,0),(1,0),(0,1)$ and $(1,1)$. We actually have to deal with sixteen different types of integrals. By asymptotes of Whittaker function on the torus, we have that, for $t_1 = h(ab,a,b^{-1},1)$,

$$W(t_1) = \sum_{\chi} \phi_\chi(a,b)\chi(a,b).$$

where the summation is taken over a finite set of finite functions χ and ϕ_χ are Schwartz-Bruhat functions over F^2. For the proof of the absolute convergence and the rationality of those integrals, we may assume that $W(t_1) = \phi_\chi(a,b)\chi(a,b)$. In other words, $W_1(t_1) = \phi_{\chi_1}(a,b)\chi_1(a,b)$ and $W_2(t_2) = \phi_{\chi_2}(c^2a^{-1},d)\chi_2(c^2a^{-1},d)$. Therefore we have to show the rationality for the integrals of following sixteen types:

(1) The integrals

$$\int \psi_1^{1-\epsilon_1}(\alpha)\psi_2^{1-\epsilon_2}(\beta)|ab\alpha^{\epsilon_1}\beta^{\epsilon_2-\frac{1}{2}}|^{s+1}\phi_1(a,b)\chi_1(a,b)\phi_2(\beta,a^{\frac{1}{2}}\alpha\beta^{-\frac{1}{2}})\chi_2(\beta,a^{\frac{1}{2}}\alpha\beta^{-\frac{1}{2}})dt_1dt_2 \tag{235}$$

where the integration is taken over the domain: $(a,b,\alpha,\beta) \in F^{\times 4}$ satisfying the following conditions: $|\alpha|,|\beta| \leq \varepsilon_2$, $|a|,|b| \leq \varepsilon_3$, $|b\alpha| \leq \varepsilon_1$, $|a^{\frac{1}{2}}\beta^{-\frac{1}{2}}| \leq \varepsilon_1$, and $|a^{\frac{1}{2}}\alpha\beta^{-\frac{1}{2}}| \leq \varepsilon_3$.

(2) The integrals

$$\int \psi_1^{1-\epsilon_1}(d)\psi_2^{1-\epsilon_2}(a)|a^{\epsilon_2-\frac{1}{2}}bcd^{\epsilon_1}|^{s+1}\phi_1(a,b)\chi_1(a,b)\phi_2(c^2a^{-1},d)\chi_2(c^2a^{-1},d)dt_1dt_2 \tag{236}$$

where the integration is taken over the domain: $(a,b,c,d) \in F^{\times 4}$ satisfying the following conditions: $|ab| < \varepsilon_1|cd|$, $|c| < \varepsilon_1|a|$, $|c^2| < \varepsilon_3|a|$, and $|a|,|b|,|d| < \varepsilon_3$.

(3) The integrals

$$\int \psi_1^{1-\epsilon_1}(b)\psi_2^{1-\epsilon_2}(a^{-1}c^2)|a^{\frac{1}{2}-\epsilon_2}b^{\epsilon_1}c^{2\epsilon_2}d|^{s+1}\phi_1(a,b)\chi_1(a,b)\phi_2(c^2a^{-1},d)\chi_2(c^2a^{-1},d)dt_1dt_2 \tag{237}$$

where the integration is taken over the domain: $(a,b,c,d) \in F^{\times 4}$ satisfying the following conditions: $|cd| < \varepsilon_1|ab|$, $|a| < \varepsilon_1|c|$, $|c^2| < \varepsilon_3|a|$, and $|a|,|b|,|d| < \varepsilon_3$.

(4) The integrals

$$\int \psi_1^{1-\epsilon_1}(\beta)\psi_2^{1-\epsilon_2}(a)|a^{\epsilon_2-\frac{1}{2}}\beta^{\epsilon_1}\alpha d|^{s+1}\phi_1(a,\beta\alpha^{\frac{1}{2}}a^{-\frac{1}{2}})\chi_1(a,\beta\alpha^{\frac{1}{2}}a^{-\frac{1}{2}})\phi_2(\alpha,d)\chi_2(\alpha,d)dt_1dt_2 \tag{238}$$

where the integration is taken over the domain: $(a,\beta,\alpha,d) \in F^{\times 4}$ satisfying the following conditions: $|\alpha|,|\beta| \leq \varepsilon_2$, $|a|,|d| \leq \varepsilon_3$, $|d\beta^{-1}| \leq \varepsilon_1$, $|\alpha^{\frac{1}{2}}a^{-\frac{1}{2}}| \leq \varepsilon_1$, and $|\beta\alpha^{\frac{1}{2}}a^{-\frac{1}{2}}| \leq \epsilon_3$.

Note that, in general, the finite function of two variables can written as

$$\chi(a,b) = \lambda_1(a)\lambda_2(b)|a|^{u_1}|b|^{u_2}(\log|a|^{n_1})(\log|b|^{n_2})$$

where λ_i, $i = 1,2$, are ramified characters of $F^\times$, u_i are real numbers, and n_i are nonnegative integers.

From now on, s can be any complex number with $Re(s)$ large so that the local zeta integral converges absolutely. The proof will again break into many cases to discuss as in the proof of the absolute convergence of the local integral. We will pick up case (2) to compute and the other cases will follow in the same way.

We set, for simplification of our notations, that

$$f_t(\alpha;\lambda;\psi;x) := |t|^{\alpha s+*}\lambda_t(t)\log|t|^*ch_{x+\mathcal{P}^p}(t)\psi(t). \tag{239}$$

Then the integral of case (2), which is denoted by I_2, can be written as

$$\int_a f_a(\epsilon_2; \lambda_a; \psi_a; x_a) \int_c f_c(\tfrac{1}{2}; \lambda_c; x_c) \int_d f_d(\epsilon_1; \lambda_d; \psi_d; x_d) \int_b f_b(1; \lambda_b; x_b) d(b, d, c, a). \tag{240}$$

The domain of the integration is $0 < |b| \leq min\{q^{-e_3}, q^{-e_1}|a^{-\frac{1}{2}}c^{\frac{1}{2}}d|\}$, $0 < |d| \leq q^{-e_3}$, $0 < |c| \leq min\{q^{-e_3}, q^{-e_1}|a|\}$, and $0 < |a| \leq q^{-e_3}$.

We first consider the integration with respect to the variable b, that is,

$$B_2 := \int_b f_b(1; \lambda_b; x_b) db \tag{241}$$

and have to separate the case $x_b \in \mathcal{P}^{p_b}$ and the case $x_b \notin \mathcal{P}^{p_b}$. When $x_b \in \mathcal{P}^{p_b}$, by lemma 4.1.2, if the character λ_b is not trivial, $B_2 = 0$; and if λ_b is trivial, one has

$$B_2 = \begin{cases} P_1(q^{-s}, 1) & \text{if } |a^{-\frac{1}{2}}c^{\frac{1}{2}}d| \leq q^{e_1 - p_b}, \\ Q_1(q^{-s}) & \text{if } |a^{-\frac{1}{2}}c^{\frac{1}{2}}d| > q^{e_1 - p_b}, \\ R_1(q^{-s}, 1) & \text{if } |a^{-\frac{1}{2}}c^{\frac{1}{2}}d| \geq q^{e_1 - e_3}, \end{cases} \tag{242}$$

where $P_1(q^{-s}, \nu)$ is a linear combination with coefficients in $\mathbb{C}(q^{-s})$ of $|a^{-\frac{1}{2}}c^{\frac{1}{2}}d|^{\nu s + *}$ and $|a^{-\frac{1}{2}}c^{\frac{1}{2}}d|^{\nu s + *} \log |a^{-\frac{1}{2}}c^{\frac{1}{2}}d|$. When $x_b \notin \mathcal{P}^{p_b}$, by Lemma 4.1.1, we have

$$B_2 = \begin{cases} 0 & \text{if } min\{q^{-e_3}, q^{-e_1}|a^{-\frac{1}{2}}c^{\frac{1}{2}}d|\} < |x_b|, \\ G_1(q^{-s}) & \text{if } min\{q^{-e_3}, q^{-e_1}|a^{-\frac{1}{2}}c^{\frac{1}{2}}d|\} \geq |x_b|. \end{cases} \tag{243}$$

Note that $min\{q^{-e_3}, q^{-e_1}|a^{-\frac{1}{2}}c^{\frac{1}{2}}d|\} \geq |x_b|$ implies that $q^{e_1 - p_b} \leq |a^{-\frac{1}{2}}c^{\frac{1}{2}}d|$. Without loss of generality, we can reformulate the results on the integral B_2 as follows:

$$B_2 = \begin{cases} P_1(q^{-s}, 1) & \text{if } |a^{-\frac{1}{2}}c^{\frac{1}{2}}d| \leq q^{e_1 - p_b}, \\ Q_1(q^{-s}) & \text{if } |a^{-\frac{1}{2}}c^{\frac{1}{2}}d| > q^{e_1 - p_b}. \end{cases} \tag{244}$$

Plugging B_2 into the integral I_2, we obtain that

$$\begin{aligned} I_2 &= \int_a f_a(\epsilon_2 - \tfrac{1}{2}; \cdots) \int_c f_c(1; \cdots) \int_d f_d(\epsilon_1 + 1; \lambda_d; \psi_d; x_d) d(d, c, a) \\ &\quad + \int_a f_a(\cdots) \int_c f_c(\cdots) \int_d f_d(\epsilon_1; \lambda_d; \psi_d; x_d) d(d, c, a) \\ &= I_{21} + I_{22} \end{aligned} \tag{245}$$

where the domain of the integration with respect to the variable d in I_{21} (or I_{22}) is $|d| \leq min\{q^{-e_3}, q^{e_1 - p_b}|a^{\frac{1}{2}}c^{-\frac{1}{2}}|\}$ (or resp. $q^{e_1 - p_b}|a^{\frac{1}{2}}c^{-\frac{1}{2}}| < |d| \leq q^{-e_3}$).

The computation of the integral D_{21} will go in the same way as that of the integral B_2 did. So we obtain that

$$D_{21} = \begin{cases} P_{21}(q^{-s}, \epsilon_1 + 1) & \text{if } |c^{-\frac{1}{2}}a^{\frac{1}{2}}| \leq q^{p_b - e_1 - p_d}, \\ Q_{21}(q^{-s}) & \text{if } |c^{-\frac{1}{2}}a^{\frac{1}{2}}| > q^{p_b - e_1 - p_d}, \end{cases} \tag{246}$$

where $P_2(q^{-s}, \epsilon_1+1)$ is a linear combination with coefficients in $\mathbb{C}(q^{-s})$ of $|c^{-\frac{1}{2}}a^{\frac{1}{2}}|^{(\epsilon_1+1)s+*}$ and $|c^{-\frac{1}{2}}a^{\frac{1}{2}}|^{(\epsilon_1+1)s+*} \log|c^{-\frac{1}{2}}a^{\frac{1}{2}}|$.

The integral D_{22} can be computed as follows. We need to distinguish two cases according to $x_d \in \mathcal{P}_d^p$ and $x_d \notin \mathcal{P}_d^p$. When $x_d \in \mathcal{P}_d^p$, D_{22} will vanish if λ_d is not trivial. If λ_d is trivial, one has, from Lemma 4.1.2, that

$$D_{22} = \begin{cases} P_2(q^{-s}, \epsilon_1) & \text{if } |c^{-\frac{1}{2}}a^{\frac{1}{2}}| \leq q^{p_b-e_1-p_d}, \\ 0 & \text{if } |c^{-\frac{1}{2}}a^{\frac{1}{2}}| > q^{p_b-e_1-p_d}, \end{cases} \tag{247}$$

where $P_2(q^{-s}, \epsilon_1)$ is a linear combination with coefficients in $\mathbb{C}(q^{-s})$ of $|c^{-\frac{1}{2}}a^{\frac{1}{2}}|^{\epsilon_1 s+*}$ and $|c^{-\frac{1}{2}}a^{\frac{1}{2}}|^{\epsilon_1 s+*} \log|c^{-\frac{1}{2}}a^{\frac{1}{2}}|$. When $x_d \notin \mathcal{P}_d^p$, we have, by the same reason, that

$$D_{22} = \begin{cases} R_2(q^{-s}) & \text{if } q^{e_1-p_b}|c^{-\frac{1}{2}}a^{\frac{1}{2}}| \leq |x| \leq q^{-e_3}, \\ 0 & \text{if } |x| < q^{e_1-p_b}|c^{-\frac{1}{2}}a^{\frac{1}{2}}|. \end{cases} \tag{248}$$

Hence we can put the results on the integral D_{22} in form:

$$D_{22} = \begin{cases} P_{22}(q^{-s}, \epsilon_1) & \text{if } |c| \geq |a|q^{2(p_d+e_1-p_b)}, \\ 0 & \text{if } |c| < |a|q^{2(p_d+e_1-p_b)}, \end{cases} \tag{249}$$

where $P_{22}(q^{-s}, \epsilon_1)$ is a linear combination with coefficients in $\mathbb{C}(q^{-s})$ of 1, $|c^{-\frac{1}{2}}a^{\frac{1}{2}}|^{\epsilon_1 s+*}$, and $|c^{-\frac{1}{2}}a^{\frac{1}{2}}|^{\epsilon_1 s+*} \log|c^{-\frac{1}{2}}a^{\frac{1}{2}}|$.

According to those computations, the integral I_{21} is a linear combination with coefficients in $\mathbb{C}(q^{-s})$ of integrals of following types:

$$I_{211} := \int_a f_a\left(\frac{2\epsilon_2+\epsilon_1}{2}; \cdots\right) \int_{|a|q^{2(p_d+e_1-p_b)} \leq |c| \leq \min\{q^{-e_3}, q^{-e_1}|a|\}} f_c\left(\frac{\epsilon_1+1}{2}; \cdots\right) d(c,a)$$

$$I_{212} := \int_a f_a\left(\frac{2\epsilon_2-1}{2}; \cdots\right) \int_{|c| \leq \min\{q^{-e_3}, q^{-e_1}|a|, |a|q^{2(p_d+e_1-p_b)}\}} f_c(1; \cdots) d(c,a)$$

and the integral I_{22} is a linear combination with coefficients in $\mathbb{C}(q^{-s})$ of integrals of following types:

$$I_{221} := \int_a f_a\left(\frac{2\epsilon_2+1}{2}; \cdots\right) \int_{|a|q^{2(p_d+e_1-p_b)} \leq |c| \leq \min\{q^{-e_3}, q^{-e_1}|a|\}} f_c\left(\frac{1-\epsilon_1}{2}; \cdots\right) d(c,a)$$

$$I_{222} := \int_a f_a(\cdots) \int_{|a|q^{2(p_d+e_1-p_b)} \leq |c| \leq \min\{q^{-e_3}, q^{-e_1}|a|\}} f_c\left(\frac{1}{2}; \cdots\right) d(c,a)$$

Since the positive integer p_d can be chosen so large that $q^{2(p_d+e_1-p_b)} > q^{-e_1}$ (choose p_d after p_b was chosen), there are no c's which satisfy the condition $|a|q^{2(p_d+e_1-p_b)} \leq |c| \leq \min\{q^{-e_3}, q^{-e_1}|a|\}$. In other words, the integrals I_{211}, I_{221}, and I_{222} are equal to zero. On the other hand, the integral I_{212} is, following similar argument, a linear combination with coefficients in $\mathbb{C}(q^{-s})$ of integrals of following types:

$$I_{2121} \quad := \quad \int_{|a| \leq q^{e_1 - p_c}} f_a\left(\frac{2\epsilon_2 + 1}{2}; \cdots\right) da,$$

$$I_{2122} \quad := \quad \int_{q^{e_1 - p_c} < |a| \leq q^{-e_3}} f_a(\epsilon_2; \cdots) da.$$

By Lemma 4.1.1 and 4.1.2 again, we conclude that the integrals I_{2121} and I_{2122} are rational functions in q^{-s}. This prove the rationality of the integral I_2. The rationality of other three cases can be proved by following the same argument.   □

Bibliography

[Adm]   J. Adams, *L-functoriality for dual pairs*, Asterisque, Vol. 171-172 (1989), 85-129.

[Art]   J. Arthur, *Eisenstein series and the trace formula*, in Proceedings of Symposia in Pure Math., Vol. 33, Part 1, Amer. Math. Soc., Providence, R.I., 1979, 253-274.

[Art1]   J. Arthur, *On some problems suggested by the trace formula*, Springer Lecture Notes in Math. 1041, 1-49, 1984.

[Art2]   J. Arthur, *Unipotent automorphic representations: Conjectures*, Asterisque, Vol. 171-172(1989), 13-71.

[Art3]   J. Arthur, *Automorphic representations and number theory*, in Canadian Math. Soc. Conf. Proc., Providence, RI. 1981, 3-51.

[ArGe]   J. Arthur and S. Gelbart, *Lectures on automorphic L-functions,* in L-functions and Arithmetic, 1-59, London Mathematical Society Lecture Note Series. 153, 1991.

[Bar]   L. Barthel, *Local Howe correspondence for groups of similitudes*, J. reine angew. Math., 414(1991), 207-220.

[Ber]   J. Bernstein, (redige par P. Deligne), *Le "centrre" de Bernstein,* dans: J. Bernstein et al. (eds.), Representations des Groupes Reductifs sur Corps Local, Hermann, Pairs, 1984.

[BGG]   I. N. Bernstein, I. M. Gelfand, and S. I. Gelfand, *Models of representations of Lie groups*, Sel. Math. Sov. 1(2)(1981), 121-142.

[BeZe1]   I. N. Bernstein and A. V. Zelevinski, *Induced representations of reductive p-adic groups* I, Ann. scient. Ec. Norm. Sup. 10(1977), 441-472.

[BeZe2]   I. N. Bernstein and A. V. Zelevinski, *Representations of the group GL(n,F) where F is a non-archimedean field*, Russian Math. Surveys 31, 1976.

[Bor]   A. Borel, *Automorphic L-functions*, in Proceedings of Symposia in Pure Math., Vol. 33, Part 2, Amer. Math. Soc., Providence, R.I., 1979, 27-61.

[Bor1]   A. Borel, *Admissible representations of a semi-simple group over a local field with vectors fixed under an Iwahori subgroup*, Invent. Math., 35(1976), 233-259.

[Bor2]   A. Borel, *Linear Algebraic Groups*, New York, Benjamin 1969.

[Bou]   N. Bourbaki, *Gropes et Algebres de Lie*, Paris, Hermann, Ch. 4-6, 1968, Ch. 7-8, 1975.

[BoJa]   A. Borel and H. Jacquet, *Automorphic forms and automorphic representations*, in Proceedings of Symposia in Pure Math., Vol. 33, Part 1, Amer. Math. Soc., Providence, R.I., 1979, 189-207.

[BoWa]   A. Borel and N. Wallach, *Continuous Cohomology, Discrete Subgroups, and Representations of Reductive Groups*, Ann. Math. Stud., Princeton Univ. Press, 1980.

[Bri]   M. Brion, *Invariants d'un sous-groupe unipotent maximal d'un groupe semisimple*, Ann. Inst. Fourier, Grenoble, 33(1983)(1), 1-27.

[Bum]   D. Bump, *The Rankin-Selberg method: a survey,* in Proceedings of Symposium in honor of A. Selberg, Oslo, 1987.

[BFG]   D. Bump, S. Friedberg, and D. Ginzburg, *Whittaker-orthogonal models, functoriality, and the Rankin-Selberg method*, Invent. Math., 109 (1992), 55-96.

[BuGi]   D. Bump and D. Ginzburg, *Symmettric square L-functions on GL(r)*, (preprint)

[BuGi1]   D. Bump and D. Ginzburg, *Spin L-functions of GSp(2n)*, Internat. Math. Res. Notices, (1992).

[Car]   P. Cartier, *Representations of p-adic groups: a survey*, in Proceedings of Symposia in Pure Math., Vol. 33, Part 1, Amer. Math. Soc., Providence, R.I., 1979, 111-155.

[Cas]   W. Casselman, *Introduction to the theory of admissible representations of p-adic reductive groups* (preprint).

[Cas1]   W. Casselman, *The unramified principal series of p-adic groups I*, Compositio Math., 40(1980), 367-406.

[Cas2]   W. Casselman, *Canonical extensions of Harish-Chandra modules to representations of G*, Can. J. Math., 41(1989).

[CaSh]   W. Casselman and J. Shalika, *The unramified principal series of p-adic groups II: the Whittaker function*, Comp. Math., Vol. 41, Fasc. 2(1980), 207-231.

[CaFr]   J. W. S. Cassels and A. Prohlich, *Algebraic Number Theory*, Academic Press 1967.

[Cop]   M. Copper, *The Fourier transform and intertwining operators for certain degenerate principal series representations of Sp(n, F)*, $F = \mathbb{R}$ or $\mathbb{C}$, (preprint)

[DHL]   W. Duke, R. Howe, and J.-S. Li, *Estimating Hecke eigenvalues of Siegel modular forms*, Duke J. Math., 67(1992)(1), 219-240.

[Fla]   D. Flath, *Decomposition of representations into tensor products*, in Proceedings of Symposia in Pure Math., Vol. 33, Part 1, Amer. Math. Soc., Providence, R.I., 1979, 179-184.

[Fur]   M. Furusawa, *On L-functions for GSp(4) $\times$ GL(2) aand their special values*, J. reine angew. Math., 438(1993), 187-218.

[Gar]   P. B. Garrett, *Decomposition of Eisenstein series: Rankin triple products.* Ann. Math., 125(1987), 209-235.

[Gel]   S. Gelbart, *An elementary introduction to the Langlands program*, Bull. Amer. Math. Soc., 10(1984)(2), 177-219.

[Gel1]   S. Gelbart, *Automorphic Forms on Adele Groups*, Ann. Math. Stud. Vol. 83, Princeton Univ. Press, Princceton, NJ, 1975.

[Gel2]   S. Gelbart, *Recent results on automrphic L-functions*, in Proc. of Symp. in honor of A. Selberg, Oslo, 1987.

[Gel3]   S. Gelbart, *On theta liftings for unitary groups*, CRM Proc. and Lect. Notes, 1(1993).

[GePS]   S. Gelbart and I. Piatetski-Shapiro, *L-functions for G $\times$ GL(n)*, in Lecture Notes in Math., Vol. 1254, Springer-Verlag, 1985.

[GeSh]   S. Gelbart and F. Shahidi, *Analytic properties of automorphic L-functions*, in Perspectives in Mathematics, Academic Press, 1988.

[GGPS]   I. M. Gelfand, M. I. Graev, and I. Piatetski-Shapiro, *Representation Theory and Automorphic Functions*, Academic Press, 1990.

[GeKa]   I. M. Gelfand aand D. Kazhdan, *Represssentations of GL(n, K) where K is a local field*, in Lie Groups and Their Representations, J. Wiley and Sons, 1975, 95-118.

[Gin]   D. Ginzberg, *L-functions for $SO_n \times GL_k$*, J. reine angew. Math., 405(1990), 156-180.

[Gin1]   D. Ginzberg, *A Rankin-Selberg integral for the ajoint representation of GL(3)*, Invent. Math., 105(1991), 571-588.

[GiRa]   D. Ginzburg and S. Rallis, *A Tower of Rankin-Selberg Integrals*, Intern. Math. Research Notices, (1994)(5), 201-208.

[GoJa]   R. Godement and H. Jacquet, *Zate Functions of Simple Algebras*, Lecture Notes in Math., Vol. 260, Springer-Verlag, 1972

[Gus]   R. Gustefson, *The degenerate principal series for Sp(2n)*, Mem. of Amer. Math. Soc., 248(1981).

[Har]   M. Harris, *L-functions of 2 × 2unitary groups and factorization of periods of Hilbert modular forms*, J. Amer Math. Soc. 6(1993)3, 637-719.

[HaKu]   M. Harris and S. Kudla, *The central critical value of a triple product L-function*, Ann. of Math., 133(1991), 605-672.

[HaKu]   M. Harris and S. Kudla, *Arithmetic automorphic forms for the nonholomorphic discrete series of GSp(2)*, Duke Math. J., Vol. 66, No. 1 (1992), 59-121.

[HST]   M. Harris, D. Soudry, and R. Taylor, *l-adic representations associated to modular forms over imaginary quadratic fieldds, I. Lifting to $GSp_4(\mathbb{Q})$*, Invent. Math., 112(1993), 377-411.

[Hos]   G. Hochschild, *Basic Theory of Algebraic Groups and Lie Algebras*, Graduate Texts in Math. 75, Springer-Verlag, New York, Heidelberg, Berlin, 1981.

[How]   R. Howe, *θ-series and invariant theory*, in Proceedings of Symposia in Pure Math., Vol. 33, Part 1, Amer. Math. Soc., Providence, R.I., 1979, 275-286.

[How1]   R. Howe, *On a notion of rank for unitary representations of classical groups*, C. I. M. E. summer school on harmonic analysis, Cortona, 1980.

[How2]   R. Howe, *Automorphic forms of low rank*, in Non-commutative Harmonic Analysis, Lecture Notes in Math., Vol. 880, Springer-Verlag, 1980.

[Hum]   J. E. Humphreys, *Introduction to Lie Aalgebras and Representation Theory*, Springer-Verlag, New York, Heidelberg, Berlin, 1972.

[Hum1]   J. E. Humphreys, *Linear Algebraic Groups*, GTM 21, Springer-Verlag, New York, 1981.

[IwMa]   N. Iwahori and H. Matsumoto, *On some Bruhat decomposition and the structure of the Hecke ring of p-adic Chevalley groups*, I.H.E.S., 25(1965), 5-48.

[Jac]   H. Jacquet, *On the residual spectrum of GL(n)*, Lecture Notes in Math., Vol. 1041, Springer-Verlag, 1984, 185-208.

[Jac1]   H. Jacquet, *Functions de Whittaker associees aux groupes de Chevalley*, Bull. Soc. Math. France, t. 95, 1967, 243-309.

[Jac2]   H. Jacquet, *Principal L-functions of the linear group*, in Proceedings of Symposia in Pure Math., Vol. 33, Part 2, Amer. Math. Soc., Providence, R.I., 1979, 63-86.

[JaLa]   H. Jacquet and R. Langlands, *Automorphic Forms on GL(2)*, Lecture Notes in Math., Vol. 114, Springer-Verlag, 1970.

[JPSS]   H. Jacquet, I. Piatetski-Shapiro and J. Shalika, *Automorphic forms on GL(3) I, II*, Ann. Math., 109(1979), 169-212 and 213-258.

[JPSS1]   H. Jacquet, I. Piatetski-Shapiro and J. Shalika, *Rankin-Selberg convolutions*, Amer. J. Math. 105(1983), 367-464.

[JaSh]   H. Jacquet and J. Shalika, *On Euler products and the classification of automorphic representations I, II*, Amer. J. Math., vol. 103, (1981)3, 499-558, 777-815.

[JaSh1]   H. Jacquet and J. Shalika, *Exterior square L-functions*, in "Automorphic Forms, Shimura Varieties and L-functions", Perspectives in Math., vol. 11, 1990.

[Jan]   C. Jantzen, *Degenerate Principal Series Representations for Symplectic Groups*, Mem. of Amer. Math. Soc., 488(1993).

[Jan1]   C. Jantzen, *Degeneratte principal series for symplectic and odd-orthogonal groups*, 1994 (preprint).

[Jia]   D. Jiang, *L-function for the standard tensor product representation of GSp(2) × GSp(2)*, Ph. D. dissertation at The Ohio State University, 1994.

[Kac]   V. G. Kac, *Some remarks on nilpotent orbits*, J. algebra 64(1980), 190-213.

[Kim]   B. Kimelfeld, *Homogeneous domains on flag manifolds*, J. Math. Anal. and Appl., 121(1987), 506-588.

[Kna]   A. W. Knapp, *Representation Theory of Semisimple Groups*, Princeton Univ. Press, 1986.

[Kos]   B. Kostant, *Lie group representations on polynomial rings*, Amer. J. Math. 85(1963), 327-404.

[Kos1]   B. Kostant, *On Whittaker vectors and representation theory*, Invent. Math. 48(1978), 101-184.

[KoRa]   B. Kostant and S. Rallis, *Orbits and Representations Associated with Symmetric Spaces*, Amer. J. Math. 93, No. 3 (1971), 753-809.

[Kud]   S. Kudla, *See-saw dual reductive pairs*, in "Automorphic Forms in Several Variables", Birkhauser Progress in Math. 46, (1984), 244-268.

[Kud1]   S. Kudla, *The local Langlands correspondence: the non-archimedean case*, (preprint)

[KuRa]   S. Kudla and S. Rallis, *Ramified degenerate principal series representations for $Sp(n)$*, Israel J. of Math. 78 (1992), 209-256.

[KuRa1]   S. Kudla and S. Rallis, *Poles of Eisenstein series and L-functions*, in Festschrift in honor of I. Piatetski-Shapiro, Israel Mathematical Conference Proceedings 3(1990), 81-110.

[KuRa2]   S. Kudla and S. Rallis, *A regularized Siegel-Weil formula: the first term identity*, Ann. of Math., 140(1994), 1-80.

[KuRa3]   S. Kudla and S. Rallis, *On the Weil-Siegel formula*, J. reine angew. Math., 387(1988), 1-68.

[KuRa4]   S. Kudla and S. Rallis, *On the Weil-Siegel formula, II: the isotropic convergent case*, J. reine angew. Math., 391(1988), 65-84.

[KuRa5]   S. Kudla and S. Rallis, *Degenerate principal series and invariant distributions*, Isr. J. Math., 69(1990), 25-45.

[KRS]   S. Kudla, S. Rallis, and D. Soudry, *On the degree 5 L-function for $Sp(2)$*, Invent. Math. 107(1992), 483-541.

[LaLa]   J. Labesse and R. Langlands, *L-indistinguishability for $SL(2)$*, Canad. J. Math. 31(1979), 726-785.

[Lan]   R. Langlands, *Euler Products*, Yale Math. Monographs, 1971.

[Lan1]   R. Langlands, *On the functional equation satisfied by Eisenstein series*, Lecture Notes in Math., Vol. 544, Springer-Verlag, 1976.

[Lan2]   R. Langlands, *Problems in the theory of automorphic forms*, Lecture Notes in Math., Vol. 170, Springer-Verlag, 1970.

[Lan3]   R. Langlands, *On the notion of an automorphic representation, a supplement to the proceding papeer*, in Proceedings of Symposia in Pure Math., Vol. 33, Part 1, Amer. Math. Soc., Providence, R.I., 1979.

[Lan4]   R. Langlands, *Automorphic representations, Shimura varieties, and motives, Ein Marchen*, in Proceedings of Symposia in Pure Math., Vol. 33, Part 2, Amer. Math. Soc., Providence, R.I., 1979.

[Lan5]   R. Langlands, *Eisenstein series, the trace formula, and the modern theory of automorphic forms*, in Proc. of Symp. in honor of A. Selberg, Oslo, 1987.

[LaSh]   R. Langlands and D. Shelstad, *On the definition of the transfer factors*, Math. Ann., 278(1987), 219-271.

[Li]   J.-S. Li, *Distinguished cusp forms are theta series*, Duke Math. J. 59(1989)(1), 175-189.

[Li1]   J.-S. Li, *Non-existence of singular cusp forms*, Compos. Math., 83(1992), 135-166.

[Lus]   G. Lusztig, *Representations of affine Hecke algebras*, Asterisque, Vol. 171-172, 1989.

[Mat]   T. Matsuki, *Orbits on flag manifolds*, in Proceedings of the International Congress of Mathematicians, Kyoto, Japan, 1990, 807-813.

[Mas]   H. Matsumoto, *Sur les sous-groupes arithmetiques des groupes semi-simples deployes*, Ann. scient. Ec. Norm. Sup., 4^e series, t. 2 (1969), 1-62.

[Moe]   C. Moeglin, *Representations unipotentes et formes automorphes de carre integrable*, (preprint)

[Moe1]   C. Moeglin, *Sur les formes automorphes de carre integrable*, in Proceedings of the International Congress of Mathematicians, Kyoto, Japan, 1990, 815-819.

[MVW]   C. Moeglin, M.-F.Vigneras, and J. L. Waldspurger, *Correspondences de Howe sur un corps p-adique*, Lecture Notes in Math., Vol. 1291, Springer-Verlag, 1987.

[MoWa]   C. Moeglin and J. L. Waldspurger, *Decomposition Spectrale et Series d'Eisenstein*, Progess in Math. 113, Birkhauser, 1994.

[Moy]   A. Moy, *Representations of GSp_4 over a p-adic field, I.* Compos. Math. 66(1988), 237-284.

[Nov]   M. Novodorsky, *Automorphic L-functions for the symplectic group GSP_4*, in Proceedings of Symposia in Pure Math., Vol. 33, Part 2, Amer. Math. Soc., Providence, R.I., 1979.

[PS]   I. Piatetski-Shapiro, *Multiplicity one theorems*, in Proceedings of Symposia in Pure Math., Vol. 33, Part 1, Amer. Math. Soc., Providence, R.I., 1979.

[PS1]   I. Piatetski-Shapiro, *On the Saito-Kurokawa lifting*, Invent. Math., 17(1983).

[PSRa]   I. Piatetski-Shapiro and S. Ralis, *L-functions for the Classical Groups*, Lecture Notes in Math., Vol. 1254, Springer-Verlag, 1987.

[PSRa1]   I. Piatetski-Shapiro and S. Rallis, *Rankin triple L-functions,* Compositio Mathematica 64 (1987),31-115.

[PSRa2]   I. Piatetski-Shapiro and S. Rallis, *A new way to get Euler products*, J. reine angew. Math., 392(1988), 110-124.

[Pop]   V. L. Popov, *Groups, Generators, Syzygies, and Orbits in Invariant Theory*, Translations of Math. Monographs, vol. 100, Amer. Math. Soc., 1992.

[Prz]   T. Przebinda, *On Howe duality theorem*, J. of Funct. Anal., 81(1988), 160-183.

[Ral]   S. Rallis, *Poles of standard L-functions*, in Proceedings of the International Congress of Mathematicians, Kyoto, Japan, 1990, 833-845.

[Ral1]   S. Rallis, *On Howe duality conjecture*, Comp. Math. 51(1984), 333-399.

[Ral2]   S. Rallis, *L-functions and the Oscillator Repressentations*, Lecture Notes in Math., Vol. 1254, Springer-Verlag, 1987.

[Ral3]   S. Rallis, *Langlands' functoriality and the Weil representation*, Amer. J. Math., 104(1982), 469-515.

[Ral4]   S. Rallis, in preparation.

[RaSc]   S. Rallis and G. Schiffman, *Weil Representation I. Intertwining distributions and discrete spectrum*, Mem. Amer. Math. Soc. vol. 231, 1980.

[RRS]   R. Richardson, G. Rohrle, and R. Steinberg, *Parabolic subgroups with abelian unipotent radical*, Invent. Math. 110(1992), 649-671.

[Sch]   G. Schiffman, *Integrales D'entrelacement et fonctions de Whittaker*, Bull. Soc. math. France, 99(1971), 3-72.

[gSch]   G.W. Schwarz, *Representations of simple Lie groups with regular rings of invarieants*, Invent. Math., (1978), 167-191.

[lSch]   L. Schwarz, *Theorie des distributions*, vol. I, II, Hermann, Paris, 1957.

[Sek]   J. Sekiguchi, *The nilpotent subvariety of the vector space associated to a symmetric pair*, Publ. RIMS. Kyoto Univ., 20(1984), 155-212.

[Shh]   F. Shahidi, *On certain L-functions*, Amer. J. Math., 103(1981)(2), 297-355.

[Shh1]   F. Shahidi, *On the Ramanujan conjecture and the finiteness of poles of certain L-functions*, Ann. of Math. (2)(1988)(127), 547-584.

[Shl]   J. Shalika, *The multiplicity one theorem for GL_n*, Ann. Math., 100(1974), 171-193.

[Sil]   A. Silberger, *Introdution to Harmonic Analysis on Reductive p-adic Groups*, Princeton University Press, Princeton, 1979.

[Spe]   B. Speh, *Unitary representations of $GL(n, \mathbb{R})$ with non-trivial $(\mathfrak{g}, K)$-cohomology*, Invent. Math. 71)1983), 95-104.

[Spr]   T.A. Springer, *Reductive groups*, in Proceedings of Symposia in Pure Math., Vol. 33, Part 1, Amer. Math. Soc., Providence, R.I., 1979, 3-27.

[Sou]   D. Soudry, *Rankin-Selberg convolutions for $SO_{2l+1} \times GL_n$:local theory*, Mem. of Amer. Math. Soc..

[Sou1]   D. Soudry, *On the archimedean theory of Rankin-Selberg convolutions for $SO_{2l+1} \times GL_n$*, (preprint).

[Sou2]   D. Soudry, *Automorphic forms on $GSp(4)$*, in Festschrift in honor of I. Piatetski-Shapiro, Israel Mathematical Conference Proceedings 3(1990), 291-303.

[Sou3]   D. Soudry, *The CAP representations of $GSp(4, \mathbb{A})$*, J. reine angew. Math., 383(1988), 87-108.

[Sou4]   D. Soudry, *A uniquness theorem for representations of $GSO(6)$ and the strong multiplicity one theorem for generic representations of $GSp(4)$*, Isr. J. Math., 58(1987)(3), 257-287.

[Sug]   M. Sugiura, *Conjugate classes of Cartan subalgebrass in real semisimple Lie Algebras*, J. Math. Soc. Jaapan, 11(1959)(4).

[Tat]   J. Tate, *Number theoretic background*, in Proceedings of Symposia in Pure Math., Vol. 33, Part 2, Amer. Math. Soc., Providence, R.I., 1979, 3-26.

[Vin]   E. B. Vinberg, *The Weyl group of a graded Lie algebra*, Izv. Akad. Nauk SSSR ser. Mat. 10(1976), 488-526.

[Vog]   D. Vogan, *Representations of Real Reductive Groups*, in Progress in Math., Vol. 15, Birkhauser, 1981.

[Vog1]   D. Vogan, *Unitary Representations of Reductive Lie Groups*, Ann. of Math. Stud., Vol. 118, Princeton Univ Press, 1987.

[Vog2]   D. Vogan, *Gelfand-Kirillov dimension for Harish-Chandra modules*, Invent. Math., 48(1978), 75-98.

[VoWa]   D. Vogan, and N. Wallach, *Intertwining operators for reaal reductive groups*, Adv. in Math., 82(1990), 203-243.

[Wald]   J. L. Waldspurger, *Correspondence de Shimura*, J. Math. Pures et Appl. 59 (1980), 1-133.

[Wal]   N. Wallach, *Real Reductive Groups* I, II, Academic Press, Boston, 1988, 1992.

[Wal1]   N. Wallach, *Asymptotic expansions of generalized matrix entries of representaions of real reductivr groups*, Lecture Notes in Math., Vol. 1024, Springer-Verlag, 1983.

[Wald2]   N. Wallach, *On the constant term of a square integrable automorphic forms*, in: Arsene, Gr. et al. (eds.) Operator Algebras and Group Representations, Proc. Conf. Romaania, 1984, 227-237,

[Wal3]   N. Wallach, *Lie algebra cohomology and holomorphic continuation of generalized Jacquet integrals*, In: K. Okamoto, T. Oshima (eds.) Rpresentation of Lie Grooups, Kyoto, (Adv. Stud. Pure Math., Vol. 14, 123-151) 1988.

[War]   G. Warner, *Harmonic Analysis on Semi-simple Lie Groups* I, Grund. Math. Wiss., 188, Springer-Verlog (1972).

[Wei]   A. Weil, *Basic Number Theory*, Berlin/Heidelbeerg/New York, 1974.

[Wei1]   A. Weil, *Sur certains groupes des operateurs unitaires*, Acta Math., 111(1964), 143-211.

[Zhe]   D. P. Zhelobenko, *The Classical Groups. spectral Analysis of Their Finite-Dimensional Representations*, Russian Mathematical Surveys.

[Zhe1]   D. P. Zhelobenko, *On Gelfand-Zetlin bases for classical Lie algebras*, in "Representations of Lie groups and Lie algebras", edited by A. A. Kirillov, 1985.

[Zel]   A. Zelevinski, *Induced representations of reductive p-adic groups* II, Ann. Sci. Ecole Norm. Sup. 13(1980), 165-210.

Index

194

Editorial Information

To be published in the *Memoirs*, a paper must be correct, new, nontrivial, and significant. Further, it must be well written and of interest to a substantial number of mathematicians. Piecemeal results, such as an inconclusive step toward an unproved major theorem or a minor variation on a known result, are in general not acceptable for publication. *Transactions* Editors shall solicit and encourage publication of worthy papers. Papers appearing in *Memoirs* are generally longer than those appearing in *Transactions* with which it shares an editorial committee.

As of May 31, 1996, the backlog for this journal was approximately 7 volumes. This estimate is the result of dividing the number of manuscripts for this journal in the Providence office that have not yet gone to the printer on the above date by the average number of monographs per volume over the previous twelve months, reduced by the number of issues published in four months (the time necessary for preparing an issue for the printer). (There are 6 volumes per year, each containing at least 4 numbers.)

A Copyright Transfer Agreement is required before a paper will be published in this journal. By submitting a paper to this journal, authors certify that the manuscript has not been submitted to nor is it under consideration for publication by another journal, conference proceedings, or similar publication.

Information for Authors and Editors

Memoirs are printed by photo-offset from camera copy fully prepared by the author. This means that the finished book will look exactly like the copy submitted.

The paper must contain a *descriptive title* and an *abstract* that summarizes the article in language suitable for workers in the general field (algebra, analysis, etc.). The *descriptive title* should be short, but informative; useless or vague phrases such as "some remarks about" or "concerning" should be avoided. The *abstract* should be at least one complete sentence, and at most 300 words. Included with the footnotes to the paper, there should be the 1991 *Mathematics Subject Classification* representing the primary and secondary subjects of the article. This may be followed by a list of *key words and phrases* describing the subject matter of the article and taken from it. A list of the numbers may be found in the annual index of *Mathematical Reviews*, published with the December issue starting in 1990, as well as from the electronic service e-MATH [**telnet e-MATH.ams.org** (or **telnet 130.44.1.100**). Login and password are **e-math**]. For journal abbreviations used in bibliographies, see the list of serials in the latest *Mathematical Reviews* annual index. When the manuscript is submitted, authors should supply the editor with electronic addresses if available. These will be printed after the postal address at the end of each article.

Electronically prepared papers. The AMS encourages submission of electronically prepared papers in $\mathcal{A}_{\mathcal{M}}\mathcal{S}$-TEX or $\mathcal{A}_{\mathcal{M}}\mathcal{S}$-LATEX. The Society has prepared author packages for each AMS publication. Author packages include instructions for preparing electronic papers, the *AMS Author Handbook*, samples, and a style file that generates the particular design specifications of that publication series for both $\mathcal{A}_{\mathcal{M}}\mathcal{S}$-TEX and $\mathcal{A}_{\mathcal{M}}\mathcal{S}$-LATEX.

Authors with FTP access may retrieve an author package from the Society's Internet node **e-MATH.ams.org** (130.44.1.100). For those without FTP access, the author package can be obtained free of charge by sending e-mail to `pub@math.ams.org` (Internet) or from the Publication Division, American Mathematical Society, P.O. Box 6248, Providence, RI 02940-6248. When requesting an author package, please specify $\mathcal{AMS}$-TEX or $\mathcal{AMS}$-LATEX, Macintosh or IBM (3.5) format, and the publication in which your paper will appear. Please be sure to include your complete mailing address.

Submission of electronic files. At the time of submission, the source file(s) should be sent to the Providence office (this includes any TEX source file, any graphics files, and the DVI or PostScript file).

Before sending the source file, be sure you have proofread your paper carefully. The files you send must be the EXACT files used to generate the proof copy that was accepted for publication. For all publications, authors are required to send a printed copy of their paper, which exactly matches the copy approved for publication, along with any graphics that will appear in the paper.

TEX files may be submitted by email, FTP, or on diskette. The DVI file(s) and PostScript files should be submitted only by FTP or on diskette unless they are encoded properly to submit through e-mail. (DVI files are binary and PostScript files tend to be very large.)

Files sent by electronic mail should be addressed to the Internet address `pub-submit@math.ams.org`. The subject line of the message should include the publication code to identify it as a Memoir. TEX source files, DVI files, and PostScript files can be transferred over the Internet by FTP to the Internet node `e-math.ams.org` (130.44.1.100).

Electronic graphics. Figures may be submitted to the AMS in an electronic format. The AMS recommends that graphics created electronically be saved in Encapsulated PostScript (EPS) format. This includes graphics originated via a graphics application as well as scanned photographs or other computer-generated images.

If the graphics package used does not support EPS output, the graphics file should be saved in one of the standard graphics formats—such as TIFF, PICT, GIF, etc.—rather than in an application-dependent format. Graphics files submitted in an application-dependent format are not likely to be used. No matter what method was used to produce the graphic, it is necessary to provide a paper copy to the AMS.

Authors using graphics packages for the creation of electronic art should also avoid the use of any lines thinner than 0.5 points in width. Many graphics packages allow the user to specify a "hairline" for a very thin line. Hairlines often look acceptable when proofed on a typical laser printer. However, when produced on a high-resolution laser imagesetter, hairlines become nearly invisible and will be lost entirely in the final printing process.

Screens should be set to values between 15% and 85%. Screens which fall outside of this range are too light or too dark to print correctly.

Any inquiries concerning a paper that has been accepted for publication should be sent directly to the Editorial Department, American Mathematical Society, P. O. Box 6248, Providence, RI 02940-6248.

Selected Titles in This Series

(Continued from the front of this publication)

(See the AMS catalog for earlier titles)